# *Correlations and Multiparticle Production* (CAMP) (LESIP IV)

International Workshop on

# Correlations and Multiparticle Production (CAMP) (LESIP IV)

Marburg, F R Germany May 14 – 16, 1990

Edited by

**M. Plümer, S. Raha and R. M. Weiner**
Department of Physics
Philipps-Universität Marburg
F R Germany

**World Scientific**
*Singapore • New Jersey • London • Hong Kong*

*Published by*

World Scientific Publishing Co Pte Ltd
5 Toh Tuck Link, Singapore 596224
*USA office:* 27 Warren Street, Suite 401-402, Hackensack, NJ 07601
*UK office:* 57 Shelton Street, Covent Garden, London WC2H 9HE

**British Library Cataloguing-in-Publication Data**
A catalogue record for this book is available from the British Library.

**CORRELATIONS AND MULTIPARTICLE PRODUCTION**



ISBN-13 978-981-02-0331-3
ISBN-10 981-02-0331-4

## Organizing Committee

P. Carruthers (Tucson)

S. Raha (Marburg), Convener

A. Webel (Marburg), Organizational Secretary

U. Ornik (Marburg), Scientific Secretary

C. C. Shih (Knoxville)

R. Weiner (Marburg), Chairman

## Advisory Committee

R. Bock (Darmstadt)
G. Fowler (Exeter)
G. Goldhaber (Berkeley)
W. Hofmann (Heidelberg)
I. Otterlund (Lund)
J. Rafelski (Tucson)
H. Satz (CERN)
R. Stock (Frankfurt)
E. C. G. Sudarshan (Austin)
W. Willis (CERN)

I. Dremin (Moscow)
E. Friedlander (Berkeley)
H. Gutbrod (Darmstadt)
W. Kittel (Nijmegen)
M. Podgoretsky (Dubna)
R. Salmeron (Palaiseau)
B. Sinha (Calcutta)
D. Strottman (Los Alamos)
D. Wegener (Dortmund)
G. Zinovjev (Kiev)

# Acknowledgement

This workshop was made possible through the generous support and funding provided by Deutsche Forschungsgemeinschaft (DFG), Hessisches Ministerium für Wissenschaft und Kunst, Philipps-Universität Marburg, Bundesminister für Forschung und Technologie (BMFT) and Marburger Universitätsbund.

# Preface

This volume contains the proceedings of the fourth meeting in the series "LESIP", which took place in May 1990 in Marburg. This series started in 1984 in Bad Honnef, reflecting the feeling of many high energy physicists that the field of 'soft' strong interactions was rapidly developing and therefore needed a special forum where the specific problems arising in this field could be debated. In 1984 one of the most actual problems of this kind was the issue of local equilibrium in strong interaction physics and herefrom the name of the series–LESIP.

The following two meetings in Santa Fe in 1986 and Tucson in 1988 were devoted to the more general topic of "Hadronic Matter in Collision." In 1990 we felt that this time the most critical issues at the frontline of research were correlations, fluctuations and multiparticle production. As a matter of fact, practitioners of the methods of quantum statistics have advocated for some time that these topics must be studied together, given the intimate link between these subjects, and this meeting gave this link the first formal status. The fact that the number of people interested in attending this meeting exceeded by far the initial upper limit of 50 the organizers had thought to impose in order to maintain the format of a workshop, proves that at least what concerns the choice of the topic, we were on the right track. As the list of participants shows, a distinctive feature of this meeting was the fact that a number of those who have done pioneering work (and are still active at the forefront of research) in Bose-Einstein correlations, multiplicity distributions and intermittency were brought together with young and new entrants. As a result, the atmosphere was very productive, the discussions very intensive and this is hopefully also reflected in the proceedings.

Among the new results on Bose-Einstein correlations presented at CAMP, one should quote, on the experimental side, the results on three and four particle correlations presented by Kageya (Na23 and Na27 collaborations) as well as the $p_T$-dependence of two body correlations seen by the Na35 collaboration at CERN in the heavy ion reactions at 200 GeV/nucleon. A strong $p_T$-dependence was also predicted in the quantum statistical approach to Bose-Einstein correlations by Andreev, and is the hydrodynamical approach by Sinyukov *et al.*

The study of multiplicity distributions in narrow rapidity windows was already suggested in 1978 as a useful method for obtaining information about the correlation length and chaoticity which control both Bose-Einstein correlations and multiplicity distributions of identical particles. The strength of this strategy has been much enhanced, among other things, through the generalization of this formalism, presented at CAMP by Shih, to charged multiplicities. This topic has become very popular, partly due to the observation of the "spike events" and also because of its possible relation with the phenomenon of intermittency. This subject was of course, as reflected in these proceedings, debated at CAMP. The question whether the growth of the moments of multiplicity distributions with narrowing width of the rapidity window continues "for ever" or saturates is, from the experimental point of view, still unsettled. From the quantum statistical standpoint, it is, however, clear that there exist two regimes in the $y$ space: as long as the window $\delta y$ is larger than the intrinsic rapidity correlation length $\xi$, the behavior of the moments as a function of $\delta y$ is controlled by $\xi$ and there can be no "intermittency." Once $\delta y \ll \xi$, it is conceivable that there is no intrinsic scale anymore, and then "intermittency" may arise. So far, there is no evidence that we have reached that second regime. As a matter of fact, the emulsion data of the EMU-01 collaboration presented by Stenlund show that the increase of moments with decreasing width $\delta y$ stops at very small $\delta y$. The Na22 data presented by Kittel, on the other hand, address the question of fluctuations in a given event, and find there an increase of the correlations with decreasing correlation distance, as predicted by Bialas and Peschanski, although the increase is much larger than expected. However, the Na22 collaboration data for hadron-nucleus reactions analyzed in terms of $\delta y$ by Verbeure show no evidence for "intermittency."

There are important methodological aspects which have to be considered in the analysis of the data and this was made clear in the talks given by Bialas, Carruthers, Friedlander, and others. Finally the proposal by Dremin to study fluctuations of moments of distributions is certainly worth mentioning, although it is not yet understood where it leads to.

All participants apparently agreed that this meeting was timely, stimulating and useful. We are confident that the reader will have the same impression from the present proceedings.

# Table of Contents

## I. Sessions on Correlations

Chairmen: *G. N. Fowler, D. Ardouin, H Gutbrod, and B. Sinha*

# ARE THE OBSERVED BOSE-EINSTEIN CORRELATIONS POSSIBLE?

M.G. BOWLER
*Department of Nuclear Physics, Oxford University*
*Keble Road, Oxford OX1 3RH, England*

ABSTRACT

**Both resonance production and final state interactions within like pion pairs narrow and suppress Bose-Einstein correlation, to the point where it becomes difficult to understand observations. For extended sources it is possible to distinguish between direct production and rescattering and I argue that in multipion production from extended sources final state interactions average to yield negligible effects. It is — just — possible to understand the correlations observed in $e^+e^-$ annihilation and in hadron-hadron collisions.**

## 1. The Observed Correlations

I am concerned with the understanding of Bose-Einstein correlations observed in $e^+e^-$ annihilation and in hadron-hadron collisions — very similar features are observed in these two cases. Those features which are both important and well established are, to my mind, the following.

1. A substantial Bose-Einstein correlation, second order or intensity interference, EXISTS.
2. The enhancement factor exceeds 1.5 as the momentum difference of two identical pions drops below 0.1 GeV/c. We can be confident that the amplitudes for radiating pions are substantially chaotic, even for a clean quark current source.
3. Within the limits of experimental accuracy, the enhancement is a function only of the single variable $Q$, where

$$Q^2 = M^2_{\pi\pi} - 4m^2_\pi \tag{1}$$

and this result is NON-TRIVIAL.

$$\begin{aligned} Q^2 = (\Delta p_\mu)^2 = (\Delta p_T)^2 &\quad + [(\Delta p_L)^2 - (\Delta E)^2] \\ = Q^2_T &\quad + Q^2_L \end{aligned} \tag{2}$$

and is an invariant. The quantities $Q^2_T$ and $Q^2_L$ are invariant under boosts along the longitudinal (jet) direction.

If the enhancement is represented by the form

$$\sim e^{-\beta^2_T Q^2_T} e^{-\beta^2_L Q^2_L} = e^{-\beta^2_T \Delta p^2_T} e^{-\beta^2_L \Delta p^2_L} e^{+\beta^2_L \Delta E^2} \tag{3}$$

it is found that $\beta_T \sim \beta_L$ (there is no difference within the errors). The enhancement GROWS as $Q^2_L$ goes negative (which is possible for non-zero $Q^2_T$). This is illustrated in fig.1.

THEREFORE the space–time structure of the (chaotic) amplitudes is a function of the invariant

$$x^2_\mu = t^2 - \mathbf{r}^2 \tag{4}$$

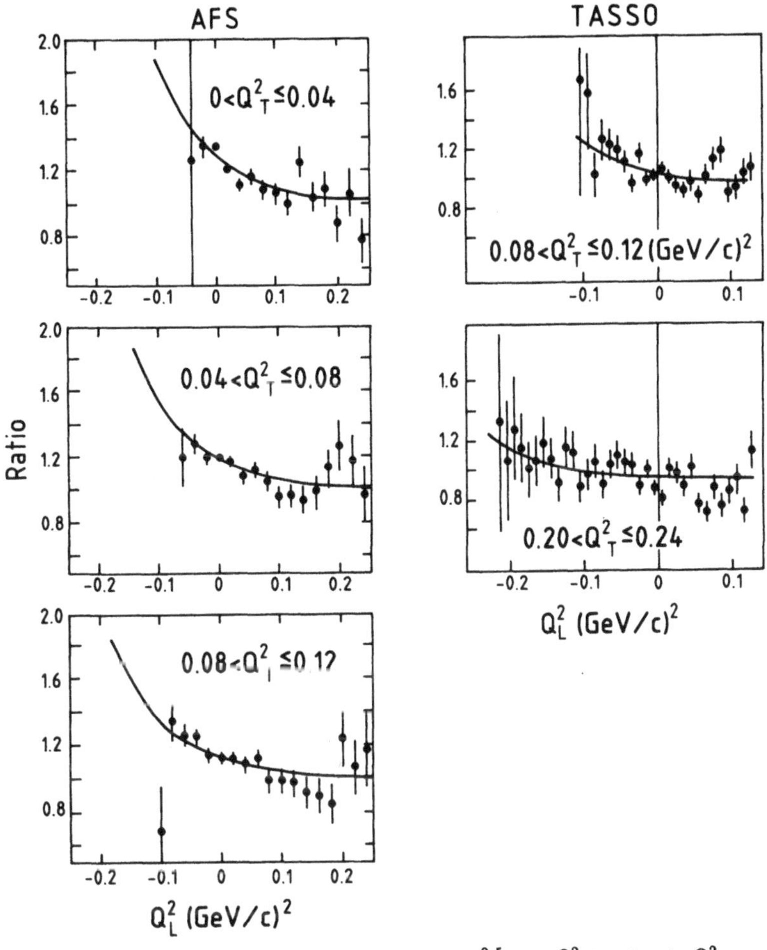


Fig.1. Variation of the like sign correlation function[3,5] with $Q_L^2$ for slices in $Q_T^2$. The correlation increases with the magnitude of negative $Q_L^2$.

and is, in the centre of mass of the pair, (approximately) spherically symmetric.

4. Three particle correlations are consistent with those features of the source extracted from two particle correlations — if the two particle source intensity is

$$f^2(r_1)f^2(r_2)$$

the three particle effect is consistent with a three particle source intensity

$$f^2(r_1)f^2(r_2)f^2(r_3)$$

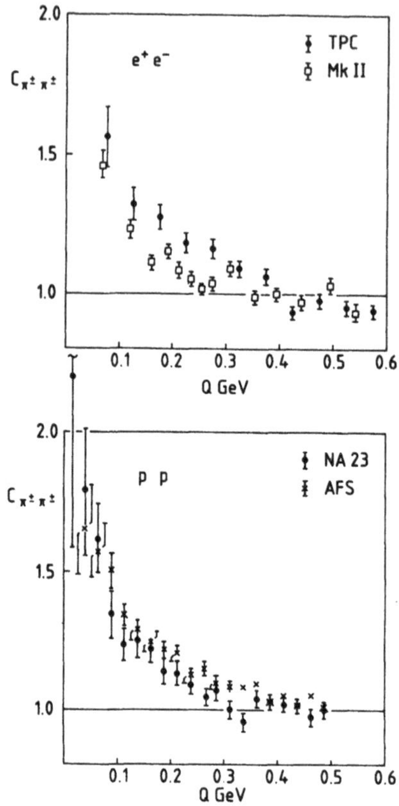


Fig.2. Examples of the like sign correlation[1–4] plotted as a function of $Q$.

The RANGE, SHAPE and ABSOLUTE MAGNITUDE of the correlations are, in contrast, less well established. Fig.2 shows data drawn from.[1–4] At first sight the similarities are more striking than the differences, but this in itself is a cause for disquiet. In $e^+e^-$ annihilation c and b quarks are created in some profusion and in events initiated by c or b quarks a lot of pions will be daughters of charmed or bottomed hadrons. A back of the envelope calculation yields an estimated suppression of the Bose-Einstein enhancement by between 30% and 40%. A similar result has been obtained by the Mark II collaboration, using Monte Carlo methods.[1] The absolute magnitude of the enhancements shown in fig.2(a) is expected to be less than the magnitude in the p-p interaction data of fig.2(b).

Although the two sets of $e^+e^-$ annihilation data and the two sets of p-p data agree well for $Q \sim 0.1$ GeV/c, there are differences within each category which amount to a factor of $\sim 2$ in the enhancement for $Q \sim 0.25$ GeV/c. This is a crucial region, as will emerge later. The significance of the experimental uncertainty is illustrated by fig.3, in which calculations[10] are compared with the Mark II and TPC data. The point at issue is not the merit of the (upper) histograms, but that the histogram for a string tension $\kappa$ of

0.2 GeV$^2$ is in good agreement with the Mark II data whereas even for a tension of 0.3 GeV$^2$ the histogram falls below the TPC data.

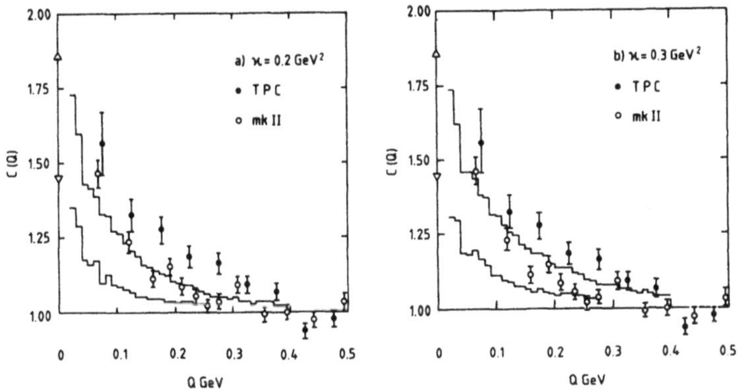


Fig.3. Comparison of Mark II[1] and TPC[2] data with calculations.[10] The figure illustrates the significance of the experimental uncertainty in the correlation function. (The upper histograms are for no $\eta'$; the lower for an $\eta'$ rate as in ref.7).

If the data of fig.2 are fitted with a single exponential in $Q^2$, using reference samples constructed using pions from different events, typical results for the range parameters $\beta$ are

| | $e^+e^-$ | | p-p | |
|---|---|---|---|---|
| | Mk II[1] | TPC[2] | AFS[3] | NA23[4] |
| $\sqrt{s}$ | 29 GeV | 29 GeV | 63 GeV | 26 GeV |
| $\beta$ | $1.0 \pm 0.1$ fm | $0.65 \pm 0.07$ fm | $0.82 \pm 0.05$ fm | $0.99 \pm 0.17$ fm |

There is substantial uncertainty on the range, and also on the shape. It is however clear from fig.2 that the correlation functions are SHARPER than gaussian in $Q$. There is no reason to expect gaussian shapes and every reason to expect shapes sharper than gaussian because:

(i) the string model, which contains the space–time structure indicated by the data, predicts a sharper function

(ii) the effects of resonance propagation sharpen a gaussian like production distribution.

If it is assumed that all pions are produced directly from the string, resonance production and final state interactions being ignored, the string model of Bose-Einstein correlations[6–9] is in effortless agreement with the data. However, both resonance production and decay and (strong) final state interactions within each like pion pair NARROW the shape of the correlation and suppress the absolute magnitude. I have been concerned to investigate whether (i) the string model can survive these effects and (ii) whether ANY source, even if much smaller than the (local) string source can yield correlations which match the data after resonances and rescattering are taken into account. The problem is not that the observed correlations are too small, but that they may be too large.

Fig.4 shows the TPC data[2] and calculations from Andersson and Hofmann.[7] The solid line is for directly produced pions (with, I believe, kinematic configurations determined by resonance production and decay) and the dotted line shows the effect of including resonance production, as in the Lund string, according to a recipe of Grassberger. The

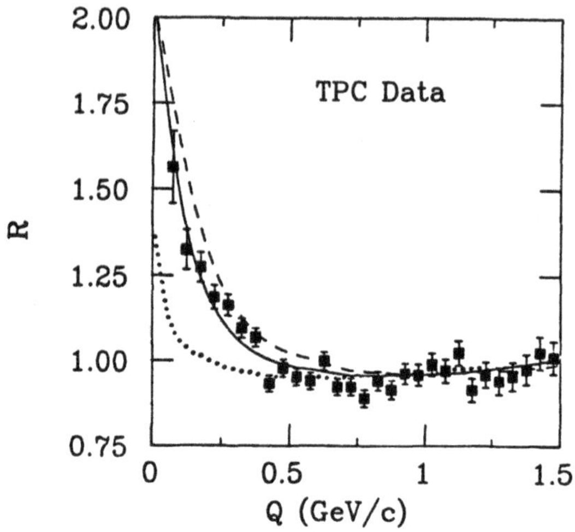


Fig.4. TPC data compared with calculations of Andersson and Hofmann.[7] The dotted curve represents their estimate of the effect of resonance propagation, but is unduly suppressed because of over-production of $\eta'$ in the generator.

effects of resonance production can be fairly cleanly separated into two: a short lived resonance such as the (predominant) $\rho$ [$c\tau = 1.3$ fm] will NARROW the distribution. Very long lived states, such as $\eta, \eta'$, will not show up at all; pairs containing a daughter are expected to show no Bose-Einstein correlation accessible to experiment. The effect of $\omega^0$ [$c\tau = 20$ fm] is in between. A range of 1.3 fm corresponds with $Q \sim 0.15$ GeV/c and 20 fm with $Q \sim 0.01$ GeV/c.

I showed[10] that the major part of the discrepancy between the resonance calculations of Andersson and Hofmann and the data may be attributed to an unrealistically large rate of $\eta'$ production in the Lund generator. Nonetheless, the TPC results still seemed broader than my calculations (fig.3). If the effects of rescattering within like pion pairs are also included, the discrepancy becomes too large to ignore. The (TPC) data are certainly consistent with a source SMALLER than the (local) string source and like pion rescattering BUT

(i) I do not WANT the source to be smaller than the string, and
(ii) since MOST pairs contain daughters of resonance decay, it seems likely that resonance propagation will not PERMIT an effective source size smaller than $\sim 1$ fm, 5 GeV$^{-1}$.

## 2. Calculations

I have investigated: (i) Effects of resonance propagation and decay; (ii) Effects of rescattering within the like sign pairs; (iii) Both together — rescattering after decay; (iv) Multiple particle rescattering. Illuminating and relatively SIMPLE calculations are made possible by the assumptions, supported by the data, that

(1) The source may be treated as spherically symmetric in frames where the pion momenta are small

(2) That the source is EXTENDED. The means that a rather clean separation can be made between pion (or resonance) direct production and the effects of rescattering AND that rescattering is easy to include. Particles are produced with separations $\sim 1$ fm but the range of the $\pi\pi$ interaction is $\sim 0.2$ fm $- 5^3 = 125$ so that the vast majority of particle pairs are produced with separations exceeding the range of $\pi\pi$ interaction (even for particles close in momentum). This is to be CONTRASTED with Coulomb final state interactions, which are long range.

Here are five steps in my calculations, of increasing complexity.

**STEP 1 Direct production of like pion pairs**

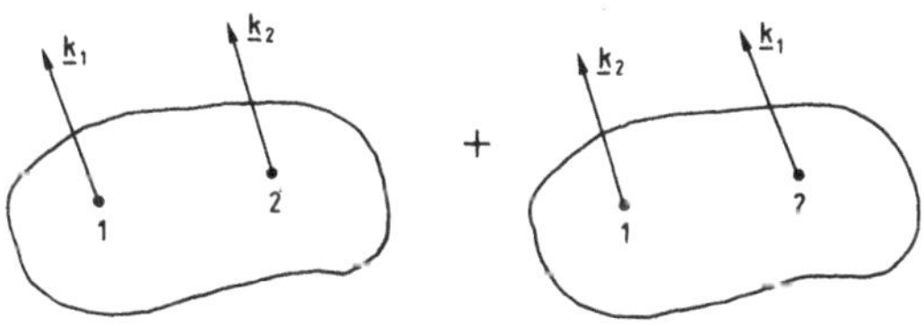


Fig.5. Bose symmetrised amplitude for direct production of two identical pions.

The Bose symmetrised amplitude corresponding to fig.5 is

$$A(\mathbf{k}_1,\mathbf{k}_2) = \frac{1}{\sqrt{2}} f(r_1) f(r_2) \left[ e^{-i\mathbf{k}_1.\mathbf{r}_1} e^{-i\mathbf{k}_2.\mathbf{r}_2} + e^{-i\mathbf{k}_2.\mathbf{r}_1} e^{-i\mathbf{k}_1.\mathbf{r}_2} \right] \tag{5}$$

For COHERENCE this amplitude is to be integrated over $r_1, r_2$ BEFORE squaring to obtain the intensity. For CHAOTIC amplitudes, square BEFORE integration. (Results such as those Weiner has recently emphasised are obtained if the function $f$ contains additive coherent and randomly phased parts). For CHAOS

$$P(\mathbf{k}_1,\mathbf{k}_2) = \left[ \int f^2(r) d^3r \right]^2 + \left[ \int f^2(r) e^{i\Delta\mathbf{k}.\mathbf{r}} d^3r \right] \left[ \int f^2(r) e^{-i\Delta\mathbf{k}.\mathbf{r}} d^3r \right] \tag{6}$$

$$C(\mathbf{k}_1,\mathbf{k}_2) = 1 + \tilde{\rho}^2(\Delta k) \tag{7}$$

where $\tilde{\rho}(\Delta k)$ is the normalised Fourier transform of the source INTENSITY.

**STEP 2 One pion directly produced, one resonance**

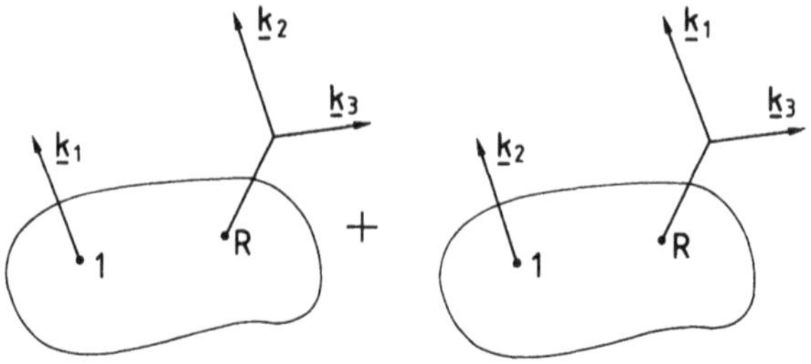


Fig.6. Bose symmetrised amplitude for direct production of a pion and a second identical pion the daughter of resonance decay.

The Bose symmetrised amplitude is shown in fig.6. Resonance propagation and decay is a COHERENT process; it is the pion and resonance direct production amplitudes which are taken to be chaotic. The three body amplitude corresponding to fig.6 is written as

$$A = f(x_1)f(X)e^{-ik_3x_2}\frac{e^{\,ik_Rx}}{r}\frac{1}{\sqrt{2}}\left[e^{-ik_1x_1}e^{-ik_2x_2} + e^{-ik_2x_1}e^{-ik_1x_2}\right] \tag{8}$$

$$x = (\mathbf{r}, t) \quad \mathbf{r} = \mathbf{r}_2 - \mathbf{R} \quad t = t_2 - t_R$$

The spherical wave

$$\frac{e^{\,ik_Rx}}{r}$$

is the propagation function for the resonance, a solution of the Klein-Gordon equation with a complex mass

$$k_R^2 = -(M_R - i\frac{\Gamma}{2})^2 \tag{9}$$

The variables $t$ and $r$ must be integrated over BEFORE squaring, and this integration selects the appropriate Breit-Wigner factors. After SQUARING and THEN integrating over the remaining variables $\mathbf{r}_1$ and $\mathbf{R}$ one obtains the obvious answer

$$P(\mathbf{k}_1, \mathbf{k}_2, (\mathbf{k}_3)) \propto BW(23)^2 + BW(13)^2 + \tilde{\rho}^2(\Delta k)2Re\left\{BW(23)^* BW(13)\right\} \tag{10}$$

which also emerges naturally from the string model.[9]

The SHAPE as a function of $\Delta k$ ($Q$) alone depends on the details of the three particle mass spectrum. For reasonable three particle mass spectra an approximately gaussian dependence due to the square of the Fourier transform of the source intensity is sharpened by the Breit-Wigner terms as illustrated in fig.7.

**STEP 3 Rescattering within a pair of directly produced particles**

If only a small fraction of pairs are produced within range of the pion-pion interaction, the effects of (s-wave) final state interactions may be evaluated very simply. Replace the plane waves which overlap the source distribution with the solution of the Schrödinger equation in the presence of the potential, but outside its range, which is asymptotic to a plane wave as $t \to +\infty$; that is, reverse the familiar phase shift treatment of scattering

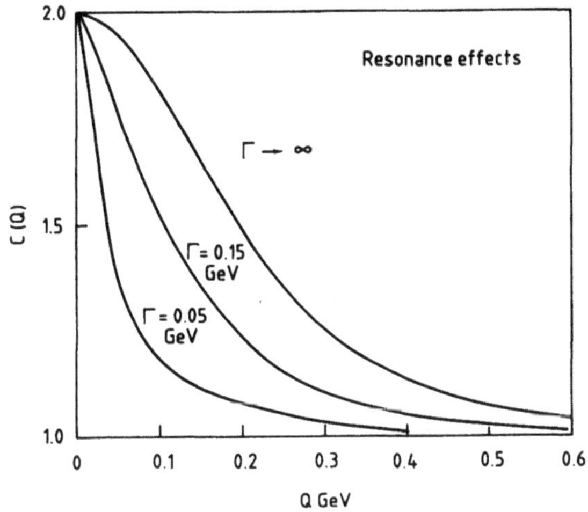


**Fig.7. Examples of the narrowing and sharpening of the correlation function as a result of resonance propagation and decay.**

by applying a phase shift of $-2\delta$ to the INCOMING (s-wave) part of the plane wave:

$$e^{-i\mathbf{k}_i\cdot\mathbf{r}_1}e^{-i\mathbf{k}_j\cdot\mathbf{r}_2} = e^{-i\mathbf{k}^+_{ij}\cdot\mathbf{r}^+_{12}}e^{-i\mathbf{k}^-_{ij}\cdot\mathbf{r}^-_{12}}$$

$$\Rightarrow e^{-i\mathbf{k}^+_{ij}\cdot\mathbf{r}^+_{12}}\left\{e^{-i\mathbf{k}^-_{ij}\cdot\mathbf{r}^-_{12}} + \frac{e^{i\delta_{ij}}\sin\delta_{ij}}{k^-_{ij}r^-_{12}}e^{ik_{ij}r_{12}}\right\} \tag{11}$$

$$\mathbf{k}^+_{ij} = \mathbf{k}_i + \mathbf{k}_j \qquad \mathbf{r}^+_{12} = \frac{\mathbf{r}_1 + \mathbf{r}_2}{2}$$

$$\mathbf{k}^-_{ij} = \frac{\mathbf{k}_i - \mathbf{k}_j}{2} \qquad \mathbf{r}^-_{12} = \mathbf{r}_1 - \mathbf{r}_2$$

Note that for SMALL $\delta_{ij}$

$$\sin\delta_{ij}/k^-_{ij} \;\rightarrow\; a_{ij} \quad \text{The SCATTERING LENGTH.}$$

With the above substitution and the assumption that the production intensity takes the form

$$f^2(r^-) \propto \exp\{-\alpha r^-\}$$

I found[11]

$$C(\mathbf{k}_1,\mathbf{k}_2) = 1 + \frac{1}{(1+\frac{Q^2}{\alpha^2})^2} + \frac{2\cos\delta\sin\delta}{k^-_{12}}\frac{\alpha}{1+\frac{Q^2}{\alpha^2}} + \frac{\sin^2\delta}{k^{-2}_{12}}\frac{\alpha^2}{1+\frac{Q^2}{\alpha^2}} \tag{12}$$

$$Q = |\mathbf{k}_1 - \mathbf{k}_2| = 2k^-_{12}$$

The second term is pure Bose-Einstein; the third and fourth contain the effects of rescattering within the like sign pair.

It is important to note that the rescattering term

$$\sim \frac{2a\alpha}{1+\frac{Q^2}{\alpha^2}}$$

is

(i) Proportional to the product of the SCATTERING LENGTH and the source size parameter $\alpha$

(ii) Of LONGER RANGE in $Q$ than the Bose-Einstein correlation.

An example of the results of these calculations is shown in fig.8. A correlation function of this kind fits the TPC data admirably, but requires $\alpha \sim 0.5$ GeV. This is roughly TWICE the size required for a match to the string model WITHOUT resonance narrowing and resonance narrowing must be present.

Rescattering within like sign pairs is particularly deadly for a reasonable source size, because it induces a zero in the overall enhancement in the sensitive region between $Q = 0.2$ and $Q = 0.3$ GeV/c.

**STEP 4 One resonance together with like pion rescattering.**

Steps 2 and 3 can be folded together.[12] I find the results, illustrated in fig.9, match well those obtained for an appropriately larger source WITHOUT resonance production. This result was not unexpected, but was not entirely obvious because of the coherent nature of resonance propagation and decay.

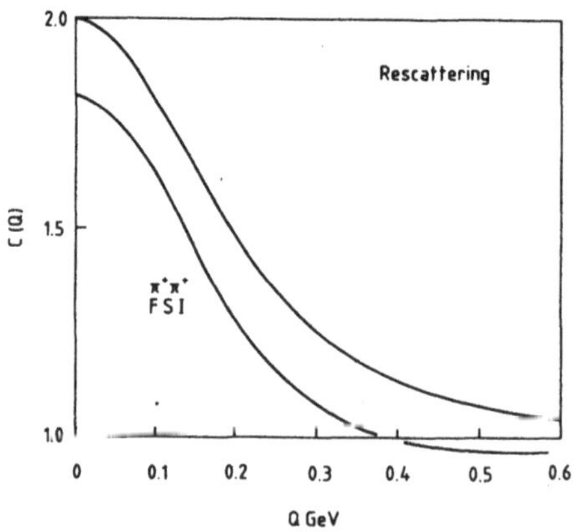


Fig.8. Example of the effect of like pion final state interactions on the correlation function

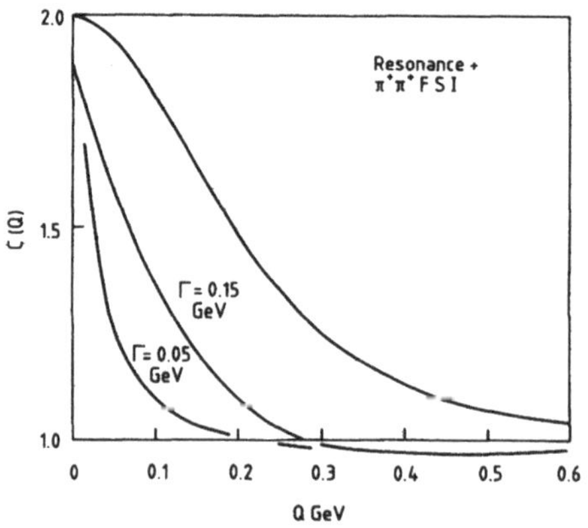


Fig.9. Example of the effects of both resonance production and rescattering of the like sign pions.

WITH BOTH RESONANCE NARROWING AND RESCATTERING WITHIN THE LIKE PION PAIR I HAVE NOT FOUND IT POSSIBLE TO MATCH PLAUSIBLY THE DATA.

A slightly optimistic estimate[12] is shown in fig.10. (The composition was chosen to match the upper histograms in fig.3 for $\alpha = 0.2$ and 0.3 GeV respectively). Increasing the parameter $\alpha$ does little good, because the effective size of the source is then controlled by the propagation of resonances.

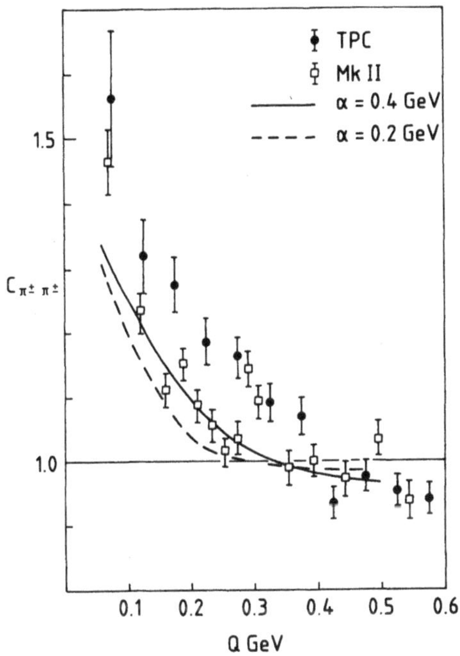


Fig.10. Estimates of the like sign correlation functions to be expected, when both resonance production and like sign final state interactions are taken into account, compared with Mark II and TPC data. The underlying source size (the potatoes of figs.5 and 6) is compatible with string model calculations for $\alpha = 0.2$ GeV but $\alpha = 0.4$ GeV is unreasonably large.

This is the reason for the title of this talk

"Are the observed Bose-Einstein Correlations possible?"

I think there is good reason to believe that in multiparticle production the effects of final state interactions among all pairs largely CANCEL and only the propagation and decay of resonances modifies the underlying source structure.

**STEP 5 Rescattering in all pairs of a three particle cluster**

Consider the waves

$$e^{-i\mathbf{k}_i.\mathbf{r}_1}e^{-i\mathbf{k}_j.\mathbf{r}_2}$$

which are overlapped with the source for the case of no rescattering. Factor into mean and relative coordinates, as in step 3. For rescattering within this pair

$$e^{-i\mathbf{k}^-.\mathbf{r}^-} \rightleftarrows e^{-i\mathbf{k}^-.\mathbf{r}^-} + e^{ik^-(r^-+a)}\frac{a}{r^-} \tag{13}$$

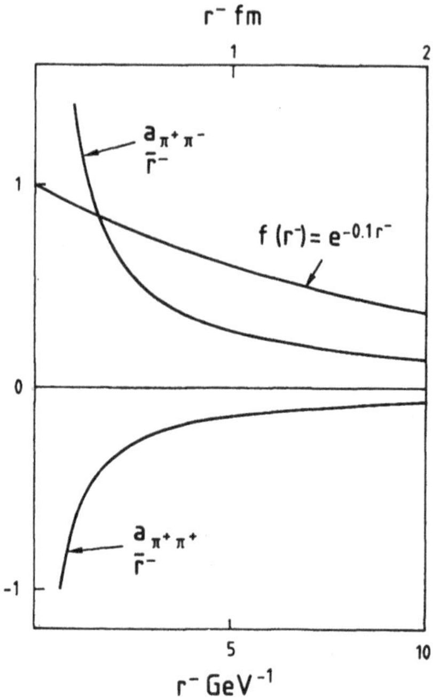


**Fig.11. The modulus of the correction to the two particle wave function due to s-wave rescattering, as a function of the relative coordinate.**

where $a$ is the (s-wave) scattering length. The correction term, sketched in fig.11, DILUTES the wave function for $r^-$ small if $a$ is negative (as for identical pions) but CONCENTRATES the wave function for $r^-$ small if $a$ is positive. If, then, we have several particles produced close enough for the long range correction factors to bite in ALL PAIRS, we have the possibility that in $\pi^+\pi^-\pi^+$ clusters the attractive $\pi^+\pi^-$ interaction will concentrate the (three particle) wave function more than the repulsive $\pi^+\pi^+$ interaction dilutes it, and thus obtain through MULTIPLE rescattering effects a net ENHANCEMENT at small $Q_{\pi^+\pi^+}$.

I have attempted to estimate the effect in the following way. If there is no rescattering, the product of three plane waves

$$e^{-i\mathbf{k}_i.\mathbf{r}_1}e^{-i\mathbf{k}_j.\mathbf{r}_2}e^{-i\mathbf{k}_k.\mathbf{r}_3}$$

is overlapped with the three particle source. I have replaced this with a modification for pairwise rescattering only, which is approximately

$$
\begin{aligned}
& e^{-i\mathbf{k}_i\cdot\mathbf{r}_1} e^{-i\mathbf{k}_j\cdot\mathbf{r}_2} e^{-i\mathbf{k}_k\cdot\mathbf{r}_3} \\
& + e^{ik_{ij}^-(r_{12}^- + a_{ij})} \frac{a_{ij}}{r_{12}^-} e^{-i\mathbf{k}_{ij}^+\cdot\mathbf{r}_{12}^+} e^{-i\mathbf{k}_k\cdot\mathbf{r}_3} \\
& + e^{ik_{ik}^-(r_{13}^- + a_{ik})} \frac{a_{ik}}{r_{13}^-} e^{-i\mathbf{k}_{ik}^+\cdot\mathbf{r}_{13}^+} e^{-i\mathbf{k}_j\cdot\mathbf{r}_2} \\
& + e^{ik_{jk}^-(r_{23}^- + a_{jk})} \frac{a_{jk}}{r_{23}^-} e^{-i\mathbf{k}_{jk}^+\cdot\mathbf{r}_{23}} e^{-i\mathbf{k}_i\cdot\mathbf{r}_1}
\end{aligned}
\qquad (14)
$$

I have evaluated, with Bose symmetrised amplitudes, the correlation function which results from this substitution,[12] employing a source intensity

$$f^2(r) \propto e^{-\alpha_0 r}$$

The vectors $\mathbf{k}_i$ were uniformly distributed over Dalitz plots drawn from a three pion mass distribution which rose from threshold to a maximum at a mass of $\sim$ 0.7 GeV/$^2$c and thereafter tailed off. Some results are shown in fig.12, which illustrates the importance of $\pi^+\pi^-$ final state interactions in determining the $\pi^+\pi^+$ correlation function and suggests that the repulsive $\pi^+\pi^+$ final state interaction may be cancelled or even overcompensated by attractive $\pi^+\pi^-$ interactions also present.

The odd pion may of course be a $\pi^\circ$, in which case all three pair interactions are repulsive. If equal numbers of $\pi^+\pi^-\pi^+$ and $\pi^+\pi^\circ\pi^+$ are considered, the effects of (s-wave) final state interactions cancel out in the two particle correlation functions. The deadly zero in the enhancement for $Q \sim 0.2 - 0.3$ GeV/c is removed. If it is indeed the case that this result obtains in the real world, then the observed Bose-Einstein correlations CAN be accounted for and the underlying source size is not obviously at variance with estimates from the string model, in the sense that the upper histograms of fig.3 are not unreasonable estimates.

Is there independent evidence that final state interactions cancel? There is, but it is not strong enough to be conclusive.

1) Correlations among like sign pairs extracted by normalising with unlike sign pairs AGREE with results extracted by normalising with artificial pairs in which the pions are drawn from different events (after trivial corrections).
2) I have seen data from pion production in proton — Aluminium interactions in which the correlation function for UNLIKE pions, normalised with pions drawn from different events, shows NO SIGN of a threshold ENHANCEMENT.[13] If any remain active in the experimental study of Bose-Einstein correlations in $e^+e^-$ annihilation or hadron-hadron collisions I commend to their attention the extraction of the correlation function for UNLIKE pairs.

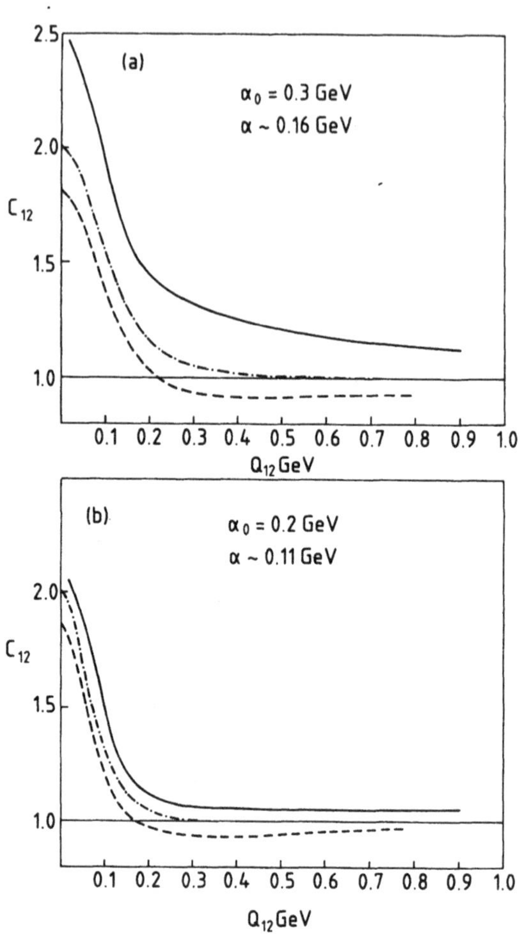


Fig.12. Estimates of the like sign correlation functions to be expected for effective source sizes characterised by $\alpha = 0.16$ GeV and $\alpha = 0.11$ GeV. The central curves are for no final state interactions, the broken curve for like sign pion rescattering only, the solid curves for rescattering within a like sign pair and the two unlike sign pairs of a triplet $\pi^+\pi^-\pi^+$.

## 3. Conclusion

The Bose-Einstein correlations observed in $e^+e^-$ annihilation and in proton-proton interactions are extremely complicated to interpret. Provided that final state interactions largely cancel, I conclude that they CAN be understood in terms of an underlying source for the generation of pions and resonances which has an effective size of the order of 0.5 — 1.0 fm. The very simple string model can account quite well for the observations and provides as fundamental an explanation for this phenomenon as exists. Further progress depends on acquisition of very high quality data, with large statistics, good momentum resolution, excellent particle identification and excellent resolution near the interaction point to facilitate removal of strange particle decay products. Very careful work is necessary in normalising the correlation function, to ensure that the extracted correlation can truly be compared with model calculations.

**References**

1. I. Juricic *et al*, Mark II Collab., *Phys. Rev.* **D 39** (1989) 1.
2. H. Aihara *et al*, TPC Collab., *Phys. Rev.* **D 31** (1985) 996.
3. T. Akesson *et al*, AFS Collab., *Z. Phys.* **C 36** (1987) 517.
4. J.L. Bailly *et al*, NA23 Collab., *Z. Phys.* **C 43** (1989) 341.
5. M. Althoff *et al*, TASSO Collab., *Z. Phys.* **C 30** (1986) 355.
6. M.G. Bowler, *Z. Phys.* **C 29** (1985) 617.
7. B. Andersson and W. Hofmann, *Phys. Lett.* **B 169** (1986) 364.
8. M.G. Bowler, *Phys. Lett.* **B 185** (1987) 205.
9. X. Artru and M.G. Bowler, *Z. Phys.* **C 37** (1988) 293.
10. M.G. Bowler, *Phys. Lett.* **B 180** (1986) 299.
11. M.G. Bowler, *Z. Phys.* **C 39** (1988) 81.
12. M.G. Bowler, *Z. Phys.* **C 46** (1990) 305.
13. K. Tokushuku, private communication.

# Bose-Einstein Correlations in Ultrarelativistic Heavy Ion Collisions

Michael Lahanas, Fachbereich Physik, University of Frankfurt
NA35 Collaboration

J. Baechler[8], J. Bartke[5], H. Bialkowska[12], R. Bock[6], R. Brockmann[6], P. Buncic[13], S. I. Chase[3], I. Derado[10], V. Eckardt[10], J. Eschke[7], C. Favuzzi[2], D. Ferenc[13], B. Fleischmann[6], M. Fuchs[6], M. Gazdzicki[11], H. J. Gebauer[10], E. Gladysz[4], C. Guerra[6], O. Hansen[14], J. W. Harris[3], W. Heck[7], M. Hoffmann[8], T. Humanic[6], S. Kabana[7], K. Kadija[13], A. Karabarbounis[1], R. Keidel[9], J. Kosiec[7], M. Kowalski[10], A. Kühmichel[7], M. Lahanas[7], J. Y. Lee[7], M. LeVine[7,14], A. Ljubicic jr. [13], S. Margetis[7], E. Nappi[2], G. Odyniec[3], G. Paic[6,13], A. D. Panagiotou[1,4], A. Petridis[1], A. Piper[9], F. Posa[2], H. G. Pugh[3], F. Pühlhofer[9], G. Rai[3], W. Rauch[3,10], R. Renfordt[7], D. Röhrich[7], H. Rothard[7], K. Runge[8], A. Sandoval[6], E. Schmidt[7], N. Schmitz[10], E. Schmoetten[8], I. Schneider[7], L. S. Schroeder[3], P. Seyboth[10], J. Seyerlein[10], E. Skrzypczak[11], P. Spinelli[2], R. Stock[7], H. Ströbele[7], L. Teitelbaum[3], A. Thomas[7], S. Tonse[3], G. Vassileiadis[1], G. Vesztergombi[10], D. Vranic[13], S. Wenig[7]

[1]Physics Dept., Univ. of Athens; [2]Dipartimento di Fisica, Univ. di Bari; [3]Lawrence Berkeley Lab.; [4]CERN; [5]Inst. of Nuclear Physics, Cracow; [6]GSI, Darmstadt; [7]Fachbereich Physik, Univ. Frankfurt; [8]Fakultät für Physik, Univ. Freiburg; [9]Fachbereich Physik, Univ. Marburg; [10]Max-Planck-Institut für Physik, München; [11]Inst. for Experimental Physics, Univ. of Warsaw; [12]Inst. of Nuclear Studies, Warsaw; [13]Rudjer Boskovic Inst., Zagreb [14]Physics Dept. Brookhaven Nat. Laboratory, Upton N. Y.

**Abstract**

Results from the analysis of Bose-Einstein correlations of negative pions produced in central collisions of S+S, S+Ag and O+Au at 200 GeV/N in a wide rapidity range $0.5 < y < 4.0$ are presented. At mid-rapidity a transverse source size significantly larger than the projectile radius is observed. A strong dependence of the transverse and longitudinal source size with rapidity, rapidity density and total multiplicity is observed.

## 1. Introduction

Geometry determines to a large extend the dynamics of the interaction and hadronization of the system formed in the collision of two nuclei at ultrarelativistic energies[1]. Pion interferometry is a direct method to determine the space-time characteristics of the pion source at freeze-out. This method is based on the fact that the effective source size (correlation length) is inversely proportional to the width of the two particle correlation function of the momentum difference of identical particle pairs. In a previous report we have presented[2] a correlation analysis of negative pions produced in O+Au at 200 GeV/N. With 16(50) participating projectile(target) nucleons this is the most asymmetric central collision system which we studied. Here we present new results for the symmetric system S+S and for the systems S+Ag and S+Au. The mid-rapidity density for these systems ranges from 28 to 85 negative particles per unit of rapidity.

## 2. Data Analysis

The data were recorded in the NA35 Streamer Chamber using the $^{16}O$ and $^{32}S$ beams of the CERN SPS. All negative tracks were measured on digitizer tablets or digital work stations. For details of the experimental setup see ref. 2,3. Due to lack of particle identification all negative particles have been treated as pions (10% of them are expected to be $K^-$, $\bar{p}$, and $e^-$ ). Table 1 summarizes the characteristics of the event samples used in this analysis. The S+Ag data were obtained by a medium

**Table 1** Summary of the characteristics of the event samples used in this analysis. *TET*: Transverse Energy Trigger; *FET*: Forward Energy Trigger; $\sigma$: Trigger cross section; $\sigma_{inel}$: Total inelastic cross section; $N$: Number of events analysed.

| $P_{lab}$ (GeV/c) | $A_p + A_T$ | Triggermode | $N$ events | $\sigma$ mbarn | $\sigma/\sigma_{inel}$ |
|---|---|---|---|---|---|
| 200 | O+Au | *FET* | 145 | 56 | 0.016 |
| 200 | S+S | *FET* | 408 | 34 | 0.020 |
| 200 | S+Ag | *TET* | 246 | 420 | 0.143 |
| 200 | S+Au | *TET* | 105 | 20 | 0.005 |

hard transverse energy trigger. The relatively wide multiplicity distribution enabled us to study the multiplicity dependence of the source parameters. In order to select a sub-sample of central collisions 91 S+Ag events with $N^- > 110$ have been chosen, with statistics comparable to the central O+Au sample. $N^-$ is the raw negative hadron multiplicity, before efficiency corrections which amount to a factor of about 1.5. The procedure of calculating the experimental two pion correlation function is described in ref. 2. The standard Gamov correction factor has been applied. For S+S also positively charged particles have been measured. The correlation of pairs of opposite charge, where Bose-Einstein correlations are absent, can be well described by the Coulomb attraction. The corresponding Gamov function is experimentally resolved giving an estimate of the pair momentum resolution which is sufficient to resolve objects of up to 20 fm. The experimental correlation function were compared to a model of a Gaussian static source, with the following parametrization:

$$C_2(q_T, q_L) = N[1 + \lambda \exp(-0.5 \cdot (q_T \cdot R_T)^2 - 0.5 \cdot (q_L \cdot R_L)^2)] \qquad (1)$$

$R_T(R_L)$ is the transverse (longitudinal) source size with respect to the beam axis, $\lambda$ describes the deviation of $C_2(0,0)$ from the expected value of 2 for a chaotic source.

$N$ is an overall normalization parameter. $q_T(q_L)$ is the transverse(longitudinal) momentum difference calculated in the overall center of mass system assuming a *clean cut geometry* at near zero impact parameter. A suggestion was recently made by G. Bertsch[4,5] to analyse the correlations in terms of the two orthogonal directions transverse to the beam, labeled **outward** $q_{Tout}$ (along the pair momentum vector sum transverse to the beam) and **sideward** $q_{Tside}$ (perpendicular to both the beam direction and $q_{Tout}$). A measure of the transverse source size along the outward direction will reflect any prolonged emission or long lived state. As expected from cascade or hydrodynamical models the distribution of the pion emission is lengthened by the quark-gluon-plasma phase transition. The experimental correlation function was compared to a model of a Gaussian source in these two components :

$$C_2(q_{Tout}, q_{Tside}) = N[1 + \lambda \exp(-0.5 \cdot (q_{Tout} \cdot R_{Tout})^2 - 0.5 \cdot (q_{Tside} \cdot R_{Tside})^2)] \quad (2)$$

The dependence of the correlation function on $q_L$ was eliminated by requiring $q_L < 40$ MeV/c for all pairs.

## 3. Experimental Results

Within a rapidity interval of unit length ($\Delta y = 1$) the parameters $R_T$, $R_L$ and $\lambda$ have been determined by shifting this window along the rapidity axis in steps of 0.2 units. The choise $\Delta y = 1$ is essential since the rapidity coherence length (the rapidity range over which the source elements contribute to the Bose-Einstein correlations) is of this order due to space-time and energy-momentum correlations present at high energies[6]. Fig. 1 shows the parameters $R_T$, $R_L$ and $\lambda$ as a function of $y$ for central O+Au and central S+Ag collisions. The arrows indicate the position of the overall c.m. system. A transverse radius about twice as large as in the target fragmentation region is observed at mid-rapidity for both systems. The enhancement observed in S+Ag is broader than in O+Au and the transverse source size in the target fragmentation domain is sligthly larger in S+Ag than in O+Au as expected. There is a positive correlation between $\lambda$ and $R_T$ for all systems. A change of the correlation function is observed as the rapidity window is shifted towards mid-rapidity. Starting from the target fragmentation region a broad Gaussian structure is seen which corresponds to a transverse source size of 4 fm. Closer to mid-rapidity a second component developo which corresponds to a source size of 7-8 fm. Both components are then present and a clear deviation of a Gaussian shape is observed. At mid-rapidity the first component vanishes and again a single Gaussian form can describe the shape of the correlation function. $R_L(y)$ is bellshaped around mid-rapidity for all systems (except for O+Au where $R_L$ stays almost constant beyond mid-rapidity). This form is expected for longitudinally expanding sources[1]. A signature for a expanding system also is the dependence of the effective source size with the transverse momentum $p_T$ of the pairs. For S+Ag $R_L$ decreases if $p_T$ is increased (see ref. 7).

If the velocity gradients of the expanding hadronic matter are large then also damped oscillations of the correlation function due to strong space-time and momentum-energy correlations can be expected[1]. If the period of the oscillations

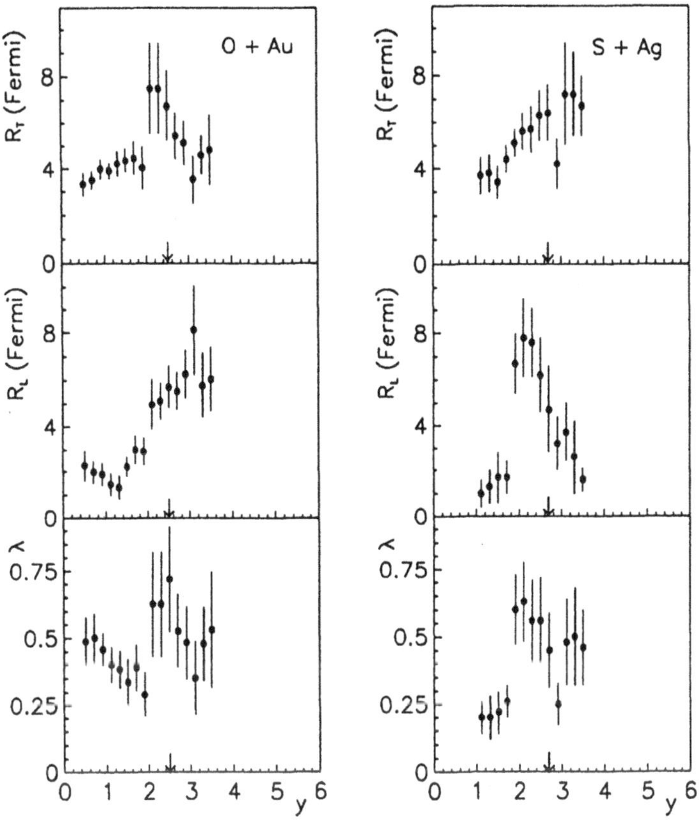


**Figure 1** Rapidity dependence of $R_T$, $R_L$, and $\lambda$ for O+Au and S+Ag. Each point corresponds to one unit of rapidity window thus the points are correlated.

depends on the total multiplicity a complicated shape could follow from the averaging of events with a variety of multiplicities. Even a shift of the maximum of the correlation function at $q_L = 0$ may be expected[8]. This is indeed seen in the S+S data. Fig. 2 shows, as an example, the correlation function $C(q_L)$ in the interval $1.6 < y < 2.6$. Also shown are two transverse correlation functions in adjacent rapidity regions. They illustrate a problem encountered in the data analysis: the single Gaussian fit accounts well for the interval $1.2 < y < 2.2$, with $R_T = 5.7$ fm whereas no fit is obtained at all in the region $2 < y < 3$ due to an upward oscillation at $q_T \approx 70$ MeV/c. A similar pattern is observed in the transverse correlation functions of S+Au (Fig.2). In the two domains, below and above the center of mass rapidity reasonable fits are obtained, with $R_T$ values of about 5 fm, whereas the superposition of a sharp maximum and broad shoulder structure prevents a successful fit in the

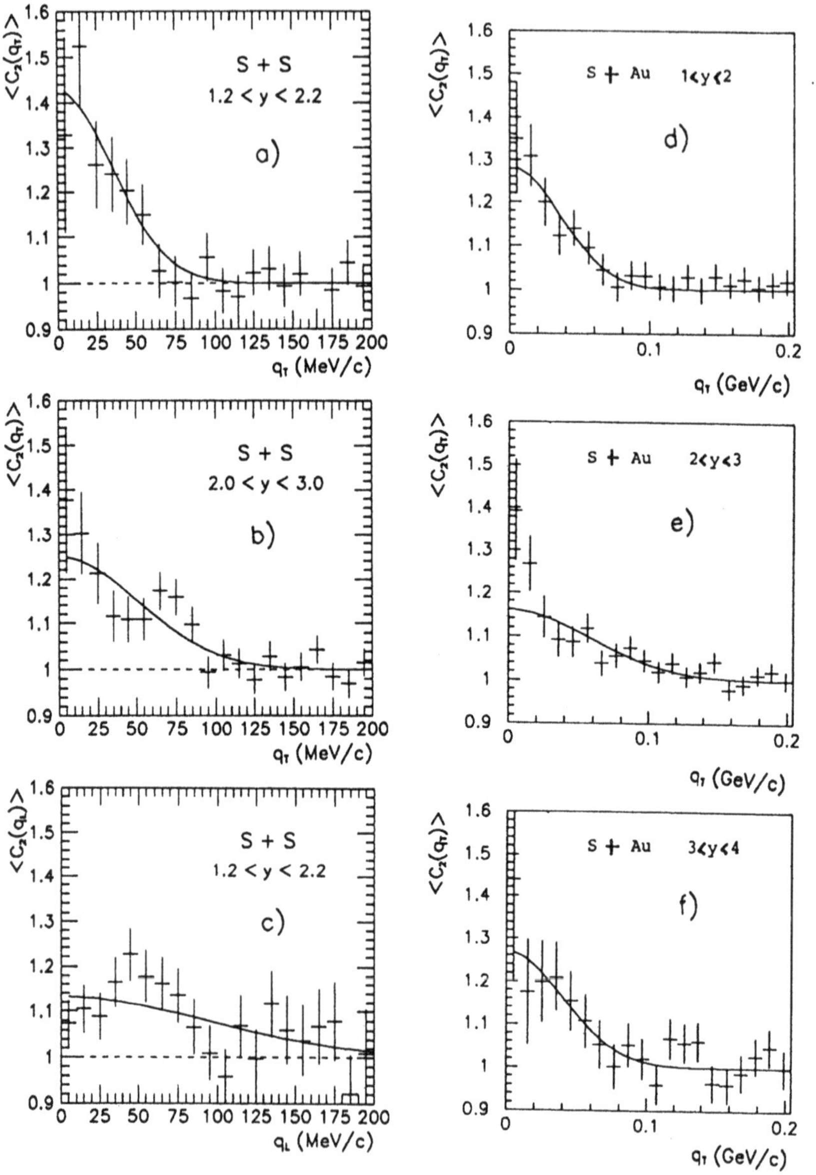


**Figure 2** a, b) Transverse- c) longitudinal correlation functions for S+S in three rapidity intervals. d, e, f) Transverse correlation functions for S+Au in three rapidity intervals.

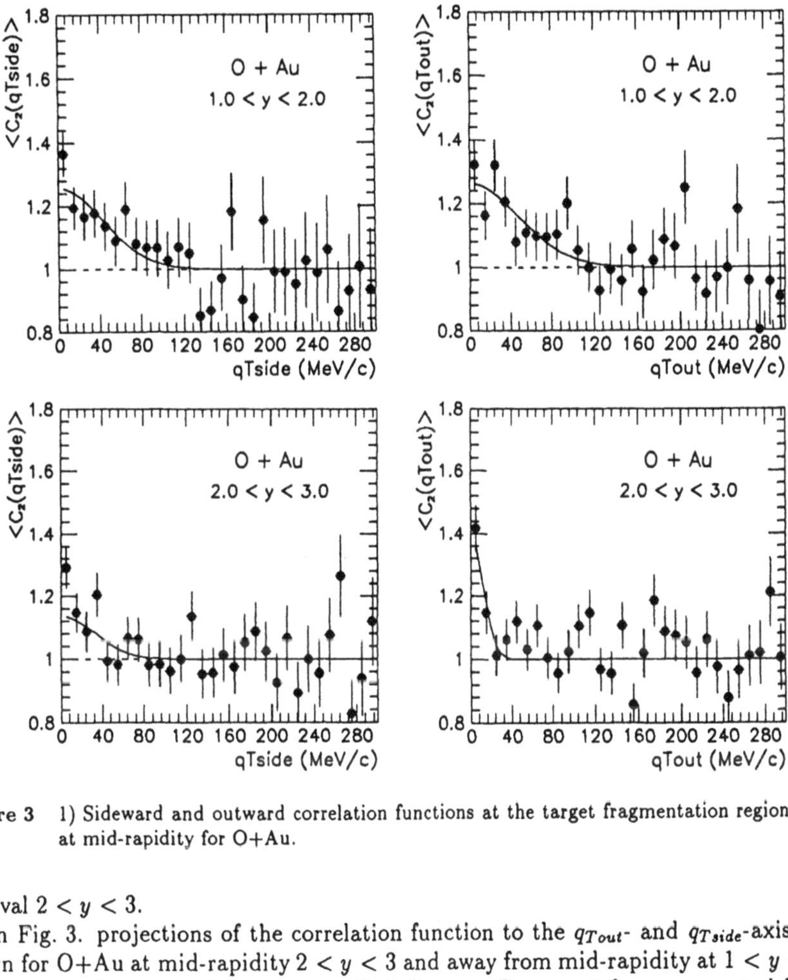


**Figure 3** 1) Sideward and outward correlation functions at the target fragmentation region and at mid-rapidity for O+Au.

interval $2 < y < 3$.

In Fig. 3. projections of the correlation function to the $q_{Tout}$- and $q_{Tside}$-axis are shown for O+Au at mid-rapidity $2 < y < 3$ and away from mid-rapidity at $1 < y < 2$. In the target fragmentation region we find $R_{Tout} \approx R_{Tside} \approx 4$ fm, as expected for a hadron gas, whereas at mid-rapidity $R_{Tout} = 16.6 \pm 3.2$ fm and $R_{Tside} = 6.2 \pm 0.8$ fm. This data suggest a long-lived state at mid-rapidity due to the large outward source size.

A comprehensive Monte Carlo investigation of possible experimental biases is underway, along with the accumulation of higher statistics. Also further theoretical studies may come up with more suitable source parametrizations. At present we can thus not report meaningful rapidity dependences of source parameters for the systems S+S and S+Au, such as given for O+Au and S+Ag reactions in Fig. 1.

## 4. Conclusions

Pion interferometry is applicable even for nonequilibrium processes. The idealized theory which is applicable only for static sources has to be extended, in order to take the nontrivial kinematical effects of the nonstatic sources in ultrarelativistic nuclear collisions into account[7]. Our analysis shows a complex rapidity dependence of the space time characteristics of the source. Deviations from a Gaussian shape of the correlation functions are seen. Geometry could explain partially the differences observed between the systems. Other differences could be of dynamical origin such as the large transverse source size at mid-rapidity in O+Au, S+Ag and S+Au in comparison to the S+S data. Higher statistics should establish the expansion mechanism of the source. The information derived from the study of the correlations of identical bosons may thus provide a space-time picture of the evolution of excited, extended and compressed hadronic matter.
**Acknowledgment:** I would like to thank Prof. R. Weiner for the invitation to the CAMP Workshop in Marburg.

## References

1. Yu. M. Sinyukov, *Nucl. Phys.* **A498** (1989)151c
2. A. Bamberger et al., *Phys. Lett.* **B203** (1989)320
3. A. Bamberger et al., *Z. Phys.* **C38** (1988)89
4. G. Bertsch and G. E. Brown, *Phys. Rev.* **C40** (1989)1830
5. G. Bertsch et al., *Phys. Rev.* **C37** (1988)1896
6. W. Hoffmann, *LBL Report* **23108** (1987)
7. D. Ferenc, *in this proceedings*
8. Y. Hama and S. S. Padula, *Phys. Rev.* **D37** (1988)3237

# PION INTERFEROMETRY IN S+Ag COLLISIONS AT 200 GeV/NUCLEON

Daniel Ferenc

Rudjer Boskovic Institute
Zagreb, Croatia, Yugoslavia
and
**NA35 COLLABORATION, C E R N**

J.Bartke[5],H.Bialkowska[12],R.Bock[6],R.Brockmann[6],P.Buncic[13],S.I.Chase[3],I.Derado[10], V.Eckardt[10],J.Eschke[7],C.Favuzzi[2],D.Ferenc[13],B.Fleischmann[6],M.Fuchs[6],M.Gazdzicki [12],H.J.Gebauer[10],E.Gladysz[5],C.Guerra[6],J.W.Harris[3],W.Heck[7],T.Humanic[6],K.Kadija[13], S.Kabana[7],A.Karabarbounis[1],R.Keidel[9],J.Kosiec[7],M.Kowalski[5,10],A.Kuhmichel[7],M. Lahanas[7],J.Y.Lee[7],A.Ljubicic[13],S.Margetis[7],E.Nappi[2],G.Odyniec[3],G.Paic[6,13],A.D. Panagiotou[1],A.Petridis[1],J.Pfennig[4],A.Piper[9],F.Posa[2],H.Pugh[3],F.Puhlhofer[9],G.Rai[3],W. Rauch[10],R.Renfordt[7],D.Rohrich[7],H.Rothard[7],K.Runge[8],A.Sandoval[6],E.Schmidt[7], N.Schmitz[10],E.Schmoetten[8],I.Schneider[7],L.S.Schroeder[3],P.Seyboth[10],J.Seyerlein[10],E. Skrypczak[11],P.Spinelli[2],R.Stock[7],H.Stoebele[7],L.Teitelbaum[3],A.Thomas[7],S.Tonse[3],G. Vassileiadis[1],Vesztergombi[10],D.Vranic[13],S.Wenig[7]

1) Phys. Dept. Univ. of Athens, 2)Dipartmento di Fisica, Univers. Bari, 3)Lawrence Berkeley Lab.,4)CERN, 5)Inst. of nucl. phus. Cracow, 6)GSI Darmstadt, 7) Fachbereich Physik, Univ. Frankfurt, 8) Fakultaet fuer Physik, Univ. Freiburg, 9)Univ. Marburg, 10)Max-Planck Inst. Munchen, 11)Inst. of Experim. Phys.Univ. Warsaw, 12)Inst. of Nucl. Studies, Warsaw, 13)Rudjer Boskovic inst. Zagreb

ABSTRACT

Dependence of the longitudinal size parameter on rapidity and transverse momentum has been investigated. Some phenomena predicted for hydrodynamically expanding systems have been observed. Few kinematical effects have been also demonstrated.

## 1. Introduction

Since different dynamical models predict different space-time evolution scenarios for the system created by ultrarelativistic nuclear collision, the information about the space-time evolution of the system is of great importance. Pion interferometry may yield interesting information about the space-time development of the system[1]. Relativistic nature of the system created by ultrarelativistic nuclear collision produces certain difficulties in measuring the "real" size of the system[2], but on the other hand enables one to learn very much about the internal motion of material in the system, as pointed out by many authors[1,2,3]. In this contribution, we applied some of those ideas to the preliminary S+Ag data[4].

*1.1. Data and triggers:* Events from the NA35 Streamer Chamber have been recorded on a 70 mm film. Tracks of negative particles have been measured by an interactive spatial reconstruction method. In our analysis, all negative particles are assumed to be pions (there is about 10% contamination coming from electrons, kaons, etc.). S+Ag data (272 events) were obtained by a medium hard transverse energy trigger, which selects semi-central collisions. Average corrected negative rapidity density at Y=2.5 is $dN/dY \cong 60$.

*1.2. Variables and Parametrizations:* Three different parametrizations of the correlation function shall be presented. In the simplest case, the correlation function is supposed to depend on a single variable, the invariant momentum difference of the pair, $Q_I$. The exponential function fits the experimental data much better than the Gaussian[4,7]. In section 3. we shall discuss some problems connected to the interpretation of the radius parameter measured in this parametrization. In the second parametrization, the correlation function is assumed to depend on two variables: the transverse and the longitudinal momentum difference, $Q_T, Q_L$:

$$C(Q_T,Q_L) = 1 + \lambda \exp(-Q_L^2 R_L^2/2) \exp(-Q_T^2 R_T^2/2) \, . \qquad (1)$$

Since the longitudinal direction is defined to be parallel to the collision axis, $Q_L$ is not Lorentz invariant to longitudinal boosts. On the other hand, $Q_T$ is not invariant to boosts along the transverse momentum of the pions. More precisely, one component of $Q_T$ vector is invariant ( the one which is perpendicular to the total transverse momentum of the pion pair), and the other is not ( the one which is parallel to the total transverse momentum of the pion pair). We shall use the nomenclature of Bertsch[5] for these two components: $Q_{Tside}$, and $Q_{Tout}$ respectively, so the correlation function may be expressed as:

$$C(Q_T,Q_L) = 1 + \lambda \exp(-Q_L^2 R_L^2/2) \exp(-Q_{Tside}^2 R_{Tside}^2/2) \exp(-Q_{Tout}^2 R_{Tout}^2/2). \quad (2)$$

*1.3. Fitting procedure:* Correlation functions Eq.1. and Eq.2. were fitted to the experimental data by the maximum likelihood method in the corresponding momentum space, and also projections of small-momentum difference slices to each momentum difference axis were fitted by the least squares method. It was systematically checked whether: (a) the results of two fitting techniques are consistent, (b) the errors are consistent, (c) chaoticity parameters from Eq.1. and Eq.2. are consistent, etc.

## 2. Longitudinal radius as a function of rapidity and transverse momentum

Since in the modern relativistic hydrodynamics theory[6] longitudinal expansion is expected to be the dominating process, longitudinal radius dependence on rapidity should be an important source of information. Makhlin and Sinyukov[3] showed that from the $R_L(y)$ distribution one may draw conclusions about the character of the source of pions. In order to demonstrate that qualitatively, let us discuss briefly two extreme scenarios for the space-time development of the source of pions in nuclear collisions. First, let us assume that the source of pions is static, without internal motion of the local elementary sources[3]. Then all the local sources contribute equally to any kinematically allowed region of phase space of emitted pions. Pions originating from distant parts of the source may have therefore similar momentum vectors, which is a necessary prerequisite for pion interferometry. Consequently, the real size of the source may be measured, and the measured value does not depend on the phase space region of the pions used in the measurement. A flat dependence of the longitudinal radius $R_L$ on rapidity is expected: $R_L$=R(real)=const(y). Rigorous calculations of Makhlin and Sinyukov[3] show that, due to relativistic effects, previous conclusion holds only when momenta are calculated in the reference frame in which the source is at rest. If, on the other hand, velocity of the "observer" with respect to the source is not equal to zero, then one measures[3]:

(a) Lorentz contracted $R_L$ values, using pions of velocities pointing to the opposite side than the velocity of the "observer" (viewed from the rest system of the source), and

(b) "Lorentz extended" $R_L$ values, using pions of velocities similar to the velocity of the "observer".

Simply speaking, the "observer" automatically assumes that he is the source himself, and therefore overcorrects or undercorrects for the relativistic effect, depending on his velocity (rapidity) relative both to the velocity of the source, and to the velocity of the particles used in the measurement. Since this interesting effect is of pure kinematical nature, it has to hold not only for the longitudinal direction, but for any direction in space. In section 3. we shall demonstrate the "Lorentz extension" effect in the transverse direction, and comment on the meaning of the radius extracted from Q-invariant parametrization, in which the "observer" is practically never in the rest frame of the source, but "flyis" together with each pair of pions.

Let us next discuss the other extreme scenario of space-time development of the source of pions, namely the Inside Outside Cascade model[6], where the source is a cylinder which rapidly expands in longitudinal direction during the emission of particles. There is a very strong correlation between the space-time position of an elementary local source and its longitudinal velocity. Like the galaxies in the Universe, all the elementary sources move (longitudinally) away from each other, the faster, the larger the distance between them. Due to a sort of the red shift effect, it is hard to find two pions with equal longitudinal momenta, which originate from two distant local sources. Pion interferometry, which needs pions of nearly equal momenta, measures therefore that short section of the whole source, which is short enough to be capable of radiating pions with similar momenta from "both ends".

In a general case of a hydrodynamic longitudinal source expansion, when the "observer" is fixed in rapidity to some arbitrary value (e.g. $y_{obs}\equiv 0$), the effective measured longitudinal size as a function of the average longitudinal momentum of the pion pair used in the measurement equals[3]:

$$R_L(P_L) = R_L(P_L = 0) / \cosh(y), \qquad (3)$$

$$R_L(P_L{=}0) = (2T_f/m_T)^{1/2}\,\tau(y)\,, \qquad P_L = (P_{L1}+P_{L2})\,/2, \qquad y = (y_1+y_2)\,/2\,,$$

where cosh(y) is the Lorentz boost parameter from the reference frame of the "observer" to the longitudinal rest frame of the pion pair, $\tau(y)$ is dependence of the proper time on rapidity ( in the Scaling model $\tau$=const(y)), $T_f$ is the freezeout temperature and $m_T$ is transverse mass of pions used in the measurement. We can see

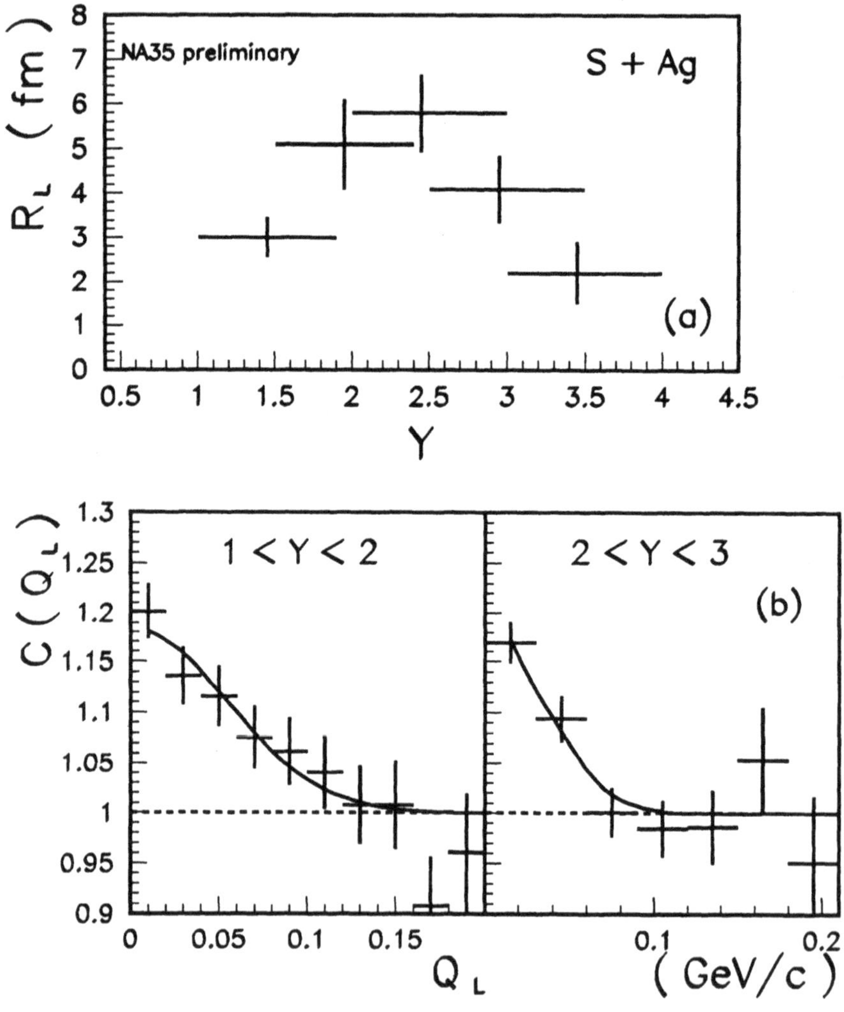


**Fig.1.** Longitudinal size as a function of rapidity

that one expects to find a bell shaped $R_L(y)$ dependence with a maximum $R_L(P_L{=}0)$ at rapidity of the "observer", ($y_{obs}{=}0$ here). Bell shaped form comes mainly from Lorentz contraction of the local source radius, measured by the "observer" which is not at rest in the frame of each local source. In order to correct for the Lorentz contraction effect, we have introduced[4] a concept of the "sliding observer frame", the reference frame which is not fixed to any particular rapidity, but adjusts itself to each individual pion pair's longitudinal center of momentum frame. By definition, we have always $P_L{=}0$, and therefore we measure $R_L(P_L{=}0)$, (see Eq.3.). If e.g. the pure Scaling hydrodynamics takes place and if we neglect the source boundaries' effect, by assuming an infinitely long source, a flat $R_L(y)$ distribution should be expected, since in the Scaling model, proper time does not depend on rapidity. At any rapidity, one should measure the same effective length, (the so called correlation length). However, since the source is not infinitely long, at high and low rapidities (and consequently at forward and backward ends of the source), the "measuring window sticks out" from the ends of the source and the measured $R_L$ values have to be smaller than the correlation length[3].
Fig.1.a. shows the $R_L(y)$ dependence for S+Ag collisions, and Fig.1.b. shows two examples of the correlation functions projected to the $Q_L$ axis ($Q_{Tside}$, $Q_{Tout}$<60 MeV/c). Events of all multiplicities were taken together, because it seems that qualitative behavior does not depend strongly on multiplicity. Fig.1 suggests that the source is probably a hydrodynamically expanding system, with the effect of the source boundaries and(or) of non-scaling hydrodynamics.

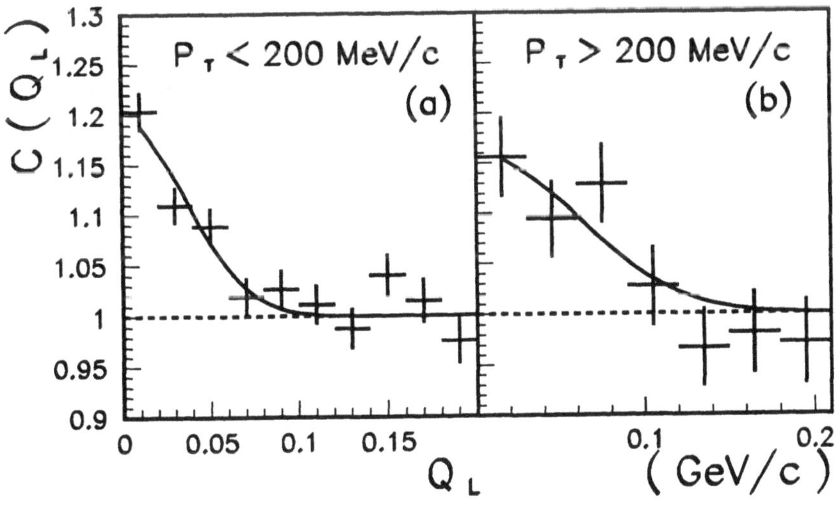


**Fig.2.** See text for explanation. The extracted radii are $R_L$=5.4±0.5 fm, and 3.2±0.6 fm for (a) and (b) respectively

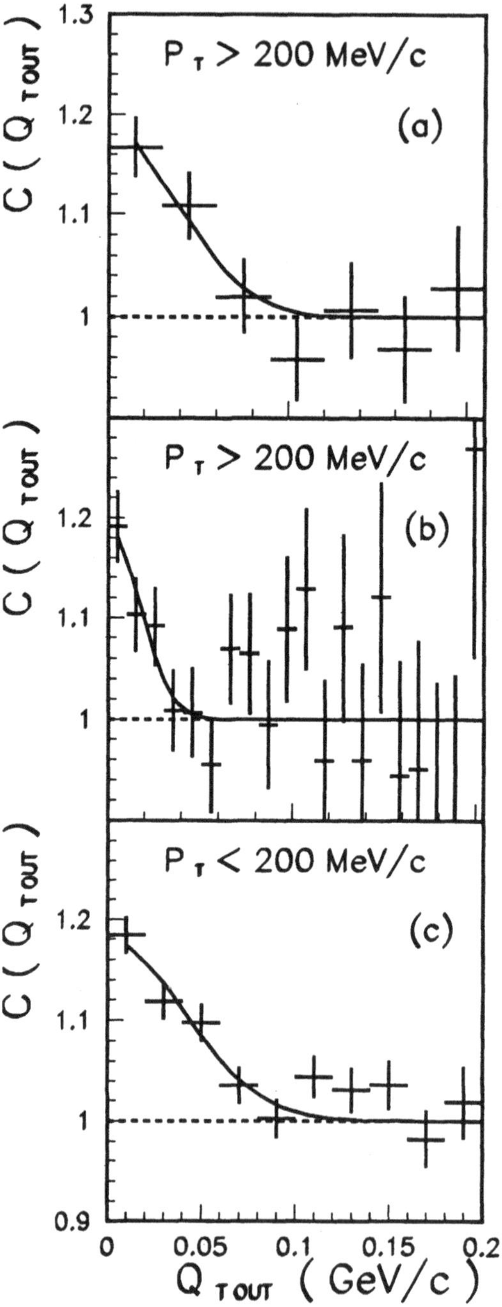


Let us now discuss the second necessary signature of any hydrodynamic system, namely dependence of the effective $R_L(P_L=0)$ on $m_T$, the transverse mass of the pions used in the measurement[3,1,2], see Eq.3.. In Fig.2. two plots are shown, both correspond to S+Ag data, in a wide rapidity range 1<y<4. In Figs. 2.a and 2.b projections of the correlation functions to the $Q_L$ axis ($Q_L$ was calculated in the "sliding observer frame") are shown, for high $p_T$ particles ($p_T$>200 MeV/c), and for low $p_T$ particles ($p_T$<200 MeV/c), respectively. Within our present statistics, we see that the effective size is smaller for higher transverse mass. Ratio of the extracted radii is 1.7, close enough to 1.5, the approximate expected value equal to the square root of the ratio of the average(!) transverse masses for the two data subsets. However, a more detailed analysis, with a careful investigation of the effects of the acceptance is still needed, as well as better statistics.

### 3. Transverse Lorentz-effects and Q-invariant

Let us first demonstrate the "Lorentz extension effect"[3], described in section 2., but in transverse (outward) direction.

**Fig.3.** See text for explanation. $R_{Tout}$=5.0±0.8 fm, 11.5±2.0 fm, and 5.1±0.5 fm for (a),(b) and (c)

If the transverse velocity of the observer is zero with respect to the source, and if we assume that the source of pions is not expanding rapidly in transverse direction, then the measured transverse radius should be independent of the transverse momentum of the particles used in the measurement, and equal to the real radius of the source $R_T(p_T)$= $R_T$(real)= const($p_T$). On the other hand, the "transverse Lorentz extension" should be present when the observer travels away from the source, together with each pair used in the measurement. The effect should be weak when low $p_T$ pions are used (because transverse boost parameter is almost equal to one), and stronger when high $p_T$ pions are used. In Fig.3. three projections of correlation functions to the $Q_{Tout}$ axis are presented, for S+Ag data in a wide rapidity interval $1<y<4$. Fig.3.a shows the correlation function for high $p_T$ particles ( $p_T$>200 MeV/c). $Q_{Tout}$ was calculated in the LAB frame, or the rest frame of the source. Fig.3.b. shows projected correlation function for the same data as Fig.3.a., but here $Q_{Tout}$ was calculated in the transverse center of momentum frame for each individual pair of particles. The "observer" flyes away from the source at the speed of the pair. The "transverse Lorentz extension" effect is present now: the OUT-transverse radius is much larger than in Fig.3.a. (the correlation function is narrower). Ratio of the measured radii is 2.3, which is relatively close to 3., the approximate Lorentz boost factor calculated from the mean(!) $p_T$ values. Finally, Fig.3.c. shows the low $p_T$ selection, and $Q_{Tout}$ was again calculated like for Fig.3.b.. The extension effect is weak, as we expect, because the transverse boost is weak.
Let us stress that the effect is very important for understanding the Q-invariant parametrization, where the "observer" is always in the center of momentum frame of each pion pair. So, depending on the orientation of the momentum difference vector with respect to the total momentum vector of the pair, and also on the magnitude of the total momentum, one measures different sizes of the same source, at the same time.

**4. Conclusion:** Dependence of the longitudinal size parameter on rapidity and transverse momentum for S+Ag collisions has been investigated. Although better statistics and deeper investigation of acceptance effect is still needed, it seems that some predicted phenomena[1,2,3], which characterize hydrodynamic systems, have been observed. Dependence of results on the chosen reference frame has been also discussed.

**Acknowledgements**
I would like to thank Prof.R.Weiner for the invitation to CAMP. Also, I would like to thank my NA35-colleagues, particularly G. Paic, for many helpful discussions. This work was supported in part by the Commission of European Communities, contract No CI-0250 Yu(A).

**References**
1. S.Pratt, Phys.Rev. D33,(1986)1314
2. Y.Hama and S.Padula, Phys.Rev. D37,(1988)3237
3. A.N.Makhlin and Yu.M.Sinyukov, Z.Phys. C39,(1988)69, and references therein
4. D.Ferenc NA35 collaboration, XXV Rencontres de Moriond, Les Arcs11.-17.3.1990., to be published
5. G.Bertsch, M.Gong, M.Tohyama, Phys.Rev. C37(1988)1896
6. J.D.Bjorken, Phys.Rev. D27,(1983)140
7. T.Csorgo et al., SPACER event generator, private communication

# PROTON PROTON CORRELATIONS AT SMALL RELATIVE MOMENTUM IN NEON AND ARGON COLLISION WITH NUCLEI AT SATURNE

J.P. ALARD, P. DUPIEUX, P. CHARMENSAT, J. AUGERAT, N. BASTID, F. BIAGI, L. FRAYSSE, J. MARRONCLE, G. MONTAROU, P. MOREL, M.J. PARIZET, D. QASSOUD, A. RAHMANI,

*LPC Clermont-Ferrand, IN2P3/CNRS, Université Blaise Pascal F-63177 AUBIERE CEDEX FRANCE*

and

R. BABINET, C. CAVATA, M. DEMOULINS, J. GOSSET, D. L'HOTE, B. LUCAS, J. POITOU, Y. TERRIEN, O. VALETTE,

*DPHN, CEN Saclay, F-91191 GIF-SUR-YVETTE CEDEX*

and

M.C. LEMAIRE

*LNS, F-91191 GIF-SUR-YVETTE CEDEX*

and

F. BROCHARD, P. GORODETZKY, C. RACCA

*CRN, BP 20CR, F-67037 STRASBOURG CEDEX*

## ABSTRACT

Proton proton correlations at small relative momentum have been studied using the Diogene $4\pi-$ Facility at Saturne. The incident energies vary from 200 to 1000 A.MeV for Ne and Ar projectiles on various targets. Experimental correlation functions are compared to theoretical calculations of S. Koonin, corrected for experimental biases. We give the values of the source size as a function of incident energy, target-projectile system and centrality of the collisions. It is shown that the method is mainly sensitive to the final stage of the interaction ; nevertheless, we have some indications of a highly compressed phase at the beginning of the collision.

## Introduction

Two particle correlations at small relative momentum can be an useful tool for the extraction of the size and lifetime of the source in relativistic nuclear collisions. If we can measure the number of participant nucleons in the source, we also can extract the value of the density of nuclear matter ; such an information is of fundamental interest for our knowledge of the equation of state of hadronic matter.

Correlations between two particles with nearly the same momentum are sensitive to the space time structure of the source, through Coulomb and nuclear interactions and statistical effects. Interesting results have been obtained showing evidence for a great difference between low incident energy collisions[1,2] (hot spot) and very high energy ones[3] (possible signature of QGP). In the range of Saturne energies (200 – 1 000 A.MeV) many results[4,5] confirm the validity of the participant spectator model and the existence of a highly excited fireball. Such studies are important both for our knowledge of the properties of hadronic matter (Equation of state) and for an approach of in medium effects.

In such analyses, $4\pi$ detectors allow to study the correlation function versus other observables (centrality for instance). But due to the complexity of such detectors, it is necessary to make a lot of corrections in order to take into account the efficiency of the experimental set up. The Diogene $4\pi$ facility[6] has been used for correlation measurements at Saturne[7].

## I – Proton-proton correlations.

The correlation function $C_2$ is calculated as the ratio of the spectra of "correlated protons" (belonging to the same event) to "uncorrelated protons" (belonging to different events) for each channel of relative momentum $\Delta p$ $[\vec{\Delta_p} = (\vec{p_1} - \vec{p_2})/2]$. Experimental correlation functions are then compared to Koonin calculations[8] distorted by experimental biases in order to extract the proton source radius $r_0$. Experimental and theoretical curves are normalized to each other for the three points $\Delta p = 80 - 90 - 100$ MeV/c.

*Results for the source size* $r_0$ :

(i) – *Discussion about the lifetime of the source*

When extracting $r_0$ values from the comparison of experimental and theoretical curves, we have to take into account the lifetime $\tau$ of the source. It is possible, in principle, to evaluate the lifetime of the source too from p-p correlations. In practice, and looking at the case $\tau \neq 0$ in the Koonin calculations[8], it becomes very difficult, due to necessary drastic experimental selections.

We have evaluated lifetime values using the intra-nuclear cascade code of J. Cugnon[9]. For each event, we define time origin at the first collision of any particle of the system and we draw the histogram of the last collision time (emission) for each proton inside the Diogene acceptance. The lifetime of the source is simply defined as the rms parameter of this spectrum, following the Koonin assumption. Lifetime values from 4 – 7 fm/c are obtained for the various analysed systems and energies. Running the Koonin code with these values, we found very small deviation $\Delta r_0/r_0$ of the order of 3% for the NaF target and 5% for the Pb one. Thus, in the following, we neglect the lifetime of the source.

(ii) – $r_0$ *values without any selection of the data*

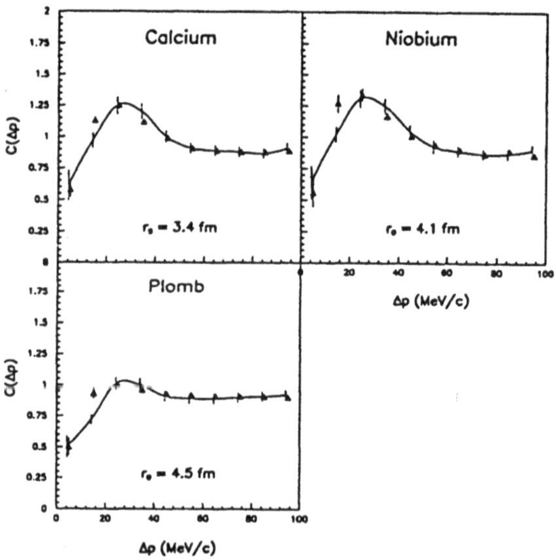


**Fig. 1 :** Proton–Proton correlation functions for Argon projectiles at 400 A.MeV.

We show on Fig. 1 some example of experimental correlation functions, with the associated theoretical ones, concerning the systems Ar+(Ca – Nb – Pb) at 400

MeV per nucleon. All the $r_0$ values are summarized on Fig. 2 and 3, respectively for Ne + (NaF – Nb – Pb) and Ar + (Ca – Nb – Pb) versus incident energy. It appears clearly that the obtained values are independent of incident energy, for a given system. It is also clear that the heavier the system, the greater $r_0$. Such results agree with a geometrical description of the collision, like in participant–spectator concept, although all obtained $r_0$ values correspond to equivalent spherical radii ($R_s = \sqrt{\frac{5}{2} * r_0}$) bigger than the classical hard sphere radii of the colliding system.

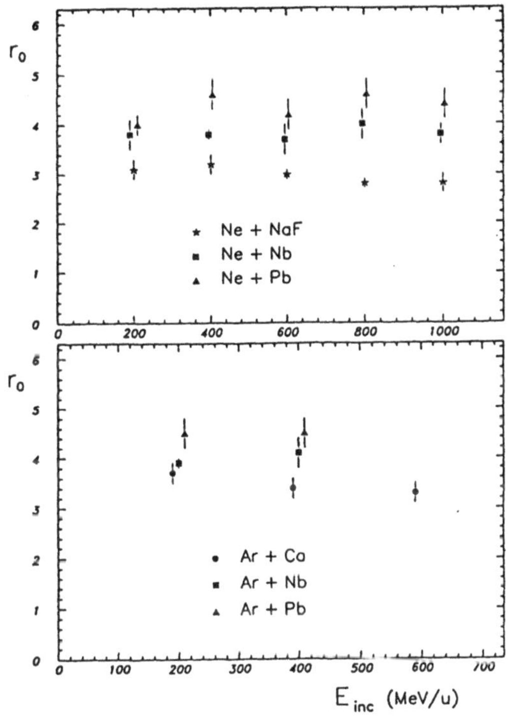


**Fig. 2–3** : Extracted $r_0$ values vs incident energy.

In order to go further in these observations we have compared $R_s$ to the radius $R_g$ ("geometrical radius") corresponding to the clean cut geometry when considering a staight–line collision. $R_g$ is calculated by a simulation program taking the Diogene trigger into account and is weighted on the impact parameter. We show on Fig. 4 the values of $R_s/R_g$ for various systems in case of the 400 MeV per nucleon incident energy. We see that we measure always the same fraction of the geometrical radius $R_g$. No compressionnal effects are pointed out. An absolute value of $R_s/R_g \simeq 2$ is in good agreement with freeze-out values.

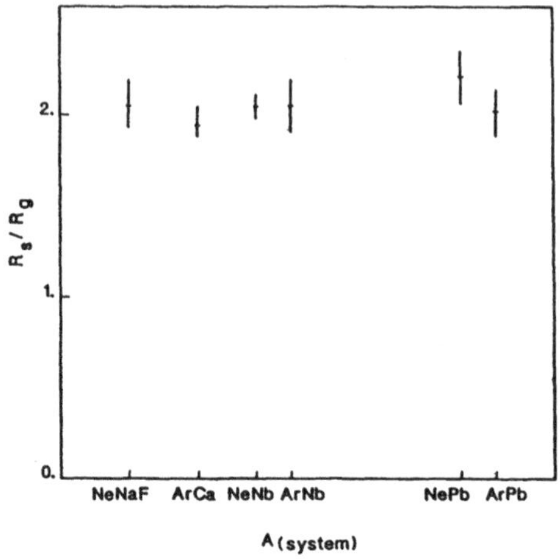


**Fig. 4 :** values of the $R_s/R_g$ ratio for different projectile-target systems.

(iii) – $r_0$ *values in case of impact parameter selections*

The impact parameter is evaluated from the proton-like multiplicity assuming a monotonous correlation between these two quantities[10]. For each system, the ratio of the proton-like cross section to the geometrical one is considered as a measurement of the square of the reduced impact parameter $(b/b_{max})^2$.

For all systems, we obtain greater $r_0$ values in case of more central collisions. We give on Fig. 5 $r_0$ values corresponding to four impact parameter classes for the system Ar + Ca and various incident energies. These results are also in good agreement with a geometrical description of the collision.

We have calculated the ration $R_s/R_g$, always in case of four classes of impact parameter, for the system Ar + Ca at 600 MeV per nucleon incident energy. The results are given on Fig. 6. In contrary to what we observe when no impact parameter selections are done, $R_s/R_g$ is not constant. It increases when going to more peripheral collisions. It can be interpreted as a signature of a bigger compression in the earlier stages of the interaction, for the more central collisions.

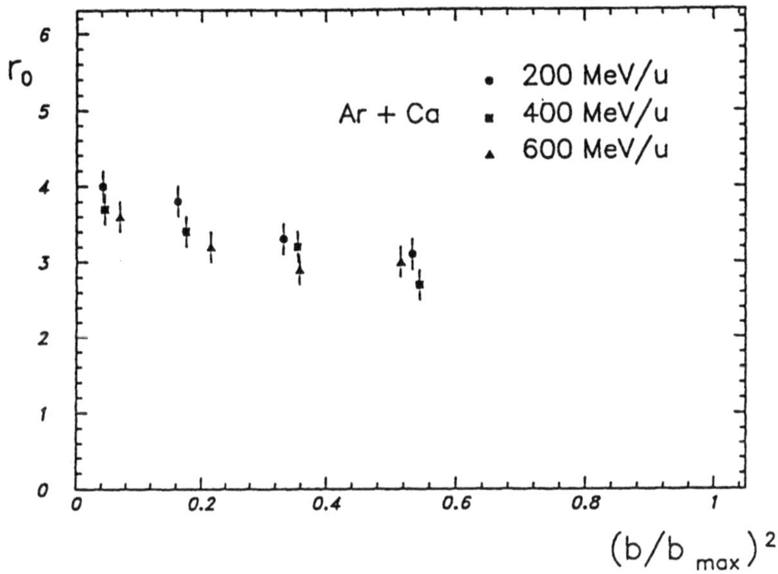


**Fig. 5 :** Extracted $r_0$ values vs impact parameter.

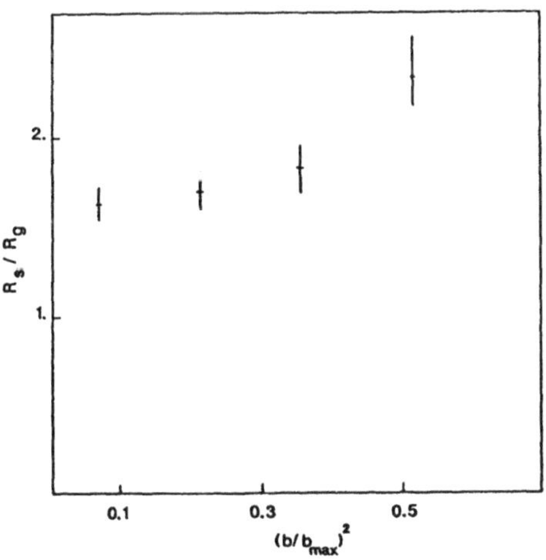


**Fig. 6 :** values of the $R_s/R_g$ ratio vs impact parameter for Ar + Ca at 600 A.MeV.

(iv) – $r_0$ *values in case of high momentum protons selection*

Results from Dubna[11] have shown a very significant decrease of the measured source radii when considering high energy protons. In our case, we see only a very slight decrease of the $r_0$ values when doing such selections (see Table I), probably due to less sensitivity of high energy protons to final state interactions. We have to notice however that Dubna results concern lighter projectiles at higher energy and lighter targets. For instance, they don't see any increase of $r_0$ with increasing the mass of the colliding system, like in other groups[4].

**Table 1** : $r_0$ values in case of high momentum protons selection, for a Ne projectile.

| Target | Incident Energy A.MeV | Momentum selection MeV/c | $r_0$ with selet. fm | $r_0$ no select fm |
|---|---|---|---|---|
| NaF | 200 | $\geq 500$ | $3.3^{+.3}_{-.3}$ | $3.1^{+.2}_{-.2}$ |
| Nb | 200 | $\geq 550$ | $3.4^{+.3}_{-.4}$ | $3.8^{+.2}_{-.3}$ |
| Pb | 200 | $\geq 550$ | $3.3^{+.5}_{-.1}$ | $4.0^{+.2}_{-.2}$ |
| NaF | 600 | $\geq 500$ | $3.2^{+.3}_{-.2}$ | $3.0^{+.1}_{-.1}$ |
| Nb | 600 | $\geq 550$ | $3.6^{+.3}_{-.3}$ | $3.7^{+.3}_{-.3}$ |
| Pb | 600 | $\geq 550$ | $3.8^{+.6}_{-.2}$ | $4.2^{+.2}_{-.3}$ |
| NaF | 1000 | $\geq 550$ | $2.9^{+.3}_{-.2}$ | $2.8^{+.1}_{-.2}$ |
| Nb | 1000 | $\geq 600$ | $3.8^{+.5}_{-.4}$ | $3.8^{+.2}_{-.2}$ |
| Pb | 1000 | $\geq 650$ | $4.0^{+.2}_{-.4}$ | $4.4^{+.4}_{-.2}$ |

## II – Conclusions.

The results of (pp) correlation studies at Saturne are in all cases representative of the freeze out stage of the interaction.

The size of the source increases when the mass of the projectile target system increases. It also increases when the centrality of the collision increases. Such observation are in good qualitative agreement with the participant spectator concept. On the other hand, we observe no clear variation of the size with the incident energy.

The extracted values of $R_s/R_g$ can show that the system keep some information from the early stage of the collision, as the compression of the nuclear matter seems to be greater for central collisions.

Finally we note that our results for (pp) correlation studies are in good agreement with those obtained with Diogene for $(\pi^-\pi^-)$ and $(\pi^+\pi^+)$ interferometry. They are also in good agreement with the Plastic ball results. It would be interesting in future experimental studies at these energies, to focus on large angle correlations, fragment fragment analyses, and measure correlations between particles which are less sensitive to final state interactions, in order to obtain more information on the initial stage of the collision.

## References


1. W.G. Lynch, C.B. Chitwood, M.B. Tsang, D.J. Fields, D.R. Klesch, C.K. Gelbke, G.R. Young, T.C. Awes, R.L. Ferguson, F.E. Obenshain, F. Plasil, R.L. Robinson, A.D. Panagiotou, *Phys. Rev. Lett.* **51**, 1850 (1983).
2. J. Pochodzalla, C.K. Gelbke, W.G. Lynch, M. Maier, D. Ardouin, H, Delagrange, H. Doubre, C. Grégoire, A. Kyanowski, W. Mittig, A. Péghaire, J. Péter, F. Saint–Laurent, B. Zwieglinski, G. Bizard, F. Lefèbvres, B. Tamain, J. Québert, Y.P. Viyogi, W.A. Friedman, D.H. Boal, *Phys. Rev.* **C35**, 1695 (1987).
3. A. Bamberger et al, *Phys. Lett.* **203**, 320 (1988).
4. H.A. Gustafsson et al, *Phys. Rev. Lett.* **53**, 544 (1984).
5. J. Bartke, *Phys. Lett.* **174**, 32 (1986).
6. J.P. Alard, J. Arnold, J. Augerat, R. Rabinet, N. Bastid, F. Brochard, M. Crouau, N. de Marco, M. Drouet, P. Dupieux, H. Fanet, Z. Fodor, L. Fraysse, J. Girard, P. Gorodetzky, J. Gosset, C. Laspalles, M.C. Lemaire, D. L'Hôte, B. Lucas, G. Montarou, A. Papineau, M.J. Parizet, J. Poitou, C. Racca, W. Schimmerling, J.C. Tamain, Y. Terrien, J. Valero, O. Valette, *N.I.M.* **A261**, 379 (1987).
7. P. Dupieux, J.P. Alard, J. Augerat, R. Rabinet, N. Bastid, F. Brochard, P. Charmensat, N. de Marco, H. Fanet, Z. Fodor, L. Fraysse, J. Girard, P. Gorodetzky, J. Gosset, C. Laspalles, M.C. Lemaire, D. L'Hôte, B. Lucas, J. Marroncle, G. Montarou, M.J. Parizet, J. Poitou, D. Qassoud, C. Racca, A. Rahmani, W. Schimmerling, O. Valette, *Phys. Lett.* **B200**, 17 (1988).
8. S.E. Koonin, *Phys. Lett.* **B70**, 43 (1977).
9. C. Cavata. M. Demoulins, J. Gosset, M.C. Lemaire, D. L'Hôte, J. Poitou, O. Valette, *"Determination of the impact parameter in relativistic nucleus-nucleus collisions", to be published in Phys. Rev.* **C**.
10. G.N. Agakishiyev, R.R. Mekhteev, V. Boldea, S. Ditza, J. Bartke, M. Kowalski, D. Armutliyski, J. Bogdanowicz, A.P. Cheplakov, L.A. Didenko, A.P. Gasparian, V.G. Grishin, I.A. Ivanovskaya, T. Kanarek, E.N. Kladnitskaya, D.K. Kopylova, V.B. Lyubimov. K. Miller, V.F. Nikitina, J. Pluta, M.I. Soloviev, R. Togoo, G.P. Toneeva, M. Chubarian, R.N. Bekmirzayev, U.D. Sherkulov, Z.V. Metreveli, Z. Strugalski, *Z. Phys.* **A327**, 443 (1987).

## Particle Correlations from $^{28}$Si+$^{197}$Au and $^{27}$Al at E/A = 14.6 GeV

*George S. F. Stephans*
*Laboratory for Nuclear Science and Department of Physics*
*Massachusetts Institute of Technology*
*Cambridge, Massachusetts 02139*

Abstract: The E802 spectrometer at the Brookhaven AGS has been used to obtain two-particle inclusive pion distributions from E/A=14.6 GeV $^{28}$Si +$^{197}$Au and $^{27}$Al interactions. Data are presented for pions near midrapidity in the nucleon-nucleon frame from central events as defined by total charged particle multiplicity. We present preliminary results for the source parameters obtained by fitting the two-particle correlation function. The potential for studying kaon correlations is also discussed.

### I. Identical Particle Interferometry

Identical particle interferometry can be a valuable tool in the elucidation of the processes underlying strong multiparticle production. The technique is based on one property of many-body wavefunctions, namely the (anti-)symmetry of a multi(-fermion)-boson wavefunction imposed by spin statistics. This aspect of the wave function causes a distortion of the two-particle phase space distribution of the particles emerging from a multiparticle produced final state. In particular, for the Bose statistics case, one would expect an enhancement in the number of bosons lying close to each other in phase space. Such enhancements have been observed in the past, [1,2,3] and further have been used in order to infer the space-time structure of the source of pion production.

The basis for the formalism applied is as follows. The probability density, $\wp$, for the observation of a two pion event is proportional to $|A_2(\vec{p}_1,\vec{p}_2)|^2$, and hence

$$\wp(\vec{p}_1,\vec{p}_2)\sim|e^{i\vec{p}_1\cdot(\vec{r}_1-\vec{r}_a)}e^{i\vec{p}_2\cdot(\vec{r}_2-\vec{r}_b)}+e^{i\vec{p}_1\cdot(\vec{r}_1-\vec{r}_b)}e^{i\vec{p}_2\cdot(\vec{r}_2-\vec{r}_a)}|^2$$
$$\sim1+\cos[(\vec{p}_1-\vec{p}_2)\cdot(\vec{r}_a-\vec{r}_b)], \qquad (1)$$

where $\vec{r}_a,\vec{r}_b$ denote pion source positions, $\vec{r}_1,\vec{r}_2$ denote pion detection locations, and $\vec{p}_1,\vec{p}_2$ denote pion momenta. The relative positive sign of the two terms comes from the symmetrisation of the wavefunction. Hence, given the assumptions of uncorrelated emission amplitudes and a source distribution function,$\rho$, we find for the coincidence probability the form

$$P_2(\vec{p}_1,\vec{p}_2)\sim\int d\vec{r}_a d\vec{r}_b\rho(\vec{r}_a)\rho(\vec{r}_b)\wp(\vec{p}_1,\vec{p}_2)\sim1+|\tilde{\rho}(\vec{p}_1-\vec{p}_2)|^2, \qquad (2)$$

where $\tilde{\rho}$ is the Fourier transform of $\rho$. This argument serves as the motivation for the definition of the two particle correlation function

$$\frac{\mathcal{N}\dfrac{d^2n}{d\vec{p}_1d\vec{p}_2}}{\dfrac{dn}{d\vec{p}_1}\dfrac{dn}{d\vec{p}_2}}\equiv1+\lambda|\tilde{\rho}(q)|^2\equiv C_2, \qquad (3)$$

where we have included the factor $\lambda$ in order to parameterise the strength of the correlation and rewritten the argument of $\tilde{\rho}$ in terms of the relative momentum of the two pions, $q$. The form of this last equation is evocative of an analogous measurement made with photons instead of hadrons, the Hanbury-Brown–Twiss experiment, [4]

$$\frac{\langle i_1 i_2 \rangle}{\langle i_1 \rangle \langle i_2 \rangle} = 1 + \lambda |\tilde{\rho}(q)|^2, \tag{4}$$

where the $i$'s might be currents read from photomultiplier tubes.

In practice, the above formulation (3) of the correlation function is not used directly. Instead, a change of variables is made from those appropriate to single particle phase space to those approriate for two particle phase space. Then the correlation function is calculated as

$$C_2 = \frac{Actual(q)}{Background(q)}, \tag{5}$$

where *Actual* is the measured phase space distribution and *Background* is a generated phase space distribution without the distortion due to statistics.

In order to compare correlation results from different experiments, it is important to stress the following: i)what gates must the selected events pass (i.e. central, jet, etc.); ii)what particles are included (identified particles, all negatives, particles in jets); iii)over what kinematic region are particles detected and how do those particles fill two-particle phase space; iv)using what phase space variables from what frame is the correlation generated; and v)how is the background handled and what corrections are applied to the data. These topics will each be addressed separately in the sections below.

We will present data binned using two different sets of two-particle phase space variables, the one-dimensional invariant four-momentum difference, $q_{inv}$, and the two-dimensional set $(q_l, q_t)$, where $q_l$ is along the beam direction in the frame of the so-called clean-cut geometric fireball center of mass. We note that implicit in our choice for the two-dimensional phase space variables (and the conventional Gaussian source distribution) is the *fixing* of the source lifetime to zero.

## II. E802

The E802 spectrometer at the BNL AGS is a large acceptance magnetic spectrometer system. It consists of twenty planes of projective drift chambers before and after the Henry Higgins dipole magnet (for a total of forty planes of drift chamber information) and a highly segmented plastic scintillator hodoscope array (TOF) used for time-of-flight particle identification. For triggering, the Target Multiplicity Array (TMA) was used in conjunction with the TOF wall and a plastic scintillator hodoscope (F0) placed between the target and the first tracking chamber. For details concerning the beam counters, and the segmentations of the various components, the reader is referred elsewhere. [5]

The present data are extracted from events triggered on the following conditions: i) *central*, in the sense of being in the upper 12% of the charged particle multiplicity distribution seen by the TMA, ii) *spectrometer*, containing at least one hit on the F0 counter, and iii) *two-negative*, containing two or more hits on the part of the TOF wall predominately struck by negative particles. The segmentation of the F0 counter is relatively coarse so that requiring two hit slats would have eliminated many pair events. This measurement was made with the spectrometer subtending polar angles of approximately 14 – 28° and with an integrated magnetic field setting of about 0.5 Tesla-meters. The negative pion acceptance in the variables of lab rapidity and transverse momentum is shown schematically in Fig. 1. Note that $y = 0$ in the nucleon-nucleon frame is at 1.72 units in the lab while in the clean-cut geometric fireball frame for central $^{28}$Si $+^{197}$Au collisions it is at 1.25 units. In this same region of phase space, the centrally triggered single particle inclusive pion distributions are found to be well characterised as exponential in $m_t$ with slope parameters of approximately 160 MeV. [6]

For individual tracks, the spectrometer has a spatial resolution on each plane of approximately 150 $\mu m$. In addition, Monte-Carlo studies indicate that the momentum resolution of $\delta p/p \leq 1.0\%$ is dominated by multiple scattering and energy loss in the target. [7] The observed two track spatial separation capability is approximately 3-4 mm per projection. Further, the Monte-Carlo indicates that the relative momentum resolution in terms of the invariant four-momentum difference is $\delta q_{inv} \approx$6-10 MeV (cf. Fig. 2). In order to understand the consequences of the detector resolutions, we make the following calculations. Two pions having momenta of 1 GeV (typical momenta in the region of phase space measured) and a transverse momentum difference of 30 Mev (i.e., in the low q region of two particle phase space which is most important in accurately fixing source parameters) one would calculate for the pion opening angle

$$\theta_{\pi\pi} \sim 1.7°. \tag{6}$$

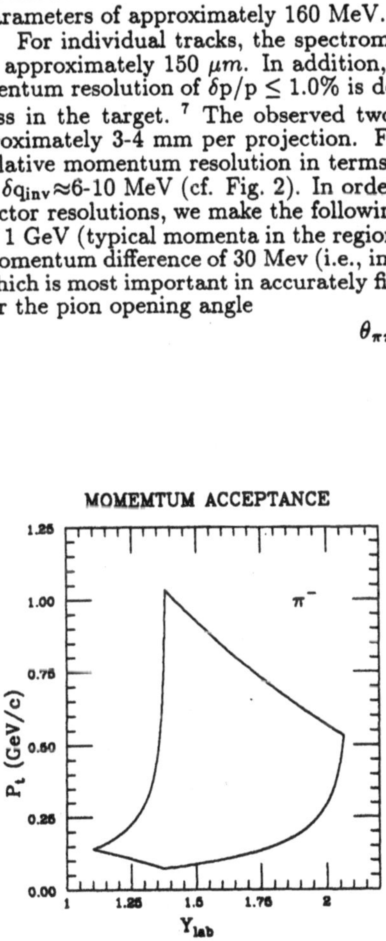


Fig. 1

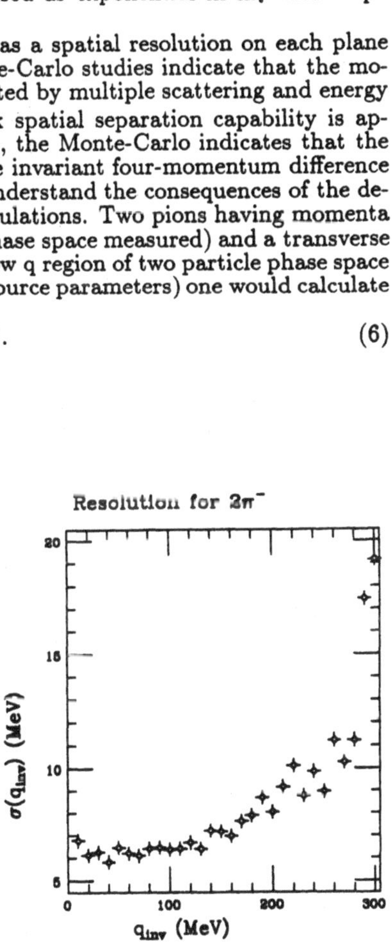


Fig. 2

This translates into a spatial separation of 4.5 cm at a distance of 1.5 m from the target, the position of the second set of ten drift planes. This is an order of magnitude larger than the two track separation resolution. Further, if we assume the Goldhaber source distribution, [1,8] i.e., a Gaussian in space and time, for which:

$$C_2 = 1 + \lambda \exp(-q_{inv}^2 R^2/2), \tag{7}$$

where $\hbar c = 1$, and if we require that the strength of the correlation has fallen to no less than half of its value in the first bin, then we find that

$$\frac{1}{2} = \exp(-r_{max}^2 \delta q_{inv}^2), \quad \text{and thus} \quad r_{max} = \sqrt{\frac{\ln 2}{\delta q_{inv}^2}}(\hbar c), \tag{8}$$

assuming $\lambda = 1$. This implies that for a resolution of 10 MeV in $q_{inv}$ we should be able to resolve up to 16 fm sources. [9]

## III. Data and Reduction

During the June 1989 running period, we collected HBT data for the reactions

$$^{28}\mathrm{Si} + {}^{197}\mathrm{Au} \rightarrow 2\pi^{\pm} + \mathrm{X} \quad \text{and} \quad ^{28}\mathrm{Si} + {}^{27}\mathrm{Al} \rightarrow 2\pi^{\pm} + \mathrm{X}$$

by bombarding $^{197}$Au and $^{27}$Al targets, of 1 and 3% interaction lengths respectively, with an E/A=14.6 GeV/c $^{28}$Si beam from the BNL Tandem-AGS complex. Following data reduction and the imposition of cuts, the yield is approximately sixty thousand negative pion pairs for the asymmetric system and approximately twenty-three thousand negative pion pairs for the nearly symmetric system. The positive pion pair yields are approximately half of the negative pion pair yields for each system. The cuts require that the tracks in pion pairs be: i) reconstructed with good status on both sides of the magnet, ii) found to hit *different* time-of-flight slats and iii) cleanly identified as pions using momentum and time-of-flight.

The data are then binned in the two-particle phase space variables as the raw *Actual* distribution. In order to account for the combined effects of pair reconstruction inefficiency, energy loss, and multiple scattering, the raw *Actual* distribution is subjected to a bin-by-bin two-particle acceptance correction. The correction factors are calculated by passing an uncorrelated set of pairs, derived from the momentum distribution of detected pions, through the E802 Geant Monte-Carlo with the physical processes of energy loss and multiple scattering activated. The result is then subjected to a data reduction process identical to that employed in the reduction of the physics data. This process was repeated in turn for the cases of negative, positive, and mixed-sign pairs.

The *Background* distribution is generated by randomly matching pion tracks from different pion-pair events in order to populate a presumably uncorrelated two-particle phase space distribution. This *Background* phase space distribution is then corrected for the di-pion Coulomb final state interaction, an effect which is presumably present in the *Actual* distribution, but absent from the *Background* distribution. In addition, the event-mixed *Background* suffers from a subtle distortion arising from its construction as a product of "single-particle" phase space distributions ($\rho^m$) formed from pions in pair events. This is effectively equivalent to an integration over an intrinsically correlated two-particle phase space distribution.

$$\rho^m(\vec{k}_1) = \int \rho(\vec{k}_1, \vec{k}_2) d\vec{k}_2 = \rho(\vec{k}_1) \int C_2(\vec{k}_1, \vec{k}_2) \rho(\vec{k}_2) d\vec{k}_2 \, . \tag{9}$$

There are two prescriptions for dealing with this distortion, it is either accounted for at the fitting stage with an empirical pre-factor such that

$$C_2^{fit} = (1 + \alpha q)C_2, \tag{10}$$

or the *Background* distribution is iteratively recalculated with weights intended to reflect the distortion. These weights amount to the average value of the correlation function over the single particle phase space cells contributing to a given two-particle phase space cell. [3,10] It is observed that for the case of one-dimensional two-particle phase space, the fitted parameters extracted using the empirical pre-factor and the iteration procedure are consistent. For the one-dimensional case, results using the empirical pre-factor prescription are presented, and for the case of two-dimensional two-particle phase space, results using the iteration procedure are presented. It is also observed that for the two-dimensional fits, the source parameters extracted before iteration can differ by as much as one sigma (68% confidence levels) from those extracted after iteration.

Finally, the source parameters are extracted by fitting the ratio of *Actual* and *Background* distributions using the principle of maximum likelihood. For the one-dimensional case we assume the so-called Goldhaber source distribution; the actual function fit is ($\hbar c = 1$):

$$C_2^{fit} = N(1 + \alpha q_{inv})(1 + \lambda \exp(-q_{inv}^2 R^2/2)) \tag{11}$$

where $N$ is for overall normalisation and the prescription $(1+\alpha q_{inv})$ is used in order to account for residual correlations in the generated background. In addition, fits have been performed using a single exponential as well as the constrained sum of two Gaussians suggested by Weiner [11]:

$$C_2^{fit} = N(1 + 2\lambda(1-\lambda)\exp(-q_{inv}^2 R^2) + \lambda^2 \exp(-2q_{inv}^2 R^2)) \,. \tag{12}$$

In the two-dimensional case we assume a static non-relativistic Gaussian in space and time (where the temporal source parameter is implicitly set to zero) and fit the form:

$$C_2^{fit} = N(1 + \lambda \exp(-({q_l}^2 R_l^2 + {q_t}^2 R_t^2)/2)) \,. \tag{13}$$

## IV. Results

The two particle correlation functions resulting from the one-dimensional analysis are displayed in the following two figures. In Fig. 3, we plot the negative and positive pion correlation functions for the $^{28}$Si + $^{197}$Au system; in Fig. 4, we likewise plot the correlation functions for the $^{28}$Si + $^{27}$Al system. In all cases, the solid curves correspond to the fits with the single Gaussian and the error bars are purely statistical. The dotted line corresponds to the single exponential fit. The dashed line showing the constrained two-Gaussian fit is mostly indistinguishable from the single Gaussian fit. The present results are inadequate to determine which of the parameterizations more closely resembles the data. Higher statistics data at very low $q_{inv}$ might discriminate between the exponential and Gaussian fits but the one- and two-Gaussians are too close to separate over the entire phase space. Below, the final fitted parameters are presented in tabular form.

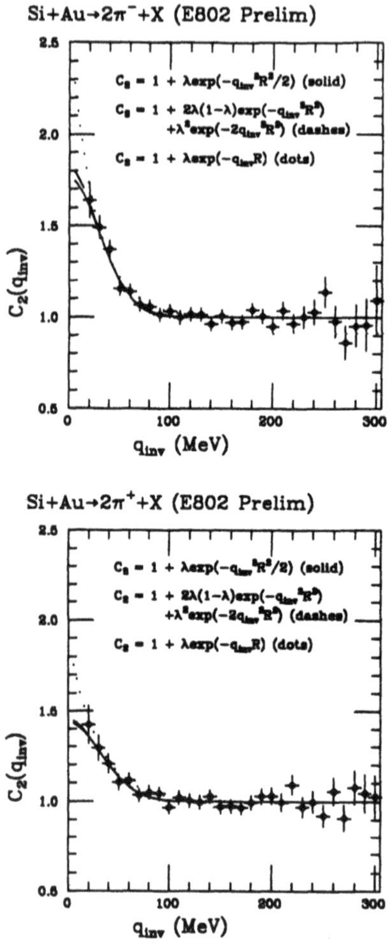


Fig. 3

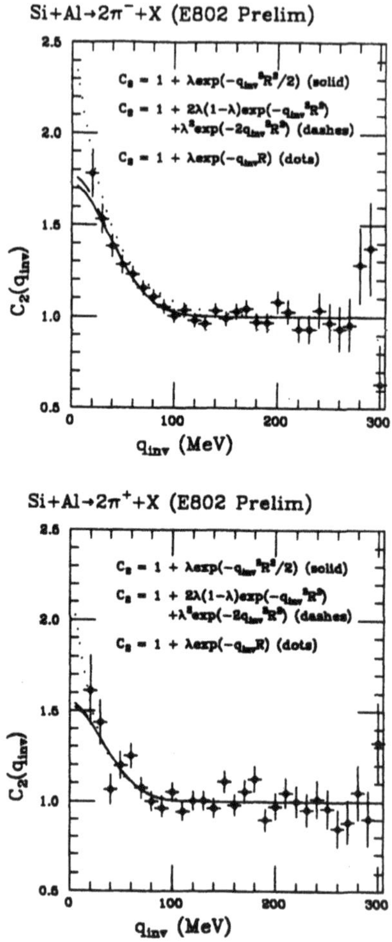


Fig. 4

In Fig. 5, we display the correlation functions generated for the unlike-signed pairs in each of the two reactions; in this case the di-pion Coulomb final state interaction approriate for unlike-signed pairs is applied. The results are consistent with no correlation, as expected. This systematic cross-check provides a sense of the internal consistency of the analysis procedure. Note that all of the above data come from a single spectrometer angle and magnetic field setting. In order to display part of the

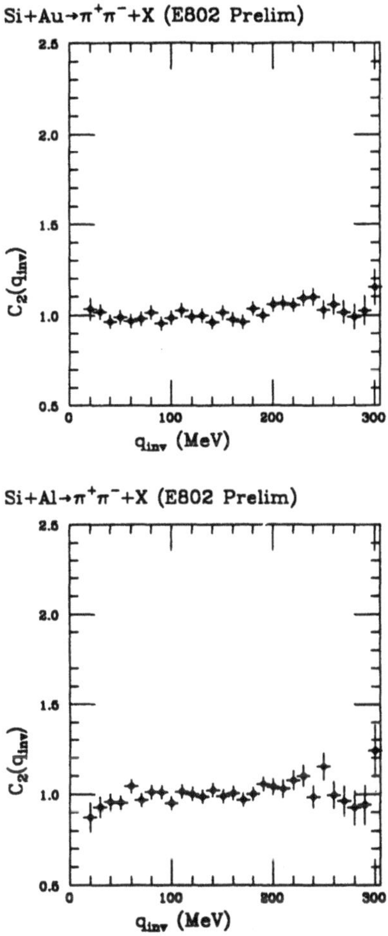


Fig. 5

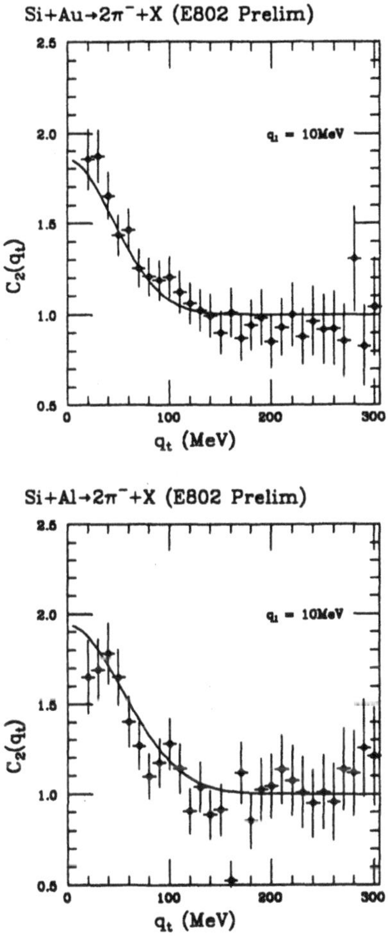


Fig. 6

two-dimensional correlation function fits, we plot a slice along the $q_l = 10$MeV axis in the variable $q_t$; in Fig. 6 are presented results for two-negative pion final states. It should be stressed that although the slices are in one dimension, the actual fits, from which come the plotted curves and the quoted source parameters, are over the entire $(q_l, q_t)$ phase space. Shown in Table III are the extracted source parameters for the two-dimensional fits.

Table I

| 1-D Single Gaussian | $R_{inv}$ | $\lambda$ | $\chi^2/dof$ |
|---|---|---|---|
| $^{28}$Si + $^{197}$Au → $2\pi^- + X$ | $6.3 \pm 0.6$ | $0.76 \pm 0.09$ | 20.3/25 |
| $^{28}$Si + $^{197}$Au → $2\pi^+ + X$ | $5.9 \pm 1.0$ | $0.44 \pm 0.10$ | 17.8/25 |
| $^{28}$Si + $^{27}$Al → $2\pi^- + X$ | $5.1 \pm 0.5$ | $0.72 \pm 0.09$ | 20.8/25 |
| $^{28}$Si + $^{27}$Al → $2\pi^+ + X$ | $5.7 \pm 1.1$ | $0.53 \pm 0.15$ | 20.8/25 |

Table II

| 1-D Double Gaussian | $R_{inv}$ | $\lambda$ | $\chi^2/dof$ |
|---|---|---|---|
| $^{28}$Si + $^{197}$Au → $2\pi^- + X$ | $4.0 \pm 0.3$ | $0.57 \pm 0.13$ | 19.5/25 |
| $^{28}$Si + $^{197}$Au → $2\pi^+ + X$ | $4.1 \pm 0.7$ | $0.26 \pm 0.07$ | 17.7/25 |
| $^{28}$Si + $^{27}$Al → $2\pi^- + X$ | $3.3 \pm 0.3$ | $0.52 \pm 0.11$ | 20.5/25 |
| $^{28}$Si + $^{27}$Al → $2\pi^+ + X$ | $3.9 \pm 0.7$ | $0.33 \pm 0.13$ | 33.2/25 |

Table III

| 2-D Gaussian | $R_t$ | $R_l$ | $\lambda$ | $\chi^2/dof$ |
|---|---|---|---|---|
| $^{28}$Si + $^{197}$Au → $2\pi^- + X$ | $4.3 \pm 0.2$ | $2.6 \pm 0.3$ | $0.86 \pm 0.06$ | 781/826 |
| $^{28}$Si + $^{197}$Au → $2\pi^+ + X$ | $3.3 \pm 0.4$ | $3.4 \pm 0.4$ | $0.55 \pm 0.07$ | 877/856 |
| $^{28}$Si + $^{27}$Al → $2\pi^- + X$ | $3.5 \pm 0.2$ | $2.9 \pm 0.3$ | $0.95 \pm 0.07$ | 555/560 |
| $^{28}$Si + $^{27}$Al → $2\pi^+ + X$ | $3.4 \pm 0.4$ | $3.2 \pm 0.8$ | $0.70 \pm 0.14$ | 233/191 |

In order to facilitate comparison with other measurements using differing conventions for source parameterisation, we convert all source radii to RMS values; for the parameterisation that we have chosen, this leads to multiplication of our extracted transverse radii by a factor of $\sqrt{3/2}$. We have done this for the four reactions and plotted the results in Fig. 7, along with similarly calculated RMS radii from other proton and heavy-ion induced reactions, as a function of the projectile mass number. The additional data come from a recent compilation by Zajc. [12] The line plotted with the data shows what would be expected for a linear variation of measured source radius with the projectile radius.

## V. Kaon Correlations

By combining the data from the two targets, a sample of roughly 1500 $K^+$ pairs was obtained. Although this sample is insufficient to perform detailed correlation analysis, it does provide a proof-of-principle that such data can be obtained in the present experimental setup. The low-statistics correlation function derived from the pairs also serves as a check of the quality of the data. A fit to the one dimensional, $q_{inv}$, correlation yields the provocative result that the radius is significantly smaller than that found for pion pairs, albeit with large uncertainty. An upgraded version of the same experiment (E859 at BNL) including a second-level trigger sensitive to the mass and momentum of particles traversing the spectrometer will shortly provide more than an order of magnitude greater statistics in order to explore this exciting result. Pion pair data over a wider range of rapidity will also be obtained.

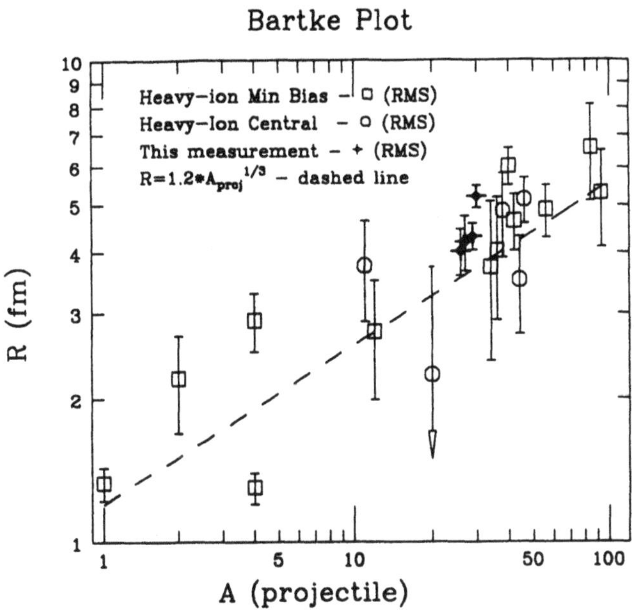


Fig. 7

## VI. Conclusion

We have reported the first use of the technique of identical particle interferometry in order to estimate the spatial extent of the source of produced pions in nucleus-nucleus collisions at AGS energies. It is found that for both negative and positive pion pairs, the observed transverse source size is comparable to the projectile radius for both the symmetric and asymmetric systems studied. It is further found that the 'chaoticity' parameter, $\lambda$, is greater than one-half. The issue of sensitivity to source lifetime is not addressed, but rather the source lifetime is implicitly set equal to zero in the spirit of other measurements. [13,14] The source extent in terms of an 'invariant' radius is extracted, and is not found to be identical to the transverse source radius as has been reported elsewhere. [13] In conclusion, more detailed studies of systematic effects in the processes of data reduction and the drawing of statistical inferences from the reduced data are in progress. Specific directions include: i) further work on multidimensional correlation functions, ii) model studies in order to estimate resonance contributions to pion production at AGS energies, and iii) the characterisation of any variations in source geometry arising from differences in the phase space regions observed, i.e. rapidity or $p_t$ variation of source geometry.

**Biblography**

1 G. Goldhaber *et al.,* Phys. Rev. Lett. 3, 181 (1959).
2 S. Y. Fung *et al.,* Phys. Rev. Lett. 41, 1592 (1978).
3 W. A. Zajc *et al.,* Phys. Rev. C29, 2173 (1984).
4 R. Hanbury-Brown and R. Q. Twiss Proc. Roy. Soc. London A242, 300 (1957), and A243, 291 (1957).
5 T. Abbott *et al.,* Nucl. Instr. and Meth. A290, 41 (1990).
6 T. Abbott *et al.,* Phys. Rev. Lett. 64, 847 (1990).
7 I. Juričić and W. A. Zajc, E-802-Mem-21 (1989).
8 G. Goldhaber *et al.,* Phys. Rev. 120, 300 (1960).
9 I. Juričić and W. A. Zajc, E-802-Mem-24 (1989).
10 I. Juričić, *Bose-Einstein Correlations in* $e^+e^-$ *Collisions*, Ph. D. Thesis, University of California, (1987). Available as LBL-24493.
11 R.M.Weiner *et al.,* Phys. Lett. 232B, 278 (1989).
12 W. A. Zajc in *Multiparticle Production,* P. Carruthers, ed. (World Scientific Publishing Co., Singapore, 1988).
13 A. Bamberger *et al.,* Phys. Lett. 203B, 320 (1988).
14 D. Beavis *et al.,* Phys. Rev. C34, 757 (1986).

# SOME FEATURES OF IDENTICAL PION CORRELATIONS FROM DUBNA EXPERIMENTS

*Jan Pluta*
*Institute of Physics, Warsaw Technical University,*
*Koszykowa 75, 00-662 WARSAW, Poland*

*V.G.Grishin, L.A.Didenko, T.B.Progulova, M.U.Sultanov, P.Tas*
*Joint Institute for Nuclear Research, Dubna, USSR*

ABSTRACT

Recent results obtained at JINR in Dubna on interference correlations of charged pions produced in $\pi^- p$ and $\pi^- C$ collisions at 40 GeV/c are presented. The size of pion emission volume for the projectile fragmentation region is close to the sizes characteristic for $e^+e^-$-annihilation. The source sizes, obtained for pions from the central region, are considerably greater. Presented results show universal properties of the pion production process in different kinds of interactions.

## 1. INTRODUCTION

The properties of multiparticle production process have been investigated at JINR in Dubna in the frame of relativistic invariant approach[1-9]. Particle characteristics have been considered in the space of four-dimensional relative velocities. This kind of investigations is related to the quark structure of hadrons and nuclei. A hypothesis was formulated that the characteristics of hadron jets produced in different kinds of interactions do not depend on the origin and properties of the colour quark system (type of interaction) but are determined by the dynamics of the hadronization process[1]. Investigations performed at the Laboratory of High Energies JINR show that many properties of hadron jets, considered in the space of four-dimensional relative velocities, have universal character and are similar even in such different reactions as $e^+e^-$- annihilation and cumulative $\pi^-C$ interactions[4-6].

Some papers on interference correlations in jets registered in $e^+e^-$ annihilation have been published so far[7-9]; interference effects were investigated also in "central jets" separated in pp-collisions at $\sqrt{s}$ = 63 GeV[10]. There are no similar data for "soft jets" in hadron-hadron collisions and nuclear reactions.

## 2. DESCRIPTION OF THE HADRON JETS IN THE SPACE OF FOUR-DIMENSIONAL RELATIVE VELOCITIES

The process of multiparticle production has been described by means of relativistic invariant variables[1-3]:

$$b_{ik} = -\left(\frac{P_i}{m_i} - \frac{P_k}{m_k}\right)^2 = -(u_i - u_k)^2 , \tag{1}$$

where $P_i$, $P_k$ - four-momenta, $m_i$, $m_k$ - masses, $u_i$, $u_k$ four- dimensional velocities of considered particles. Variables $b_{ik}$ are four-dimensional relative velocities. Jets (clusters) are defined as groups of particles with small values of $b_{ik}$. The centre of jet is defined as

$$V_\alpha = \frac{\Sigma u_i}{\sqrt{(\Sigma u_i)^2}}. \tag{2}$$

It has been shown[4-6] that in the reactions considered, two jets can be distinguished in general. In order to separate them, the following quantity is minimized:

$$A_2 = \min \left[ -\sum_k (V_\alpha - u_k^\alpha)^2 - \sum_i (V_\beta - u_i^\beta)^2 \right]. \tag{3}$$

To minimize $A_2$, all possible distributions of particles into two groups in each event are considered, and minimal value of $A_2$ is fixed. The values of $V_\alpha$ and $V_\beta$ are simultaneously determined.

Distance between jets in four-velocity space is defined by

$$b_{\alpha\beta} = -(V_\alpha - V_\beta)^2. \tag{4}$$

To characterize the region of jet (cluster) production, the invariant variables, $x_{sn}$ and $x_{sm}$, are used. These variables express the part of the four-momenta of colliding particles taken by a given jet:

$$x_{sn} = \frac{M_{\alpha(\beta)}}{m_I} \frac{(V_{\alpha(\beta)} u_{II})}{(u_I\ u_{II})}, \qquad x_{sm} = \frac{M_{\alpha(\beta)}}{m_{II}} \frac{(V_{\alpha(\beta)} u_I)}{(u_I\ u_{II})}, \quad (5)$$

where $M_{\alpha(\beta)}$ - effective mass of the cluster $\alpha$ or $\beta$, $m_I$ - mass of the projectile, $m_{II}$ - target mass.

Detailed criteria of jet selection will be discussed below.

## 3. FOUR-DIMENSIONAL JETS IN $\pi^-$p AND $\pi^-$C COLLISIONS AT 40 GEV/C

The characteristics of secondary pions, registered in the 2-meter propane bubble chamber irradiated by negative pions at 40 GeV/c, were analysed in our earlier papers[1-6]. Jet properties were investigated with a traditional method applying "sphericity" and "thrust" variables[4-6]; invariant method was used as well[1-3]. As a result, the criteria of jet selection have been worked out and applied in this paper.

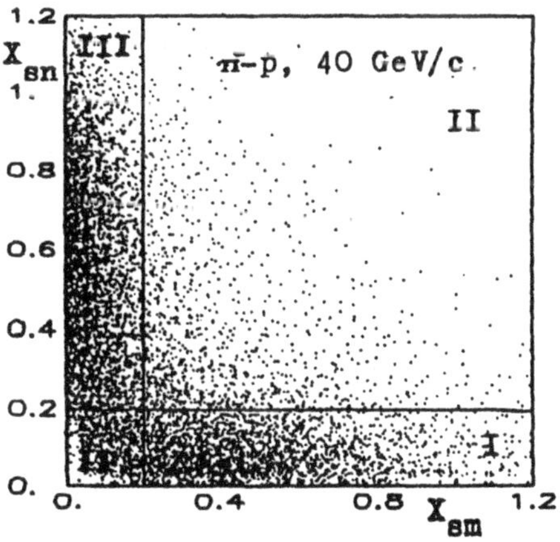


*Fig. 1. Distribution of pion jets over the $x_{sn}$ and $x_{sm}$ variables.*

Jets were selected in events with secondary pion multiplicity $n_{\pi^\pm} \geq 4$. For each event the minimization of $A_z$ was performed and the value of $b_{\alpha\beta}$ was calculated. The quantities $x_{sn}$ and $x_{sm}$ were then determined for two selected jets. Fig. 1. presents the two-dimensional distribution over $x_{sn}$ and $x_{sm}$ for $\pi^-$p interactions at 40 GeV/c.

Large $x_{sn}$ values correspond to beam fragmentation region, large $x_{sm}$ values denote the target fragmentation. Each event is represented by two points in this figure. It has been accepted that two jets are separated in the four-velocity space if $b_{\alpha\beta} \geq 10$. Different regions of pion production has been selected by the following criteria:

I. target fragmentation region: ( $x_{sm} \geq 0.2$ and $x_{sn} < 0.2$ )

II. central region: ( $x_{sn} \geq 0.2$ and $x_{sm} \geq 0.2$ ) or ( $x_{sn} < 0.2$ and $x_{sm} < 0.2$ ).

III. beam fragmentation region: ( $x_{sn} \geq 0.2$ and $x_{sm} < 0.2$ ).

These regions are indicated in Fig. 1. The groups of pions located in the regions of beam and target fragmentation were considered separately. If the x values for two groups were located in the central region, all pions were considered as a one group. Fig. 2. demonstrates a clustering effect. If one group (jet) is located in the beam fragmentation region (Fig 2.b), the second one appears in the region of target fragmentation (Fig. 2a). Central region contains events in which no clear two-jet structure is seen in the sense described above.

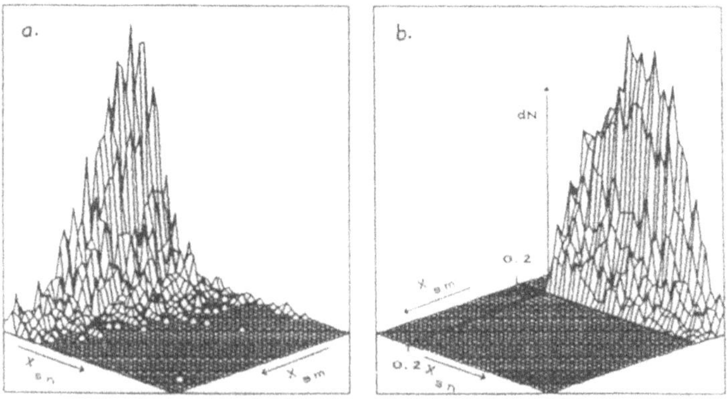


*Fig. 2. The clustering effect in $\pi^- p$ interactions at 40 GeV/c.*
*b. - $x_{sn}$ v.s. $x_{sm}$ distribution for jets in the beam fragmentation region, a - analogous distribution for the second jet in the same event.*

A physical interpretation of the origin of jets in the fragmentation regions (I, III) is connected with the hadronization of quark and diquark from pion and proton, not participating in the $\pi^- p$ interaction[1]. Pions in the central region appear as a result of hadronization of interacting quarks.

The analysis was performed at the cms reference frame of $\pi^- p$ ($P_{\pi^-}$ = 40 GeV/c). Data obtained for all events with $n_{\pi^\pm} \geq 2$ in $\pi p$ and $\pi C$ collisions are presented for comparison as well. The statistics of events and pion numbers in each considered group are listed in Table 1.

Table 1. Statistics of experimental data

| region | $\pi^- p$ | | $\pi^- C$ | |
|---|---|---|---|---|
| | $N_{events}$ | $N_{\pi^\pm}$ | $N_{events}$ | $N_{\pi^\pm}$ |
| all $\pi^\pm$ | 17352 | 88434 | 8307 | 58900 |
| beam fragmentation | 5187 | 17199 | 1483 | 5049 |
| target fragmentation | 4209 | 13865 | 2605 | 10388 |
| central region | 1896 | 11528 | 799 | 6453 |

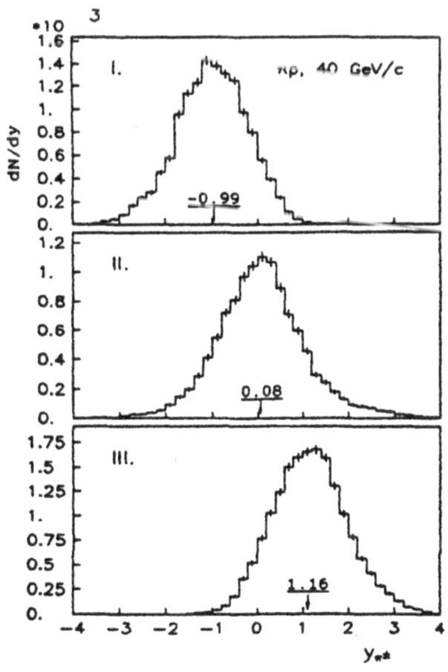


Fig. 3. presents rapidity distributions in $\pi^-$p-cms for pions from three selected regions. The difference of the mean rapidities for pions from beam and target fragmentation regions is about 2, which is more than double value of the dispersion of rapidity distributions themselves. Both the value of mean rapidity and the form of y distribution for pions from the central region correspond to their symmetric emission in the c.m. system of colliding particles.

*Fig. 3. Rapidity distributions for pions from three selected regions.*

## 4. INTERFERENCE CORRELATIONS IN DIFFERENT REGIONS OF PION PRODUCTION

Interference correlations of charged and neutral pions have been investigated at JINR in Dubna in hadron-hadron, hadron-nucleus and nucleus-nucleus collisions. The results are described elsewhere[11-20]; some of them will be used here for comparison.

Two-particle correlations in the region of small relative momenta were described by the ratio

$$R(\vec{q},q_o) = \frac{\rho(\vec{q},q_o)}{\rho_b(\vec{q},q_o)}, \qquad (6)$$

where: $\rho(\vec{q},q_o)$ - density of identical particle pairs in a phase space, $\rho_b(\vec{q},q_o)$- density of pairs in the conditions of absence of interference. Here $\vec{q} = (\vec{p}_1 - \vec{p}_2)$, $q_o = E_1 - E_2$; $\vec{p}_1,\vec{p}_2$ and $E_1,E_2$ - momenta and energies of pair particles. Assuming space-time distribution of particle sources in a form of gaussian distribution, two-particle correlation function can be expressed in the form[21]:

$$R(\vec{q},q_o;\ r_o,\tau_o) = A\ (\ 1 + \lambda\ e^{-\vec{q}^2 r_o^2 - q_o^2 \tau_o^2}), \qquad (7)$$

where $r_o$ and $\tau_o$ define the space and time characteristics of particle emission process, $\lambda$ - phenomenological parameter, A - normalization coefficient. In this form of parameterization, the rms radius of particle emission volume is: $\langle r^2\rangle^{1/2} = \sqrt{3}r_o$.

Due to limited statistics of our data we have analysed only one-dimensional distributions $R(\vec{q};\ r_o)$, $(q_o < 0.2$ GeV$)$ . Experimental distributions were fitted with the formula

$$R(\vec{q};\ r_o) = A\ (\ 1 + \lambda\ e^{-\vec{q}^2 r_o^2}). \qquad (8)$$

To construct the background distributions pairs of unlike charge pions were used. This method allows to remove interference effects but preserves other kinds of correlations due to jet structure of events.

Fig. 4. presents the results of analysis of interference correlations for the groups of pions selected in $\pi^- p$ - interactions at

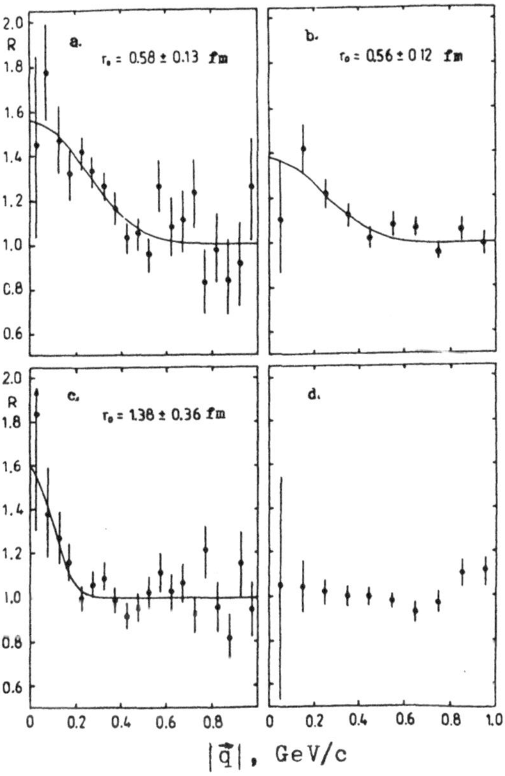


40 GeV/c. Figures (a,b,c) show correlation distributions for pions from three selected regions (III,I,II) respectively. Distribution in Fig. 4.d was obtained by correlation of pions taken from different groups selected in accordance with formula (3). In this case no correlation is seen, though pion pairs are located in the interference region.

*Fig. 4. Correlation distributions R(q) for pions from different selected regions: a - beam fragmentation, b - target fragmentation, c - central region, d - pions from different groups.*

The fitted parameters are shown in the Table 2.

Table 2. $\pi^- p$ interactions at 40 GeV/c

| region | $r_0$ , fm | $\lambda$ | $\chi^2$/NDF |
|---|---|---|---|
| all $\pi^{\pm}$ | 0.93 ± 0.14 | 0.42 ± 0.07 | 1.05 |
| beam fragmentation | 0.58 ± 0.13 | 0.56 ± 0.12 | 0.82 |
| target fragmentation | 0.56 ± 0.12 | 0.37 ± 0.16 | 1.10 |
| central region | 1.38 ± 0.36 | 0.58 ± 0.27 | 0.95 |

The sizes of pion emission volume, obtained for the regions of beam and target fragmentation, are considerably smaller than those for all pions in a given reaction, and are smaller than a half of the size, obtained for pions from the central region. Their absolute values are consistent with the sizes characteristic for $e^+e^-$-annihilation[7-9]. Interference correlations in those processes are usually described by Q-variable - a four-dimensional momentum difference; $Q^2 = -(P_1 - P_2)^2$, where $P_1$, $P_2$ are four-momenta of correlated pions.

Fig.5.a presents correlation distribution obtained by us using Q-variable for pions from beam fragmentation region. Analogous distribution for pions registered in $e^+e^-$-annihilation at $\sqrt{s}$ = 29 GeV[7], is shown in this figure as well. Both distributions were approximated with the formula, which takes into account a slow increase of correlation function beyond the interference region:

$$R(Q,r) = A\ (\ 1 + \lambda\ e^{-Q^2 r^2}\ )\ (\ 1 + \gamma\ Q\ ). \tag{9}$$

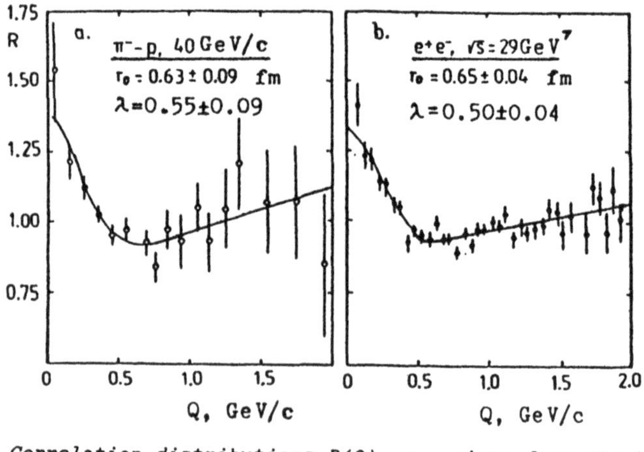


*Fig. 5. Correlation distributions R(Q). a - pions from the beam fragmentation region in $\pi^- p$ interactions at 40 GeV/c ,b - pions from the reaction of $e^+e^-$ - annihilation at $\sqrt{s}$ =29 GeV[7].*

Fig 6. and Table 3. present the results obtained for $\pi^- C$ interactions at 40 GeV/c. Here the size of pion emission volume for beam fragmentation region is also close to the sizes, characteristic for lepton processes. For other regions the results show greater values than in the case of $\pi^- p$ collisions. This effect may reflect the greater mass of the target nucleus.

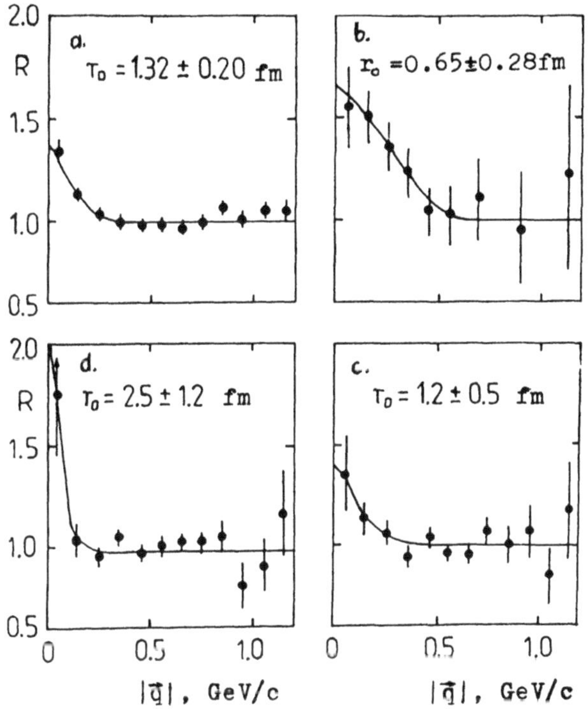


*Fig. 6. Correlation distributions R(q) for pions from different selected regions in $\pi^-C$ interactions at 40 GeV/c. a - total sample, b - beam fragmentation, c - target fragmentation, d - central region.*

Table 3.
The results of approximation with the formula (8) of the correlation distributions for different groups of pions selected in $\pi^-C$ - interactions at 40 GeV/c.

| region | $r_0$ , fm | $\lambda$ | $\chi^2$/NDF |
|---|---|---|---|
| all $\pi^{\pm}$ | 1.32 ± 0.20 | 0.34 ± 0.08 | 1.62 |
| beam fragmentation | 0.65 ± 0.28 | 0.68 ± 0.28 | 0.85 |
| target fragmentation | 1.20 ± 0.51 | 0.33 ± 0.20 | 1.15 |
| central region | 2.5 ± 1.2 | 1.3 ± 0.7 | 1.35 |

## 5. MULTIPLICITY DEPENDENCE

One can easily calculate the mean pion multiplicities in different regions of pion production using the data from Table 1. Fig. 7. presents the obtained radii of the pion emission volume as a function of mean multiplicity for different groups of pions.

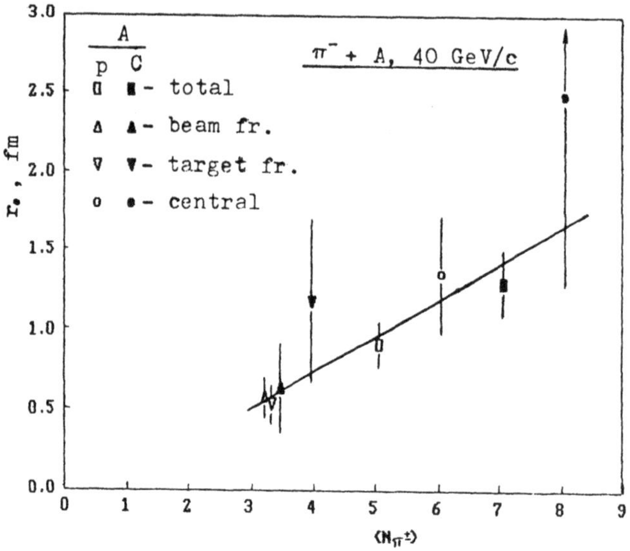


*Fig. 7. The radii of pion emission volume as a function of mean multiplicity for different groups of pions.*

The form of multiplicity dependence seems to reflect some features of the pion emission process.

1. Projectile fragmentation gives the same sizes and multiplicities in hadron-hadron and and hadron-nucleus collisions.
2. In $\pi^-p$ collisions the projectile and target fragmentation regions show the same values of sizes and multiplicities.
3. Approximately linear dependence holds between mean pion multiplicities in the selected groups and the sizes of pion emission volume.
4. The sizes obtained for central and target fragmentation regions in hadron-nucleus collisions show relatively greater values.

It should be noted however that the sizes of pion emission volume for different pion multiplicities in $\pi^- p$ collisions at 40 GeV/c were estimated earlier and no multiplicity dependence was seen[22]. Fig. 8. presents the form of $Q^2$-distribution for pions from two groups of events with different multiplicities. The fitted radii ($R_g$) shows even opposite tendency to those from Fig. 7.

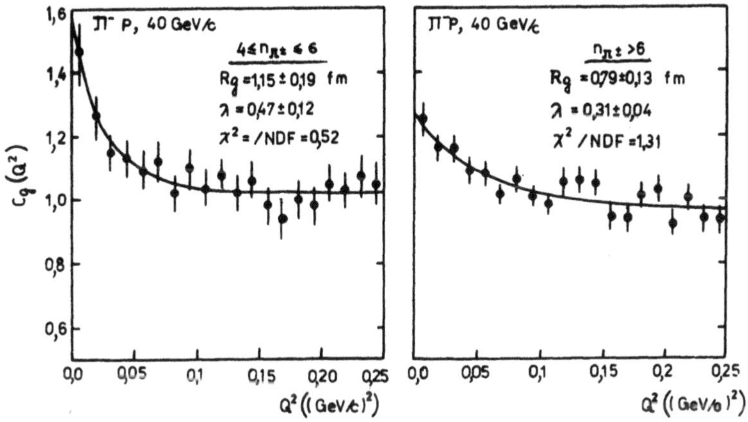


*Fig. 8. The $Q^2$-distributions for pions from two groups of events with different pion multiplicity.*

Similar result has been obtain recently by NA23, EHS-RCBC collaboration for pions from pp collisions at 360 GeV/c[23]. On the other hand, the increase of the sizes with increasing pion multiplicity has been reported for the central rapidity interval in ISR and collider data at CERN[24,25].

To better understand the observed dependence we compared the mean multiplicities in the events where two jets were distinguished, ( $\langle n_{\pi^\pm}\rangle_j$= 6.11±0.07 ) with the analogous value for events where pions are located in the central region, $\langle n_{\pi^\pm}\rangle_c$=6.08±0.13. The same values obtained for both cases show that the pion multiplicity per event is not essential here. In Fig.9, these two types of events are shown as a distribution of pions in rapidity and perpendicular momentum scale.

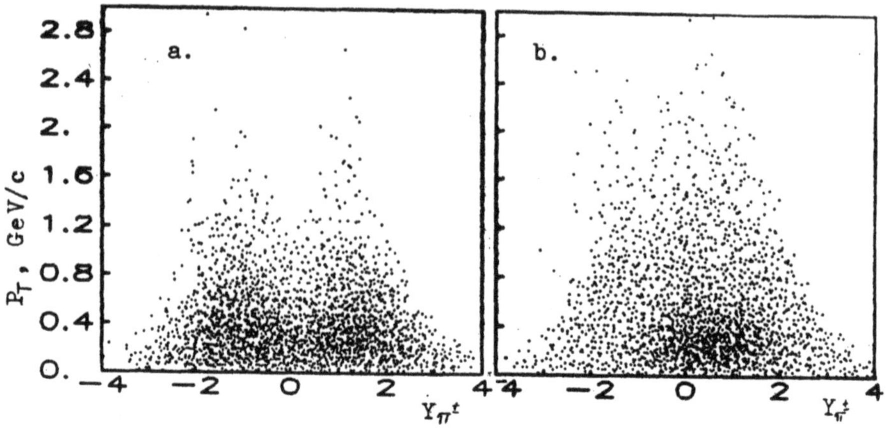


*Fig. 9. Distribution of pions in rapidity and perpendicular momentum scale. a - for two-jet events, b - for no-jet events.*

Different structure of events in rapidity scale is seen. It is presented in Fig.4d that the interference analysis shows no correlations between pions from the beam and target fragmentation regions. Fig. 10 shows the rapidity distribution for pions from the beam fragmentation region in $\pi^-p$ collisions at 40 GeV/c plotted in the rest frame of the jet. This distribution has approximately a gaussian-like shape with the dispersion, $\sigma$=0.67. It is characteristic for the isotropic decay of the particle source.

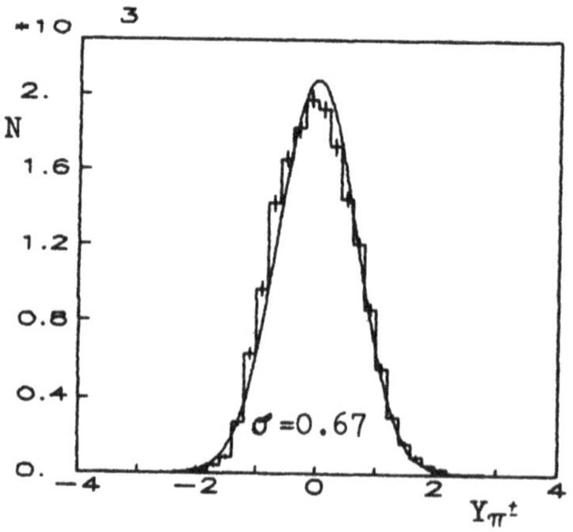


*Fig. 10. Rapidity distribution for pions from the beam fragmentation region in $\pi^-p$ interactions at 40 GeV/c plotted in the rest frame of the jets. Curve represents approximation with the Gaussian function.*

Thus our selection procedure separates events with different pion momentum structure whereas interference analysis reveals the connections between momentum and space-time characteristics of particle emission process.

## 6. SUMMARY

The analysis of interference correlations in $\pi^- p$ and $\pi^- C$ - interactions at 40 GeV/c leads to the following conclusions:

1. Interference correlations of secondary pions, produced in the beam fragmentation region in soft hadron-hadron and hadron-nucleus collisions, do not depend on the target mass (p or C-nucleus) and coincide with the analogous data for hard processes of $e^+e^-$ annihilation.

2. For the target fragmentation region, the dependence of pion emission volume on the target mass is seen . In the case of $\pi^- p$ collisions, this volume is compatible with that of beam fragmentation; in $\pi^- C$ collisions, it is considerably greater.

3. In the events where no clear two-jet structure is seen the sizes of pion emission volumes are much greater than in the case of pions from the fragmentation regions. Obtained volumes are also considerably greater than those for the total samples of registered pions. The observed effect is similar to that in NA35 experiment in nucleus-nucleus collisions at 200 GeV [26], though absolute values differ considerably.

4. The size of pion emission volume increases with the increasing multiplicity in the considered groups of pions.

Presented results show that the space-time characteristics of hadron jets, selected by the invariant method in the incident hadron fragmentation region, are the universal property of multiparticle production process not depending on the type and nature of the colliding objects.

The authors are grateful to M.I. Podgoretskii and N. Angelov for useful discussion and to the members of Propane Collaboration at JINR for the participation in the collection of experimental material. The investigations were supported in part by the Research Program RR-I-15 in Poland.

REFERENCES

1. A.M. Baldin et al. *Z.Phys.* **C33**, 363 (1987).

2. A.M. Baldin et al. JINR, P1-87-**552**, Dubna (1987).

3. A. Kuznetsov, *Proc of the XVIII Int. Symp. on Multi-particle Dynamics, Tashkent, USSR Sept. 8-12, 1987*, p.285.

4. V.G. Grishin et al. JINR, **P1-81-542,** Dubna (1981).

5. A.M. Baldin et al. JINR, E1-84-317, Dubna (1984).

6 A.M. Baldin et al. *Yad.Fiz*, **39, 1215 (1984**).

7. H. Aihara et al., *Phys. Rev.D*, **31**, 996 (1985).

8. M. Althoff et al., *Z.Phys.C*, **30**, 355 (1986).

9. I. Juricic et al., *Phys. Rev.D*, , **39**, 1 (1989).

10.T. Åkesson et al., *Z.Phys.C*, **36**, 517 (1987).

11. N. Angelov et al., *Yad.Fiz.* **26**, 796 (1977).

12. N. Angelov et al., *Yad.Fiz.* **27**, 675 (1978).

13. N. Angelov et al., *Yad.Fiz.* **31**, 411 (1980).

14. N. Angelov et al., *Yad.Fiz.* **33**, 1257 (1981).

15. N.S. Angelov et al., *Dokl.BAN* **35**, 1061 (1982).

16. N. Akhababian, et al., *Z.Phys.C*, **18**, 97 (1983).

17. G.N. Agakishiev et al., *Yad.Fiz.* **39**, 543 (1984).

18. J. Pluta, *Proc.Int.Nucl.Phys.Conf. Harrogate U.K. 25÷30 August 1986*, **1**, 463 (1986)

19. V.G. Grishin et al., *Yad.Fiz.* **47**, 439 (1988).

20. V.G. Grishin et al., JINR, P1-89-639, Dubna (1989)

21. M.I. Podgoretsky, EPAN (ЭЧАЯ), **20/3, 628 (1989**).

22. N. Angelov et al., JINR, P1-83-309, Dubna (1983).

23. J.L. Bailly et al., *Z.Phys C.* **43**, 341 (1989).

24. T. Åkesson et al., *Phys.Lett.B*, **129**, 269 (1983).

25. C. Albajar et al., *Phys.Lett.B*, **226**, 410 (1989).

26. A. Bamberger et al., *Phys.Lett.B*, **203**, 320 (1988).

# ON THE EFFICIENCY AND METHOD OF CHECKING THE RELATIVITY BASED ON STUDYING THE INTERACTION BETWEEN HIGH-ENERGY PARTICLES

I.L. ROZENTAL

Space Research Institute Academy of Sciences.

117810 Moscow,Profsojuznaya 84/32, USSR

ABSTRACT

The paper deals with the pros and cons of the hydrodynamic theory of multiple processes. It analyses one difficulty of the hydrodynamic theory which is connected with the relativity. The experiment is proposed to check the relativity based on studying the interaction between high-energy heavy ions.

## 1. PROS OF HYDRODYNAMIC THEORY

The hydrodynamic model proposed more than thirty five years ago permits unpresedented euristic approaches. Indeed, already in the early stages of model development the most important characteristics of multiple processes were predicted and then were convincingly confirmed by the latest generation collider experiment.

These characteristics are as follows:

1. Rapidity and pseudorapidity distributions[1].
2. Average, multiplicity dependence total energy [1,2].
3. Distribution over transverse momenta [3,4].

4. Weak dependence of mean transverse momentum $< p_\perp >$ on the energy $\sqrt{s}$ [5].
5. Weak dependence of average multiplicity $<N>(A)$ on the target nucleus atomic number [6].
6. Appearance of liptons in the hadronic collision [7].

Among these predictions the emphasis should be given to the phenomena which refer to the transverse motion of quark-gluonic fluid. First, these characteristics are weak-dependent on not well defined initial conditions, second they involve an extremely fine effect - very slow growth of $< p_\perp >$ with energy.

In view of great importance of this effect the theoretical data were recalculated and turned out to be in good agreement with the Collider and Tevatron results (up to energies $\sqrt{s}$ = 1800 GeV) [8,9]. The experimental data correlate with the available theoretical predictions with the dependence of mean transverse momentum on multiplicity [10,11]. In conclusion we note that in primary Landau's work [1] the thermal motion was not taken into consideration. The thermal motion was introduced into hydrodynamic theory in paper [3] and on this ground the correct value $< P_t >$ was calculated [4].

## 2. CONS OF HYDRODYNAMIC THEORY

In addition to the well-known uncertainty of initial conditions attention should be drawn to the other problem.

In the relativity there are some questions which do not have general solution. More than one hundred works were published which analyze these problems. All of them without exception, are related to extensive or macrosonic bodies and have the following grounds:

1. the boundary condition problem;
2. the averaging problem;
3. the non-local (collective) interaction.

Common reason: the extensive bodies distort the characteristics of the Minkovski space. The mentioned problems manifest themselves in the hydrodynamic theory. All elements of the relativistic fluid have a boundary and interact with the neighboring elements. Therefore it is impossible to define a single element of the fluid. The condition that the proper time of the element is equal zero does not agree with the condition of isotropy of particles motion in the element [12]. So there are the grounds to check the relativity in the condition, when perhaps the non-local interaction exists to which the well-known experiments of NA-35 collaboration indicate [13-15]. This experiment employed an approach to studying identical particle interference in different rapidity intervals. Probably in this condition the collective (non-local) interaction exists. Due to this fact a test was made to check the relativity.

## 3. THE PROPOSED EXPERIMENT

The main idea of the experiment is to measure and to compare the inclusive distribution $\frac{dN}{dy}$ in the two different reference frames (for example in C- and L-reference frames) with the similar parameter s. It is expected to put into operation heavy ions colliding beams accelerators at the BNL after which it will be possible to conduct the above mentioned measurements. A more simple way of studying the relativity is comparison of the distributions $\frac{dN}{dy}$. It should be borne in mind that the pseudorapidity is not as accurate invariant variable. There is one positive argument in favour of this proposal. For new generation of accelerators very few fundamental problems were formulated. Therefore, the study of new, even risky fundamental problem may justify the creation of great accelerators. The author is grateful to M.I. Podgoretskii for discussing the questions mentioned in the paper.

## REFERENCES

1. L.D. Landau, Izv. Acad. Nauk SSSR. 17 (1953) 51. Collected papers ed. D. Ter Haar (Pergamon, Oxford 1965) p. 569
2. S.Z. Belenkii a.ð. Uspechi Phys. Nauk 62 (1957) 1
3. I.L. Rozental. Journ. Eksp. Theor. Fiz.(JETP). 31 (1956) 278
4. G.A. Milechin, I.L. Rozental. JETP. 33 (1957) 309
5. G.A. Milechin. Proc. Int. Conf. Cosmic Ray. Moscow, vol. 1, p. 212, 1960
6. I.L. Rozental, D.S. Chernavskii. Uspechi Phys. Nauk 52 (1954) 185
7. E.L. Feinberg. Izv. Acad. Nauk SSSR, 26 (1962) 622
8. I.L. Rozental, Yu.A.Tarasov. JETP, 85 (1983) 1535
9. I.L. Rozental, Yu.A. Tarasov. Izv. Acad. Nauk SSSR, 50 (1986) 2094
10. Yu.A. Tarasov. Yadernaja Phys. 42 (1985) 411
11. Yu.A. Tarasov JETP Letters 43 (1986) 562
12. M.I. Gorenstein, Yu.M. Sinyukov. Local anisotropy effects in the hydrodynamical theory of multiparticle production. Kiev 1983. Preprint Inst. Theor. Phys. Academy of Sciences Uk. SSR; 83-128E
13. NA35 Collaboration (A. Bamberger e.a.) Phys. Lett. B 203 (1988) 320.

14. T.Y. Humanic. Zt. Phys. C38 (1988) 79

15. E.Y. Shuryak. Phys. Lett. B207 (1988) 345.

# Bose-Einstein Correlations in $e^+e^-$ Annihilations at 10 GeV Center-of-Mass Energy

G. Herrera◇
Institut für Physik V,Universität Dortmund†
4600 Dortmund 50, F R Germany

**Abstract**

Using the ARGUS detector at the DORIS II $e^+e^-$ storage ring, Bose-Einstein Correlations in $\pi$−pairs are studied. Preliminary results on the space time size and on the coherence factor are presented both, for the hadronic decay of the $\Upsilon(1S)$ resonance and for the $e^+e^-$ annihilation into two quarks. The problems in measuring the effect in K-pairs at this energy are discussed.

# 1 Introduction

The Bose-Einstein effect, 30 years after its first observation in high energy physics (GGLP effect) [1,2], continues to be an important tool for the study of the fragmentation process. It enables one to extract information about the space-time interaction region. One expects that its geometry should reflect, to a large extent, the underlying physics during the hadronization of the system. In the region of energy around 10 GeV, it is interesting to make inferences about the q$\bar{\text{q}}$- and 3-gluon-fragmentation processes by comparing the Bose-Einstein correlation in the direct decay of the $\Upsilon(1S)$ with those of the continuum nearby. The Kopilov-Podgoretsky [3] and Goldhaber et al [2] parametrizations are used. In the Kopilov-Podgoretsky parametrization the Correlation Function is constructed as a function of the component of $\vec{q} = (\vec{p}_1 - \vec{p}_2)$ (the vectorial momentum difference) transverse to the direction of the pair, while in the Goldhaber parametrization the relative squared four-momentum of the pair $Q^2 = -(p_1 - p_2) \cdot (p_1 - p_2)$ is used. A gaussian form in these variables with amplitude $\alpha$ (coherence factor) and width $\beta$ (coherence length) was taken to fit the distributions.

---

◇Representing the ARGUS collaboration, this work has been supported by a fellowship of DAAD
†Supported by the Bundesministerium für Forschung und Technologie (BMFT), under contract number 054DO51P.

# 2 Data analysis

The analysis is based on a data sample collected with the ARGUS detector at the DORIS II storage ring at DESY. The data sample comprises an integrated luminosity of 21.6 $pb^{-1}$ at 9.4 and 10.5 GeV continuum energies and 22.5 $pb^{-1}$ on the $\Upsilon(1S)$ resonance. The ARGUS detector is a $4\pi$ magnetic spectrometer. Details of the detector, the trigger and the event reconstruction are discussed elsewhere [4]. Charged particle identification was made on the basis of specific energy loss in the drift chamber (dE/dx) and time-of-flight (TOF) measurements. This information was used to calculate, for all charged tracks, a likelihood ratio for each particle hypothesis (e,$\mu$,$\pi$,K,p). All particle hypotheses with a likelihood ratio larger than 1% were accepted.

Since the detector is triggered not only by events of interest, namely multihadron events due to $e^+e^-$ annihilation into $q\bar{q}$-pairs and direct decays of the $\Upsilon$ resonances, but also by background from QED processes, beam-gas and beam-wall interactions and two-photon processes, global selection criteria are needed to suppress the unwanted background. As discussed in detail in ref. 5 these backgrounds are reduced to a negligible level ($< 2\%$) by requiring the following :

(a) 3 or more charged tracks reconstructed with a common vertex within a radius $r < 1.5$ cm and a longitudinal distance $|z| < 5$ cm of the nominal interaction point (if the three tracks have no common vertex an energy of at least 1.7 GeV has to be deposited in the shower counters).

(b) the angle between each track and the beam axis must be greater than $23^0$.

(c) in order to suppress beam-wall, beam-gas and $\gamma\gamma$-events the normalized scalar momentum sum of charged and neutral particles and the normalized momentum sum along the beam axis is constructed and a "parabola cut" in these variables is applied.

These requirements reject most background events while retaining nearly all multihadron events, i.e. only 5% of all $q\bar{q}$ events are lost by these cuts.

A residual background from radiative QED events is largely eliminated by requiring at least 2 photons to be detected in the calorimeter [5].

The radius of the source ($r = 0.197\sqrt{\beta}$ fm) is obtained from the fit to the $Q^2$-spectrum in the plots of the ratio of like- and unlike-sign boson pairs. There exist experimental limitations in the measurement of the radius that can be determined from our data. The finite resolution of the detector imposes a minimum value in $Q^2$ which can be resolved and thus the maximum value of the parameter $\beta$ to be determined.

Using Monte Carlo simulations of the detector this maximum value was determined as follows: The difference between Q as calculated from the momenta of pions given by the event generator, and Q as calculated from the reconstructed tracks was plotted for unlike-sign pairs in continuum data. The distribution has a full width at half maximum (FWHM) of $\Delta Q = 14.028$ MeV/c (similar results are obtained in resonance data and like charged pairs). We find

$$r_{max} = \hbar c \sqrt{\frac{\ln 2}{\Delta Q^2}} = 11.7 \text{ fm},$$

this is well above the expected value of 1 fm found in other $e^+e^-$ collision experiments.
The limited experimental resolution introduces a systematic error in the Bose-Einstein enhancement. With the help of Monte Carlo data and a detector simulation an estimation of this error was made. An expected systematic error of 8 % on the coherence factor $\alpha$ and 4 % on the coherence length $\beta$ is derived.

# 3 Analysis for pion pairs

After having selected multihadronic events as described in the last section, some criteria are imposed to the remaining events in order to reduce resonance effects and to eliminate events with undesirable characteristics.

- A pion hypothesis was accepted if the corresponding likelihood ratio exceeded 10 % and the particle was assumed to be produced at the main vertex if the $\chi^2$ from the vertex fit was less than 36.

- The particle momentum was required to be in the range of 0.2-2.0 GeV/c, the lower cutoff is imposed for the absorption in the detector material. The upper cutoff corresponds to $x_F = 2p/\sqrt{s} < 2/5$ and is applied to avoid phase space limitations (if a particle with large momentum appears in the event, it is unlikely for another to appear, with similar momentum and thus few pairs in this event would have small Q).

- A transverse momentum $p_t$ larger than 0.08 GeV/c was required to guarantee a good acceptance in the drift chamber.

- A cut in the opening angle for the two tracks of $cos(\theta_{\pi\pi}) < 0.998$ was required to avoid acceptance problems for the like charged pairs. It also removed from the like charged pion pairs sample all those broken tracks having being reconstructed as two tracks and the remaining background from converted photons in the unlike-sign pion pairs.

- In order to improve the detector resolution and to reduce the systematic error due to this experimental limitation, a cut on the momentum error of the track $\sigma(p) < 0.01$ was applied.

After these cuts we are left with a sample of 54412 events in the continuum and 226520 at the $\Upsilon(1S)$ center-of-mass energy.

In order to measure the Bose-Einstein effect in like-charged boson pairs, it is necessary to have a comparison sample which does not exhibit the BE correlation. An obvious choice are unlike charged bosons since there is no reason to expect preferences in the phase space distribution, although the presence of resonances must be taken into account. The combination of pions from different events - event mixing - represents another possible reference sample; it is free of resonances and of course it does not exhibit BE-Correlation. Both kinds of study samples were used in the analysis and the results were compared.
To construct the event-mixing sample we combined pions from different events using the particle momenta as expressed in an event-related coordinate system, defined for the eigenvectors $\hat{e}_1, \hat{e}_2, \hat{e}_3$ of the momentum tensor

$$M_{\alpha\beta} = \frac{\sum_i p_i^\alpha p_i^\beta}{\sum_i p_i^2}$$

where $p_i^\alpha$ is the $\alpha$-component of the momentum for the i-th particle (as was done in ref. 6 ). This method avoids the large invariant masses that appears due to the jet structure correlations. Moreover the energy-momentum violation becomes softer. In order to avoid different acceptances the sphericity axes of two events were required to agree in polar angle to within $3^0$ for the particles to be mixed and the polar angle of sphericity was required to be greater than $40^0$ to assure that the event was well contained in the detector.

Any particles which were misidentified as pions are not subject to the Bose-Einstein statistics. They appear as an uncorrelated background and a special correction is needed. With the use of MC data, the fraction of misidentified pions was determined and a misidentification correction applied.

The final state of two or three charged bosons is affected by strong and electromagnetic interactions. The effects on the Bose-Einstein enhancement due to the strong interaction have been estimated by Suzuki [7]. A correction of this effect has not been performed in this analysis.
Two unlike (like) charged bosons experience electromagnetic attraction (repulsion). This supresses the number of pairs in the region of small relative momenta. The correction factors for this effect, known as Gamov factors [8], were applied to the experimental data. The Coulomb correction was not applied to the event mixing sample.

Many of the pions used to construct the correlation distribution come from decays of resonances and long lived particles, this constitutes a background which

reduces the strength of the effect. A pair in which one of the pions comes directly from the fragmentation and the other from a resonance decay is expected to be less correlated than if both came directly from the fragmentation. A possible check of this is obtained by combining pions from $K^0$ decays with pions from the fragmentation [9].
The suppression of the coherence factor $\alpha$ is in fact caused mainly by resonance and particle decays. Among the most abundant in the region of $Q^2$ under examination are $\eta$, $\eta\prime$, $\omega$,$K^0$ and $\rho$.

The correlation function and the fitted curve for the continuum and for the $\Upsilon(1S)$ full corrected data are shown in fig. 1. To suppress the effects of resonances the histograms were fitted in the following intervals : [0.0175,0.12] (GeV$^2$), [0.7,1.0] (GeV$^2$) for the $Q^2$ parameter and [0.12,0.62] (GeV) for the $q_t$ variable. Table 1

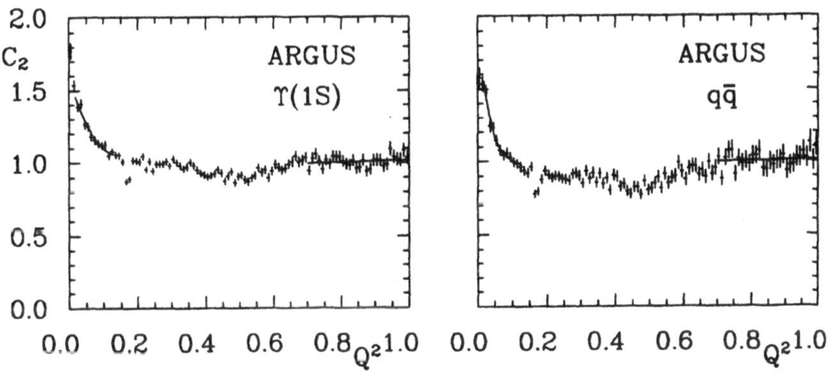


Figure 1: Correlation function for continuum and $\Upsilon(1S)$ data as a function of $Q^2$ after misidentification and Coulomb correction. The $\eta$'s were subtracted from both data samples. In the case of the $\Upsilon(1S)$ data sample, a subtraction was performed to remove the fraction of $\Upsilon(1S)$ that decays into $q\bar{q}$.

summarizes the results of the fits. The difference observed in the coherence factor ($\alpha$) between continuum and $\Upsilon(1S)$ data as a function of $Q^2$, is due to the respective difference in abundancy of $\omega$'s $\eta\prime$'s and $\eta$'s at these two energies. The LUND model predicts $0.64\frac{\omega}{event}$ and $0.31\frac{\eta\prime}{event}$ in $\Upsilon(1S)$ and only $0.47\frac{\omega}{event}$ and $0.22\frac{\eta\prime}{event}$ in continuum data, i. e. the suppression in $\alpha$ is larger at the resonance energy (see table 1). This has been observed during the subtraction of $\eta$'s which was possible from the ARGUS measurement [10] of its production rates. The difference in the coherence

factor between $\Upsilon(1S)$ and continuum data becomes smaller in the absence of $\eta$'s and would not exist if $\omega$'s and $\eta\prime$'s were also subtracted.
In terms of the transverse momentum $q_t$, the BE-effect is not strongly affected for resonance decays, since the $q_t$ distribution of pions originating from resonances is rather flat in the fitted region.

| Fitted Expression: $\gamma(1+\delta Q^2)(1+\alpha e^{-\beta Q^2})$ | | | | | |
|---|---|---|---|---|---|
| | $\gamma$ | $\delta$ | $\alpha$ | $\beta$ | $\frac{\chi^2}{DOF=36}$ |
| $\Upsilon(1S)$ | 0.941±0.062 | 0.081±0.084 | 0.718±0.092 | 15.08±3.27 | 38.5 |
| Continuum | 0.960±0.041 | 0.054±0.055 | 1.116±0.199 | 32.96±8.08 | 30.3 |
| Fitted Expression: $\gamma(1+\alpha e^{-\beta q_t^2})$ | | | | | |
| | $\gamma$ | | $\alpha$ | $\beta$ | $\frac{\chi^2}{DOF=36}$ |
| $\Upsilon(1S)$ | 0.958±0.004 | | 0.369±0.025 | 17.95±1.68 | 42.8 |
| Continuum | 0.933±0.007 | | 0.394±0.047 | 19.67±3.15 | 24.8 |

Table 1: Results of the fits in the GGLP (above) and Kopilev-Podgoretsky (below) parametrizations. The $\gamma$ and $\delta$ parameters are used to describe the background.

# 4 Analysis for Kaon pairs

It is evidently also possible to see such correlation effects in other systems of bosons. Only few data exist on Kaon correlations; a work on $K_S^0 K_S^0$ correlations has been reported [11] and another on Bose-Einstein correlations in kaons in $\alpha\alpha$, $pp$ and $p\overline{p}$ collisions [12]. The study of Bose-Einstein correlations in K pairs is particularly interesting since the production of the strange quark introduces a new element into the fragmentation dynamics. Unfortunately in $e^+e^-$ annihilations at 10 GeV the low abundancy of kaons represents a major problem in the observation of the Bose-Einstein enhancement. Only 0.908 $K^{\pm}$/event at $\Upsilon$(1S) and 0.888 $K^{\pm}$/event at continuum center-of-mass energy are produced. In particular, the charm production and its subsequent decay accounts for the most of the existing kaons which, as we know, constitutes a non correlated sample of kaons, ( the production rates u:d:c:s are 4:1:4:1 and the charm quark decays almost always into strangeness ). In addition, the presence of the $\phi$ resonance makes impossible the observation of the Bose-Einstein effect in kaon pairs at $\sqrt{s} = 10\ GeV$.

# 5 Summary

Using the ARGUS detector, we have extracted the coherence factor and the source radius of the pion source in the hadronic decay of the $\Upsilon(1S)$ resonance and in the $e^+e^-$ annihilation into two quarks. We see no difference between the hadronization of quarks and gluons when $q_t$ is used as parameter and the data agree with the results of the CLEO experiment [13] in the same energy region. The small difference between $\Upsilon(1S)$ and continuum data observed when $Q^2$ was used as parameter can be explained in terms of $\eta\prime$ and $\omega$ contributions which have not been subtracted and is expected to be larger for the $\Upsilon(1S)$ than for the Continuum data.
The results are summarized in Table 1 where the $\beta$ value corresponds to a radius of $0.765 \pm 0.166$ fm in $\Upsilon(1S)$ and to $1.131 \pm 0.277$ fm in the Continuum for the $Q^2$ parametrization and to $0.835 \pm 0.078$ fm and $0.874 \pm 0.139$ fm respectively for the $q_t$ parametrization. The event mixing method produces similar results using the parameter $q_t$ but this is not the case for the $Q^2$ one, where the values of $\alpha$ are rather small. Improvements on the event mixing technique are under study.

## References


1. G. Goldhaber et al., *Phys. Rev. Lett.* **3**, (1959)181
2. G. Goldhaber et al., *Phys. Rev.* **120**, (1960)300
3. G. Kopilev and M. Podgoretsky, *Yad. Fiz.* **18** (1973)656
   *Sov. J. Nucl. Phys.* **18**, (1974)336
4. ARGUS Collab., H.Albrecht et al., *Nucl. Inst. and Meth.* **A275** (1989) 1.
5. A. Drescher, Ph. D. thesis, Universität Dortmund 1987
   ARGUS Collab., H.Albrecht et al., *Z.Phys.* **C41** (1989)557
6. H. Aihara et al. *Phys. Rev.* **D31**, (1985)996
7. M. Suzuki, *Phys. Rev.* **D35**, (1987)3359
8. A. S. Davydov, *Quantum Mechanics*, Pergamon Press, London (1965)
9. G. Herrera, Ph.D. work University of Dortmund in preparation
10. ARGUS Collab., H.Albrecht et al., *DESY* **89**-066
11. T. Åkesson et al, *Phys. Lett.* **155B**, (1985)128
12. A. M. Cooper et al, *Nucl. Phys.* **B139**, (1978)45
13. P. Avery et al, *Phys. Rev.* **D32**, (1985)2294

# OBSERVATION OF THE COULOMB INTERACTION EFFECT IN PION PAIRS FROM THE REACTION $p+Ta \to \pi^{+}+\pi^{-}+X$ AT 70 GeV

L.G. Afanasyev, A.S. Chvyrov, V.V. Karpukhin, V.I. Komarov, A.V.Kolomyichenko, V.V.Kruglov, A.V.Kuptsov, L.L.Nemenov, M.V.Nikitin, M.A. Ivanov, Zh.P. Pustylnik, M.N. Shumakov, S.M. Frolikov
Laboratory of Nuclear Problems JINR, P.O.Box 79, Moscow I0I000 USSR

O.E. Gorchakov, A.V. Kulikov, S.V. Trusov
Moscow State Univ., Inst.for Nuclear Physics, Dubna Br., Dubna I4I980 USSR

A.P.Kurov
Institute for High Energy Physics, Serpukhov I42284 USSR

V.V.Yaskov
Moscow State University, Moscow II9899 USSR

ABSTRACT

Pion pair production from the proton-Tantal nuclear interaction at proton energy 70 GeV has been studied. Pion correlations as a function of the relative momentum in a pair CMS have been measured at the low momentum region of momenta below 40 MeV/c. The experimental data agree with the calculation that takes into account the Coulomb factor in the differential cross section of pion pairs production.

Coulomb interaction between the components of a particle pair essentially changes the pair production cross section at small relative momenta. If this interaction has a relatively weak influence on the production process itself, proceeding at small distance, then the differential cross section of particle production with momenta $\vec{k}_1$ and $\vec{k}_2$ can be factorized in the form

$$d\sigma(\vec{k}_1,\vec{k}_2) \simeq A_c(|\vec{k}_1-\vec{k}_2|)\cdot d\sigma_0(\vec{k}_1,\vec{k}_2), \qquad (I)$$

where $d\sigma_0$ means the cross section when one neglects the interaction between the particle charges $Z_1$, $Z_2$ and $A_c$ is a Coulomb factor:

$$A_c(q)=2\pi\eta/[\exp(2\pi\eta)-1], \quad \eta=z_1z_2\alpha/\beta_{12}\simeq z_1z_2\,2k_c/q \,. \tag{2}$$

Here $k_c=m\alpha$ stands for the characteristic Coulomb momentum $k_c$ and $q=|\vec{k}_1-\vec{k}_2|$ for the relative momentum in the C.M. frame of the pair.

The factorization in form (I), (2) for the pair production processes was obtained first by Sakharov /I/ for the case of $e^+e^-$-pair photoproduction in the Coulomb field of a nucleus. The Coulomb effect for the pion pair production in the nucleus-nucleus collisions was investigated theoretically in ref. /2/ and for a wider range of processes in ref. /3/. The effect is difficult to be experimentally observed because of the small width of the Coulomb peak, so the precision of its describing in the form (I), (2) was out of question. However it has recently become important in connection with the study of interference of identical particles and especially with the investigation of the pion dimesoatoms /4/.

In the present experiment $\pi^+\pi^-$-pairs were detected at the mean angle of 8.4° to the proton beam. The angular acceptance was equal to $4\cdot 10^{-5}$sr. An earlier unstudied region of small q (0–40 MeV/c) was investigated with the resolution of (q) = I.2 MeV/c. The correlation function

$$R=N\frac{d\sigma}{d\vec{p}_+\,d\vec{p}_-}\Big/\left(\frac{d\sigma}{d\vec{p}_+}\cdot\frac{d\sigma}{d\vec{p}_-}\right) \tag{3}$$

was obtained, where $\vec{p}_+$ and $\vec{p}_-$ are the $\pi^+$- and $\pi^-$-momenta respectively, and $N$ is a normalizing constant chosen so that $R(q)\to 1$ at $q\to\infty$.

The experimental two-pion correlation as a function of the momenta $q$ and its longitudinal $q_L$ and transverse $q_T$ components with respect to the mean momentum $\vec{p}=(\vec{p}_+ + \vec{p}_-)/2$ direction is shown in the Figure. The curves are the result of the calculation where the correlation is caused by the Coulomb factor $A_c$ alone. The calculation describes the experimental data somewhat better if the pairs with one pion resulting from $K^0_S$- or $\eta$-meson decay are taken into account. The subsequent experiments will probably

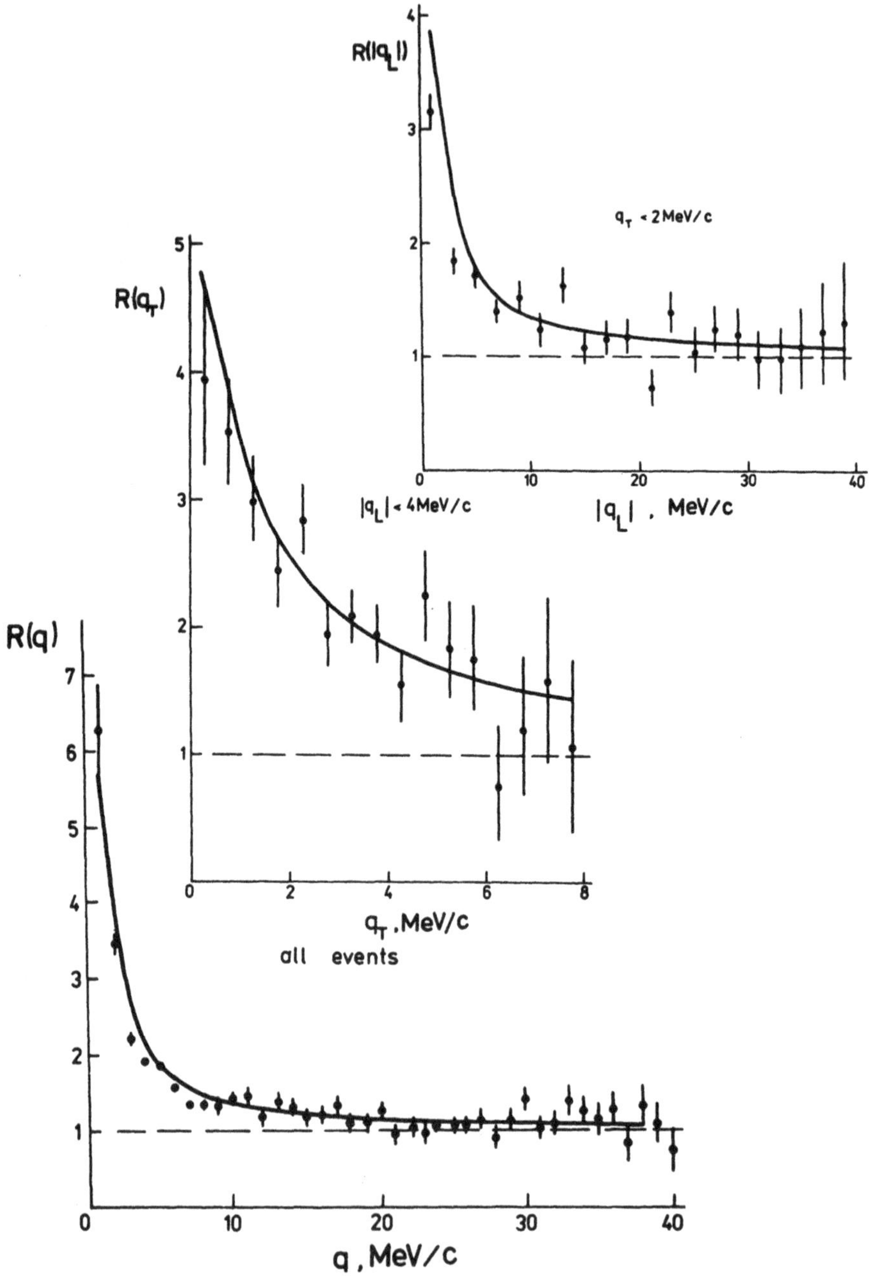
R(|q_L|)
4
3
2
1
0
10
20
30
40
q_T < 2MeV/c
|q_L| , MeV/c
R(q_T)
5
4
3
2
1
0
2
4
6
8
|q_L| < 4MeV/c
q_T , MeV/c
all events
R(q)
7
6
5
4
3
2
1
0
10
20
30
40
q , MeV/c

Figure I.

reveal the value of the contribution of pion pairs produced by dissociation of $A_{2\pi}$-dimesoatoms in the target into the spectrum region with the momenta $q_T \lesssim$ I.6 MeV/c, $|q_L| \lesssim$ I.2 MeV/c.

## REFERENCES


I. Sakharov A.D., JETPh I8, 63I (I948).

2. Gyulassy M. e.a., Phys. Rev. 20C, 2267 (I979).

3. Lednicky R. and Lyuboshitz V.L., Jad.Fis. 35, I3I6 (I982).

4. Nemenov L.L., Jad.Fis. 4I, 980 (I985).

# Soft pions and geometry in Bose-Einstein correlations

I.V. Andreev[1] and R.M.Weiner[2]

1 P.N. Lebedev Physical Institute, Moscow, USSR
2 Physics Department, Univ. of Marburg, F.R.G.

**Abstract**

The two exponential feature in Bose-Einstein correlations (BEC) due to the presence of a coherent component is rediscussed in a space-time approach. The mixed-coherent-chaotic exponent in this formalism is strongly damped unless soft hadrons are considered. The relation between the geometrical radius and the coherence length of the chaotic fields is exhibited in a simple case.

Introduction. There are various theoretical reasons why one expects that coherent states play a significant role in hadronic reactions (cf. e.g. [1],[2]). Experimentally effects of coherence should be seen among other things in Bose-Einstein correlations (BEC). Unfortunately the simplest way to look in BEC for coherence, namely in the intercept $C_2(0)$, does not work satisfactorily, because of other effects which might mimic the effect of coherence. Here $C_2$ is the two body correlation function for identical particles

$$C_2(y_1, y_2) = P_2(y_1, y_2)/[P_1(y_1) \cdot P_1(y_2)] \tag{1}$$

where $y_i$ are kinematical (multicomponent) variables which characterize the state of a particle (momenta, rapidity etc.) and $P_k$ the corresponding two particle and single particle probabilities. If $C_2$ depends only on $|y_1 - y_2|$, then the value $C_2(0)$ is strongly dependent on whether the particles are in a coherent state (in that case $C_2(0) = 1$) or in a chaotic state ($C_2(0) = 2$), but to disentangle this dependence on coherence from other effects like final state interactions, insufficient resolution on $|y_1 - y_2|$ etc. is at the present stage quite difficult. However, it has been pointed out recently [1],[2] that another "criterion" for coherence may be available, if one considers a more accurate expression for $C_2$ than that used so far in experimental analysis. Indeed, the usual empirical parametrization of $C_{(2)}$ is

$$C_{(2)} \sim 1 + \lambda exp(-|y_1 - y_2|^2 \cdot R^2) \tag{2}$$

where $\lambda$ is an ad hoc "incoherence" factor and $R$ the source "radius" defined by a given space-time distribution of the source. The correct expression for $C_{(2)}$, however, contains in the presence of a coherent component two exponentials and the detection of this two exponential structure could constitute evidence for coherence, provided the ratios of these two exponentials are those predicted by the theory.

The derivation of the two exponentials relation for BEC was given in [1],[2] in a quantum statistical (QS) treatment in momentum (rapidity) space, without reference to the space-time distribution of the source. Given the possible important implications of this result, it appears necessary to investigate its space-time aspects.

The quantum statistical formalism. In the simplified treatment of BEC the two exponential formula for the correlation functions is

$$C_{(2)} = 1 + 2p(1-p)exp(-|y_1 - y_2|^2 \cdot \tilde{R}^2) + p^2 exp(-2|y_1 - y_2|^2 \cdot \tilde{R}^2) \qquad (3)$$

where $p$ is the chaoticity:

$$p = < n_{ch} > / < n >, and < n > = < n_{ch} > + < n_c > \qquad (4)$$

$< n_{ch} >$ and $< n_c >$ are the mean numbers of particles due to the chaotic and coherent fields respectively and $< n >$ the total mean multiplicity. Here $\tilde{R}$ is the inverse coherence (correlation) length in $y$ space defined in terms of the chaotic component of the field $\pi_{ch}(y)$:

$$< \pi^*_{ch}(y_1)\pi_{ch}(y_2) > = < n_{ch,y} > exp(-|y_1 - y_2|^2 \cdot \tilde{R}^2) \qquad (5)$$

$< n_{ch,y} >$ is the mean chaotic number density related to the $< n_{ch} >$ by $< n_{ch} > = \int < n_{ch,y} > dy$. If $y$ stands for the 3 momentum $k$, e.g. then

$$< n_{ch} > = \int < n_{ch,k} > d^3k.$$

Formally eq. (3) reduces to (2) for $p = 1$, so that experimental evidence for two exponentials with the "radii" and their coefficients given by eq. (3) would be a strong indication for coherence. On the other hand eq. (3) is based on the following simplifying assumptions:

A. A QS formalism is used, in which two extremes (chaotic and coherent) correspond to particular forms of the density matrix in the $\mathcal{P}$ representation (gaussian and delta-function respectively). It is known that more general situations can occur [3],[4]. Thus e.g. $C_2(0)$ can take for squeezed coherent state values below 1 and above 2.

B. "Stationarity" in the variables $y$ was assumed (eq. (5)) and the $y$ dependence of the coherent fields was neglected. If $y$ refers to rapidity, these assumptions are suggested by the existence of the rapidity plateau and imply, among other things, that $p$ in eq. (3) is independent of $y$.

C. The central "frequency" of the chaotic field coincides with the "frequency" of the coherent field [5],[6]. We shall address below briefly assumption C, because it is particularly relevant for some conclusions of the present paper.

Consider the superposition of a coherent field $\pi_c(y) = A(y)exp(i\omega_c y)$ and a chaotic field $\pi_{ch}(y) = \pi'_{ch}(y)exp(i\omega_{ch}y)$ where $\omega_c$ is the "frequency" of the coherent signal (coherent fields are "monochromatic") and $\omega_{ch}$ is the central "frequency" of the chaotic field. Here "frequency" means the variable conjugated to $y$; in optics $y$ is usually the time variable and then $\omega$ is a proper frequency. If $y$ refers to rapidity then $\omega$ is the boost variable. The total field is then

$$\pi(y) = \pi_c(y) + \pi_{ch}(y) \tag{6}$$

and its complex amplitude can be written as $\pi'(y) = \pi'_c(y) + \pi'_{ch}(y)$ with

$$\Delta\omega = \omega_c - \omega_{ch} \,. \tag{7}$$

By repeating now the calculations of refs. [1], [2], one gets

$$C_{(2)} = 1 + 2p(1-p)\cos[\Delta\omega(y_1 - y_2)]exp(-|y_1 - y_2|\cdot\tilde{R}^2) + p^2 exp(-2|y_1 - y_2|\cdot\tilde{R}^2) \quad (8)$$

which differs from eq.(3) by the factor $\cos\Delta\omega(y_1 - y_2)$ in the second term of the r.h.s. With assumption C, i.e. $\Delta\omega = 0$, this factor reduces to unity and we recover eq. (3).

Space-time approach to Bose-Einstein correlations in the QS formalism. We start by considering a mixture of a coherent and chaotic field. We write the positive frequency part of the total (real) field as

$$\tilde{\pi}(x) = \tilde{\pi}_c(x) + \tilde{\pi}_{ch}(x) \quad (9)$$

where $\tilde{\pi}_c$ and $\tilde{\pi}_{ch}$ are the coherent and chaotic components and $x$ is the space-time coordinate. The amplitude of emission of two pions with four-momenta $k_1, k_2$ is

$$A(k_1, k_2) \sim \int_V d^4x_1 \int_V d^4x_2 \quad exp(ik_1x_1 + ik_2x_2)\tilde{\pi}(x_1)\tilde{\pi}(x_2) \equiv \pi(k_1)\pi(k_2). \quad (10)$$

The corresponding probability is

$$\begin{aligned} P(k_1, k_2) &= \int d^4x_1 \int d^4x_2 \int d^4x_3 \int d^4x_4 exp(ik_1x_1 + ik_2x_2 - ik_1x_3 - ik_2x_4)\cdot \\ &\cdot < \tilde{\pi}(x_1)\tilde{\pi}(x_2)\tilde{\pi}^*(x_3)\tilde{\pi}^*(x_4) > \equiv < \pi(k_1)\pi(k_2)\pi^*(k_1)\pi^*(k_2) > \quad (11) \end{aligned}$$

where the <> brackets denote the ensemble average over the chaotic fields. We define the correlation function of the chaotic components as

$$C(x_1 - x_2) = < \tilde{\pi}_{ch}(x_1)\tilde{\pi}_{ch}(x_2) > . \quad (12)$$

By writing $C$ as a function of $x_1 - x_2$ only, we have assumed stationarity in time and homogeneity in space, which is a valid assumption as long as $x_1, x_2$ are inside e volume of the source. Although formally this assumption is quite similar to sumption (5), the physics is quite different and leads, as will be shown below, to fferent results for the BEC.

he expectation value of products of fields is calculated as usual [6] taking into count that odd powers of the expectation values of the chaotic field vanish. This, course, is not the case for the coherent fields. From eqs. (9)- (12) we get then

$$\begin{aligned}(k_1,k_2) &= |\pi_c(k_1)|^2|\pi_c(k_2)|^2 + |\pi_c(k_1)|^2 < \pi^*_{ch}(k_2)\pi_{ch}(k_2) > + \\ &+ |\pi_c(k_2)|^2 < \pi^*_{ch}(k_1)\pi_{ch}(k_1) > +2Re[\pi_c(k_1)\pi_c(k_2) < \pi^*_{ch}(k_1)\pi^*_{ch}(k_2) > + \\ &+ \pi_c(k_1)\pi^*_c(k_2) < \pi^*_{ch}(k_1)\pi_{ch}(k_2) >] + \\ &+ < \pi^*_{ch}(k_1)\pi_{ch}(k_1) >< \pi^*_{ch}(k_2)\pi_{ch}(k_2) > + \\ &< \pi_{ch}(k_1)\pi_{ch}(k_2) >< \pi^*_{ch}(k_1)\pi^*_{ch}(k_2) > + < \pi^*_{ch}(k_1)\pi_{ch}(k_2) >^2.\end{aligned} \tag{13}$$

or "standard" QS [6] the averages of the type $< \pi_{ch}(k_1)\pi_{ch}(k_2) >$ vanish, so that . (14) simplifies becoming

$$\begin{aligned}(k_1,k_2) &= |\pi_c(k_1)|^2|\pi_c(k_2)|^2 + |\pi_c(k_1)|^2 < \pi^*_{ch}(k_2)\pi ch(k_2) > +|\pi_c(k_2)|^2 < \pi^*_{ch}(k_1)\pi_{ch}(k_1) > \\ &+ 2Re[\pi_c(k_1)\pi^*_c(k_2) < \pi^*_{ch}(k_1)\pi_{ch}(k_2) >] + \\ &+ < \pi^*_{ch}(k_1)\pi_{ch}(k_1) >< \pi^*_{ch}(k_2)\pi_{ch}(k_2) > + < \pi^*_{ch}(k_1)\pi_{ch}(k_2) >^2.\end{aligned} \tag{14}$$

his equation coincides with eq. (4.66) of ref. [7] and also with eq. (13) of ref. [2] r the particular case that the $k$ dependent chaoticity (cf. below) is replaced by ne global chaoticity.

/e proceed now to the parametrization of the space-time dependence of the fields. /e assume that the correlation function of the chaotic fields (12) reads

$$C(x_1 - x_2) = C(0)exp[-(x_1 - x_2)^2_0/L^2 - (\vec{x}_1 - \vec{x}_2)^2/L^2] \tag{15}$$

where $L$ is the coherence length in space-time. (The gaussian form is chosen for reasons of mathematical convenience.) Note that this form of $C$ leads to an Euclidian metric in space-time, so in the following we shall write $x^2 = x_0^2 + \vec{x}^2$. We have thus at least two length scales in the problem, one related to the volume of the source $V$ (if the source is not spherical, more than one radius $R$ appears), and one given by $L$. In particular the space-time distribution of the source is parametrized as

$$\rho(x) \sim \exp[-x^2/R_\perp - x_\parallel^2/R_\parallel^2 - x_0^2/R_0^2] \tag{16}$$

and all space-time integrals have to be performed with this distribution. From eqs. (10),(12),(15) we get

$$\begin{aligned} < \pi_{ch}(k_1)\pi_{ch}^*(k_2) > \; &= \int d^4(\frac{x+y}{2})d^4(x-y)exp[\frac{i(k_1-k_2)_\mu(x+y)_\mu}{2}] \cdot \\ &\cdot \; exp[-\frac{(\vec{x}+\vec{y})_\perp^2}{2R_\perp^2} - \frac{(x_\parallel+y_\parallel)^2}{2R_\parallel^2} - \frac{(x_0+y_0)^2}{2R_0^2}] \cdot \\ &\cdot \; exp[-\frac{(\vec{x}-\vec{y})_\perp^2}{2R_\perp^2} - \frac{(x_\parallel-y_\parallel)^2}{2R_\parallel^2} - \frac{(x_0-y_0)^2}{2R_\parallel^2}] \cdot \\ &\cdot \; exp[i\frac{(k_1+k_2)_\mu(x-y)_\mu}{2}] \cdot \\ &\cdot \; C(0)exp[-\frac{(x_0-y_0)^2}{2L^2} - \frac{(\vec{x}-\vec{y})_\perp^2}{2L^2} - \frac{(x_\parallel-y_\parallel)^2}{2L^2}]. \end{aligned} \tag{17}$$

Here the notation $a_\mu b_\mu \equiv a_0 b_0 - \vec{a}\vec{b}$ was used. For simplification we shall assume in the following $R_\perp = R_\parallel = R_0 \equiv R$. The integrals can then be performed immediately by using the relation

$$\int_{-\infty}^{\infty} dx e^{-ax^2+ibx} = \sqrt{\frac{\pi}{a}} exp(-\frac{b^2}{4a}) \tag{18}$$

and we obtain

$$< \pi_{ch}(k_1)\pi_{ch}^*(k_2) > \; = \frac{1}{16}(2\pi)^2 R^4 exp[-R^2(\vec{k_1}-\vec{k_2})^2/8 - R^2(k_1-k_2)_0^2/8] \cdot$$

$$\cdot \quad (2\pi)^2 R_L^4 C(0) exp[-R_L^2(\vec{k_1}+\vec{k_2})^2/8 - \tag{19}$$

$$- \quad R_L^2(k_1+k_2)_0^2/8] .$$

Here

$$R_L^2 = \frac{R^2 L^2}{R^2+L^2} \simeq \begin{cases} R^2 & if \quad L^2 >> R^2 \\ L^2 & if \quad L^2 << R^2 . \end{cases} \tag{20}$$

For the coherent fields we get

$$\begin{aligned} \pi_c(k_1)\pi_c^*(k_2) &= |\pi_0|^2 \int_V d^4x_1 \int_V d^4x_2 expi(k_1x_1 - k_2x_2) \cdot \\ &\cdot \quad exp[-(\vec{x}_1^2 + x_{10}^2)/R^2 - (\vec{x}_2 + x_{20})^2/R^2] = \\ &= |\pi_0|^2 \pi^4 R^8 exp\{-(R^2/8)[(\vec{k_1}+\vec{k_2})^2 + (k_{10}+k_{20})^2 + \\ &+ \quad (\vec{k_1}-\vec{k_2})^2 + (k_{10}-k_{20})^2]\} = \\ &= \pi_c(k_1)\pi_c(k_2) - \pi_c^*(k_1)\pi_c^*(k_2) \end{aligned} \tag{21}$$

where $\pi_0$ is the value of $\pi_c(x)$ at $x = 0$.

One could now define the $k$ dependent chaoticity in analogy to eq.(4) by writing:

$$p(k) \equiv < |\pi_{ch}(k)|^2 > /[< |\pi_{ch}(k)|^2 > + < |\pi_c(k)|^2 >] . \tag{22}$$

In the following we shall use for convenience instead of this quantity the value of $p$ at $k = 0$,

$$p(0) = R_L^4 C(0)/[R_L^4 C(0) + R^4 |\pi_0|^2] . \tag{23}$$

The single particle distribution reads

$$\begin{aligned} P_1(k) &= < \int_V d^4x_1 \int_V d^4x_2 \, exp(ikx_1 - ikx_2)[\pi_c(x_1) + \pi_{ch}(x_1)][\pi_c^*(x_2) + \pi_{ch}^* ch(x_2)] > \\ &= \pi_c(k)\pi_c^*(k) + < \pi_{ch}(k)\pi_{ch}^*(k) > . \end{aligned} \tag{24}$$

From eqs. (15,20) we obtain after some simple transformations

$$\begin{aligned} P_1(k) &= \pi^4 R^4 [R^4 |\pi_0|^2 + R_L^4 \;\; C(0)] exp[-\frac{R_L^2}{2} k^2] \cdot \\ &\cdot \; \{p + (1-p) exp[-\frac{1}{2}(R^2 - R_L^2)(2\vec{k}^2 + m_\pi^2)]\}. \end{aligned} \tag{25}$$

The two components of the field appear here under the form $p$ (for the chaotic part) and $1-p$ (for the coherent part) and the factor multiplying the coherent part $(1-p)$ is worth noticing. It is the analogon of the $\cos\Delta\omega$ factor in eq. (8). It has the effect that only for soft pions ($k \sim m_\pi$) the coherent term survives. This feature will reappear below also in the two- particle distribution which reads:

$$\begin{aligned} C_2(k_1, k_2) &= 1 + 2p(0)[1 - p(0)\exp[-\frac{1}{4}(R^2 - R_L^2)(k_1^2 + k_2^2)] \cdot \\ &\cdot \; \{\exp[-\frac{1}{8}(R^2 - R_L^2)(k_1 - k_2)^2]\}/A(k_1)A(k_2) + \\ &+ \; p^2(0)\{\exp[-\frac{1}{4}(R^2 - R_L^2)(k_1 - k_2)^2]\}/A(k_1)A(k_2) \end{aligned} \tag{26}$$

where

$$A(k_i) = p(0) + [1 - p(0)] exp[-\frac{1}{2}(R^2 - R_L^2)k_i^2]\,. \tag{27}$$

Main results. Let us compare now eq. (26) with eq.(3): We observe that the main characteristic, i.e. the two-exponential form, where the second exponent is the square of the first, is reobtained with $(y_1 - y_2)^2 \cdot \tilde{R}^2$ replaced by $\frac{1}{8}(k_1 - k_2)^2 \cdot (\tilde{R} - R_L^2)$. There are, however, also some important differences.

1. Soft pions and coherence

The second term in (26) which represents the interference between the chaotic and coherent fields contains in the numerator also the factor $exp[-\frac{1}{4}(R^2 - R_L^2)(k_1^2 + k_2^2)]$ which leads to a strong damping of this term except for the case $(R^2 - R_L^2)(k_1^2 +$

$k_2^2) << 1$. This last condition can be fulfilled either for very small values of $R$ or for soft pions. If one disregards the first possibility, it follows that unless cuts in momenta of the observed particles are introduced, or unless one considers only low energy data (e.g. $\bar{p} - p$ annihilation, heavy ion reactions in the 1-2 GeV/A regime or low energy $e^+ - e^-$ reactions), the coherent term and thus the first exponential is practically absent. This new effect can serve as a possible signal for the existence of coherent states in BEC. Intuitively it can be understood in the sense that for long wave length particles the system appears coherent. Formally this damping factor is the analogon of the $cos\Delta\omega(y_1 - y_2)$ term in eq. (8). Both factors contribute to the reduction of the coherent term although the physics may be different.

A consequence of eq. (26) is that soft pions are associated with an effective radius that is smaller than that seen with hard pions. This follows from the fact that for hard(er) pions the second exponent with the factor $-1/4$ dominates. Hints for such a phenomenon may have been seen by the Na35 collaboration [J. Bartke, talk prepared for the CORINNE workshop, Nantes 1990].

2. Correlation length versus radius

The geometrical "radius" $R$ and the coherence length $L$ are present in eq. (26) only in the combination $R^2 - R_L^2$. This appears to be a consequence of the fact that the parametrizations of the space-time extension of the source (16) and the correlation of the chaotic fields (15) were both chosen of gaussian form. However, in actual experiments, because of limited statistics, more than one length can hardly be measured, and then this kind of parametrization seems reasonable. It is important to note that the link between the radius $R$ and the coherence length $L$ is independent of the value of $p$, i.e. it is present also for purely chaotic fields ($p = 1$). It then follows that when measurements are performed in the variables $k_i$, the geometrical size $R$ of the source is underestimated, unless the coherence length $L$ is much smaller than $R$.

One may ask when eqs. (3),(8) and when eq. (26) apply. Eqs. (3) and (8) refer to point-like sources only and furthermore assume stationarity in $k(y)$ space (eq. (5)) which may or may not hold. Furthermore the damping factor in eq. (8) is unknown and amounts to an additional free parameter $\Delta\omega$. Eq. (26) is more general, because it describes finite size sources. However, it is based on another physics assumption, namely stationarity in space-time (eq. (12,(15)). As pointed out, this last condition is supposed to hold at least for $L < R$. All this seems to indicate that at least in the absence of a generalization of eqs. (3),(8) for finite size sources, and as long as there does not exist a direct test of eq. (5), eq. (26) is to be prefered.

We have benefitted from instructive discussions with G.N. Fowler and C.C. Shih.

This work was supported in part by the Gesellschaft für Schwerionenforschung, Darmstadt and the Heraeus-Stiftung. Hanau.

# References

[1] R.M. Weiner, From Two-Particle Correlations to Multiplicity Distributions, in Multiparticle Production, Proceedings of the Perugia Workshop, eds. R. Hwa, G. Pancheri and Y. Srivastava, World Scientific 1989, p. 231.

[2] R.M. Weiner, Physics Letters B232, 278 (1989); ibid. B242, 547 (1990).

[3] M. Schubert and B. Wilhelmi, Non-Linear Optics and Quantum Electronics, John Wiley and Sons, New York (1986).

[4] A. Vourdas and R.M. Weiner, Phys. Rev. D38, 2209 (1988).

[5] G.N. Fowler et al., Phys. Rev. D37, 3127 (1988).

[6] B. Saleh, Photoelectron Statistics, Springer-Berlin, Heidelberg, New York (1978).C20,2267 (1979).

[7] M. Gyulassy, S.K. Kaufman, and L.W. Wilson, Phys.Rev.

[8] E.V. Shuryak, Sov.J.Nucl.Phys. 18 667 (1974).

[9] G.N. Fowler and R. M. Weiner, Phys.Rev. D17, 3118 (1978).

# BOSE-EINSTEIN CORRELATIONS FOR PIONS PRODUCED IN p-p COLLISIONS AT 360 GeV/c AND 400 GeV/c

EHS-RCBC (NA23) COLLABORATION
LEBC-EHS (NA27) COLLABORATION

Tsuneo KAGEYA

Tokyo Metropolitan University, Department of Physics

Fukazawa 2-1-1, Setagaya-ku, Tokyo, Japan

## ABSTRACT

Correlations among identically charged pions were studied for pions produced in p-p collisions at 360 GeV/c and 400 GeV/c using the EHS spectrometer. The effective radius, $\rho$, and the chaoticity factor, $\lambda$, for pion production at 360 GeV/c were determined to be $\rho = 0.99 \pm 0.17$ fm, $\lambda = 0.37 \pm 0.08$ and those at 400 GeV/c to be $\rho = 1.09 \pm 0.06$ fm, $\lambda = 0.31 \pm 0.02$, where the latter is a preliminary result. These values are compared with those determined by other experiments. A model based on a superposition of chaotic and coherent fields for pions produced was employed to describe the correlations for two, three and four identical bosons observed in our data.

## 1. INTRODUCTION

In 1954, the second order interference between lights was discovered by Hanbury Brown and Twiss [1] and it was used to measure the size of stellar objects. Later the second order interference between identically charged pions produced in particle collisions was observed by Goldhaber, Goldhaber, Lee and Pais [2], which is known as the GGLP effect or as the Bose-Einstein correlations. They have found that identically charged pions with small relative momenta are emitted with enhanced probability as compared to uncorrelated ones [2].

We have studied the Bose-Einstein correlations as a function of the variable , $Q^2$, the negative value of the invariant four momentum transfer squared defined as

$$Q^2 = -(p_1 - p_2)^2 = M_{12}^2 - 4m_\pi^2, \tag{1}$$

where $p_1$, $p_2$ are the four momenta of the particles 1, 2, respectively, $M_{12}$ is the effective mass of the two particle system and $m_\pi$ the pion mass. If the pion source has a Gaussian form in space with the radius, $\rho$, of the interaction volume, then the correlation function, $C_{(2)}$, for identically charged pion pairs can be expressed as a function of $Q^2$ as follows,

$$C_{(2)}(Q^2) = 1 + \lambda \exp(-\rho^2 Q^2). \tag{2}$$

The factor, $\lambda$, is a phenomenological parameter introduced by Deutschmann *et al.* [3] to take account of partial interference, which is called "chaoticity". A completely chaotic source is realized for $\lambda = 1$ for which the Bose-Einstein correlation is maximal.

Up to the present, the effective radius, $\rho$, of the volume of pion production and the chaoticity, $\lambda$, have been measured in hadron-hadron interactions [3-12], electron-positron annihilations [13-18] and lepton-hadron interactions [19-20] in the energy range from $\sqrt{s}$= 2 to 900 GeV. They are mostly $\rho \sim 1$ fm and $\lambda < 1$, which implies that pions emitted consist of chaotic part and coherent one. Theoretical explanations for the Bose-Einstein correlations observed in high-energy hadronic processes are given by several authors [3][21-32].

Quantum statistical properties of a mode of propagation of light which is a superposition of thermal and coherent radiations was theoretically investigated by Lachs [33] and Mandel et al. [34]. Taking these properties into consideration, several authors describe the pion fields as a mixture of chaotic and coherent ones in hadronic production processes [23-24]. Based on the model of superposed chaotic and coherent emissions of pions, identically charged three- and four-pion correlations as well as two-particle correlations were derived theoretically and correlation functions are given in [23-29].

The correlation function $C_{(2)}$ of this model for two identical bosons is as follows:

$$C_{(2)}^{QS} = 1 + 2p(1-p)e^{-r^2Q^2} + p^2 e^{-2r^2Q^2}, \tag{3}$$

where $C_{(2)}^{QS}$ means the correlation function derived by the quantum statistics. The parameter, $r$, represents the source size of the chaotic emission of pions and $p$, the chaoticity. In this model we assume that the chaotic source of bosons is Gaussian form in space with the radius, $r$, and the coherent one doesn't have any $Q^2$ dependence.

The correlation function, $C_{(3)}$, for three-particle correlations can be written [25][28-29] as:

$$C_{(3)}^{QS} = 1 + 6p(1-p)e^{-\frac{1}{3}r^2Q_{3\pi}^2} + 3p^2(3-2p)e^{-\frac{2}{3}r^2Q_{3\pi}^2} + 2p^3 e^{-r^2Q_{3\pi}^2}, \tag{4}$$

where $Q_{3\pi}^2$ is defined as

$$Q_{3\pi}^2 = M_{3\pi}^2 - 9m_\pi^2,$$

using the effective mass of the three-boson system, $M_{3\pi}$. In this case, the following relation is asummed:

$$Q_{12}^2 = Q_{23}^2 = Q_{31}^2 = \frac{1}{3}Q_{3\pi}^2,$$

for $Q_{12}^2(Q_{23}^2, Q_{31}^2)$, which is $Q^2$ of particles 1 and 2(2 and 3, 3 and 1).

In the same way as in the equation (4), the correlation function, $C_{(4)}$, for four identical bosons is given as [29],

$$\begin{aligned} C_{(4)}^{QS} &= 1 + 12p(1-p)e^{-\frac{1}{12}r^2Q_{4\pi}^2} + 6p^2(7 - 8p + 2p^2)e^{-\frac{1}{6}r^2Q_{4\pi}^2} \\ &+ 4p^3(11 - 9p)e^{-\frac{1}{4}r^2Q_{4\pi}^2} + 9p^4e^{-\frac{1}{3}r^2Q_{4\pi}^2}, \end{aligned} \tag{5}$$

where

$$Q_{4\pi}^2 = M_{4\pi}^2 - 16m_\pi^2,$$

and $M_{4\pi}$ is the effective mass of the four-boson system. For $Q_{i,j}^2 (i = 1 \sim 4, j = 1 \sim 4)$, $Q^2$ of particles $i$ and $j$ in the four-boson system, the following assumption is made,

$$Q_{12}^2 = Q_{13}^2 = Q_{14}^2 = Q_{23}^2 = Q_{24}^2 = Q_{34}^2 = \frac{1}{6}Q_{4\pi}^2.$$

## 2. EXPERIMENTAL APPARATUS AND DATA

The NA23 data was obtained at the proton beam momentum of 360 GeV/c ($\sqrt{s}$ = 26 GeV) of CERN-SPS using European Hybrid Spectrometer (EHS) [35]. From an exposure of the hydrogen Rapid Cycling Bubble Chamber (RCBC) [36], we selected data samples consisting of 9,884 events having more than 4 prongs for which all charged tracks were measured. The experimental set up is shown in Fig.1. Pions were identified by ISIS (identification of secondaries by ionization sampling). For the tracks having no ISIS information, those which stopped in the bubble chamber were identified as protons while all other tracks were assigned as pions.

The NA27 experiment was performed at CERN-SPS using European Hybrid Spectrometer (EHS) and the small Lexan liquid hydrogen bubble chamber (LEBC) exposed to beams of 400 GeV/c ($\sqrt{s} = 27$ Gev) [37] as shown in Fig. 1. The goal of this experiment was to study the production properties of charmed hadrons [37]. For the bulk of the events where no charmed hadron candidate has been observed, the LEBC pictures were not measured, whereas the coordinates of the primary vertices and their charged multiplicities were recorded and could be used in conjugation with EHS spectrometer information. It was found that the reconstruction efficiency of tracks by the EHS spectrometer is high enough in the kinematical region $X_F > 0$ and , that is, more than 95 % . Particle identification to select pions was done by using ISIS, FC(forward Cerenkov counter) and TRD(transition radiator detector). Having 433,016 events of p-p interaction in the LEBC, we selected samples with more than 4 prongs, and used secondary tracks of these samples within interval $0 < X_F < 0.5$ for the present analysis.

## 3. DATA ANALYSIS

In order to obtain the radius, $\rho$, and the chaoticity, $\lambda$, with the equation (2), $Q^2$ distributions of pions of like charged pairs of each event were taken together with those of uncorrelated sample. The uncorrelated sample was created by mixing tracks randomly of different events of the same track configuration. After normalizing these $Q^2$ distributions for like charged pairs and uncorrelated samples by the total number of entries, we calculated ratios of the former to the latter at each $Q^2$ bin. Then the following equation is fitted to these ratios.

$$R_{(2)}(Q^2) = C(1 + \lambda e^{-\rho^2 Q^2})/(1 + \delta Q^2), \tag{6}$$

where $C$ is a normalization constant and the term $(1 + \delta Q^2)$ is introduced to take account of global behavior of the ratios at large $Q^2$ region. (See Fig.2). In this fitting procedure, the first data point $(0. < Q^2 < 0.01(GeV/c)^2)$ of the ratios was dropped because of possible contaminations due to faked tracks at small $Q^2$ region [12]. Fitted parameters obtained using the NA23 data are:

$\rho = 0.99 \pm 0.17 fm$, $\lambda = 0.37 \pm 0.08$, $\delta = 0.09 \pm 0.14 (GeV/c)^{-2}$, $\chi^2(NDF) = 31(35)$.

The same analysis was done for the NA27 data, and calculated ratios $R_{(2)}(Q^2)$ are shown in Fig.5(a) with the best fit curve to the equation (6). The fitted parameters obtained using the the NA27 data are:

$\rho = 1.09 \pm 0.06 fm$, $\lambda = 0.31 \pm 0.02$, $\delta = 0.21 \pm 0.02 (GeV/c)^{-2}$, $\chi^2(NDF) = 59(35)$.

The results for NA27 are preliminary. The results of NA23 and NA27 data are compared with those of other experiments in Fig.4 . One can see that our results are consistent with others.

From Fig.4 one can see that the chaoticity $\lambda$'s experimentally determined using data of both hadron-hadron reactions and $e^+e^-$ reactions are mostly less than one. Taking account of this observation, we tried to apply the quantum statistics to describe the Bose-Einstein effect. Like as the above analysis, the ratios of $Q^2$ distributions of two- and three-like charged pions to those of the uncorrelated ones were calculated using the NA23 data and plotted in Fig.2 and for the two- and three-particle correlations, respectively. The following equations

$$R_{(2)}(Q^2) = C\frac{C_{(2)}(Q^2)^{QS}}{1+\delta Q^2}, \tag{7}$$

for two-particle correlations and

$$R_{(3)}(Q^2_{3\pi}) = C\frac{C_{(3)}(Q^2)^{QS}}{1+\delta Q^2_{3\pi}}, \tag{8}$$

for the three-particle correlations were fitted to the calculated ratios and the resulting fitted curves are drawn in Fig.2 and Fig.3 for two- and three-particle correlations respectively. Fitted parameters obtained for two-particle correlations are:

$r_{2\pi} = 1.36 \pm 0.24 fm$, $p_{2\pi} = 0.22 \pm 0.05$, $\delta = 0.09 \pm 0.14 (GeV/c)^{-2}$, $\chi^2(NDF) = 31(35)$.

and those for three-particle correlations:

$r_{3\pi} = 1.30 \pm 0.16 fm$, $p_{3\pi} = 0.20 \pm 0.04$, $\delta = 0.16 \pm 0.12 (GeV/c)^{-2}$, $\chi^2(NDF) = 21(16)$.

We can see that the fitted values of $r$ and $p$ for two-particle correlations are consistent with those of three-particle correlations in NA23 data set.

Also analyzed are the NA27 data for two-, three- and four-particle correlations. An equation to make a fit to the ratios of $Q_{4\pi}$ distributions for correlations of four-identical bosons is

$$R_{(4)}(Q^2_{4\pi}) = C\frac{C_{(4)}(Q^2_{4\pi})^{QS}}{1+\delta Q^2_{4\pi}}. \tag{9}$$

In our fitting procedure the first data point $(0. < Q^2 < 0.01(GeV/c)^2)$ of the ratios for two-particle correlations was dropped, and correspondingly the region $(0. < Q^2_{3\pi} < 0.03(GeV/c)^2)$ and $(0. < Q^2_{4\pi} < 0.06(GeV/c)^2)$ of the ratios were dropped for three- and four-particle correlations, respectively. The ratios and the fitted curves for two-, three- and four-particle correlations are shown in Fig.5(a), Fig.6(a) and Fig.7(a), respectively.

Parameters obtained are given below.

For two-particle correlation:

$r_{2\pi} = 1.58 \pm 0.06 fm$, $p_{2\pi} = 0.16 \pm 0.01$, $\delta = 0.06 \pm 0.02 (GeV/c)^{-2}$, $\chi^2(NDF) = 43(35)$,

for three-particle correlation:

$r_{3\pi} = 1.51 \pm 0.04 fm$, $p_{3\pi} = 0.14 \pm 0.01$, $\delta = 0.07 \pm 0.01 (GeV/c)^{-2}$, $\chi^2(NDF) =$ 136(93),

and for four-particle correlation:

$r_{4\pi} = 1.85 \pm 0.10 fm$, $p_{4\pi} = 0.14 \pm 0.01$, $\delta = 0.36 \pm 0.05 (GeV/c)^{-2}$, $\chi^2(NDF) =$ 155(90).

Parameters $r$ and $p$ for two- three- and four-particle correlation of both data are summarized in Fig.8.

In order to study possible correlations caused by the procedure to create uncorrelated samples, we analyzed generated events using the Monte Carlo program, FRITIOF, with the same procedure as employed in our data analysis. The program, FRITIOF, is an event generator of production of low $p_t$ particles in high energy collisions base on the Lund fragmentation scheme. Apart from the Bose-Einstein correlations, it is known that this model reproduces various aspects of single particle distributions experimentally observed.

The ratios of $Q^2$ distributions of identical charged pions to those of uncorrelated ones were calculated using the generated events, which are shown in Fig. 5(b), 6(b) and 7(b) for two-, three- and four-particle correlations, respectively. One can see from those figures that they show almost flat distributions which suggests that there is no significant correlations to disturb our analyses for the Bose-Einstein effect.

## 4. SUMMARY

1) Analyzing for the Bose-Einstein effect of charged pions the data of p-p collisions at 360 GeV/c and 400 GeV/c, we determined the effective radius, $\rho$, and the chaoticity, $\lambda$, as follows,

$\rho = 0.99 \pm 0.17 fm$, $\lambda = 0.37 \pm 0.08$, for NA23 data,

$\rho = 1.09 \pm 0.06 fm$, $\lambda = 0.31 \pm 0.02$, for NA27 data, where the results of NA27 are preliminary.

2) We compared the above results with those obtained by other experiments of different energies. It was found that in hadron-hadron reactions the effective radius and the chaoticity obtained at different energies are consistent with those obtained at our analysis.

3) Assuming coherence and chaotic emissions of pions produced at p-p reactions, we studied the Bose-Einstein effect based on the quantum statistics. Using the NA23 data we obtained the chaotic source size, $r$, and the chaoticity, $p$, by this study as follows,

$r_{2\pi} = 1.36 \pm 0.24 fm$, $p_{2\pi} = 0.22 \pm 0.05$ for two-particle correlations and

$r_{3\pi} = 1.30 \pm 0.16 fm$, $p_{3\pi} = 0.20 \pm 0.04$ for three-particle correlations.

With NA27 data, we obtained:

$r_{2\pi} = 1.58 \pm 0.06 fm$, $p_{2\pi} = 0.16 \pm 0.01$, for two-particle correlations,
$r_{3\pi} = 1.51 \pm 0.04 fm$, $p_{3\pi} = 0.14 \pm 0.01$, for three-particle correlations and
$r_{4\pi} = 1.85 \pm 0.10 fm$, $p_{4\pi} = 0.14 \pm 0.01$, for four-particle correlations.

The results obtained using the NA27 data are preliminary.

One can see that the fitted values for $r_{2\pi}$, $r_{3\pi}$ and $r_{4\pi}$ are consistent within errors and those for $p_{2\pi}$, $p_{3\pi}$ and $p_{4\pi}$ are also consistent within errors.

After this workshop, it has been found that the results obtained by the analysis of NA27 data should be corrected by taking account of the following points.

(1) Contaminations due to the $\gamma \to e^+e^-$ converted tracks. (2) Limited resolution of the EHS for two-track separation. (3) Possible biases imminent in "mixed track" technique for data of $R_{(3)}$ and $R_{(4)}$ caused by violation of energy and momentum conservation.

Because of these reasons, we called the results obtained by the analysis of NA27 data "preliminary results" in this paper. The final results of this analysis will come soon after minor corrections.

## ACKNOWLEDGEMENTS

We are deeply indebted to the CERN-SPS, beam and EHS crews for their support during preparation and runs of our experiments. Thanks are due to the EHS collaborators for their helpful discussions with us.

## REFERENCES

[1]. R. Hanbury Brown, E.Q. Twiss: Phil. Mag. 45 (1954) 663
[2]. G. Goldhaber et al.: Phys. Rev. 120 (1960) 300
[3]. M. Deutschmann et al., Nucl. Phys. 204B (1982)333
[4]. N. Akhababian et al.: Z. Phys. C - Particles and Fields 18 (1983) 97
[5]. M. Adamus et al.: Z. Phys. C - Particles and Fields 37 (1988) 347
[6]. A. M. Cooper et al.: KEK-PREPRINT-79-13 (June 1979)
[7]. T. Åkesson et al.: Phys. Lett. 129B (1983) 269; 187B (1987) 420
[8]. T. Åkesson et al.: Z. Phys. C - Particles and Fields 36 (1987) 517
[9]. A. Breakstone et al.: Phys. Lett. 162B (1985) 400
[10]. A. Breakstone et al.: Z. Phys. C - Particles and Fields 33 (1987) 333
[11]. C. Albajar et al.: Phys. Lett. 226B (1989) 410
[12]. J.l. Bailly et al.: Z. Phys. C - Particles and Fields 43 (1989) 341

[13]. G. Goldhaber, Inter. Conf. on High Energy Physics, Lisbon, (Ed. Dias de Deus and J. Soffer) p. 767(1981)

[14]. I. Juričić: Ph. D. Thesis, Univ. of California, Berkeley, LBL-24493, (1987)

[15]. I. Juričić et al.: Phys. Rev. D39 (1989) 1

[16]. P. Avery et al.: Phys. Rev. D32 (1985) 2294

[17]. H. Aihara et al.: Phys. Rev. D31(1985) 996

[18]. M. Althoff et al.: Z. Phys. C - Particles and Fields 29 (1985) 347; 30 (1986) 355

[19]. M. Arneodo et al., Z. Phys. C - Particles and Fields 32 (1986)1

[20]. D. Allasia et al., Z. Phys. C - Particles and Fields 37 (1988)527

[21]. G. Kopylov and M. Podgoretsky: Sov. Nucl. Phys. 19 (1974) 434; 18 (1973) 656; 16 (1972) 392

[22]. G. Cocconi: Phys. Lett. 49B (1974) 459

[23]. G. N. Fowler and R. M. Weiner: Phys. Lett. 70B (1977) 201; Phys. Rev. D17 (1978) 3118

[24]. R.M. Weiner: Proc. Perugia Workshop on Multiparticle Production, Perugia(1988), p.231. R.Hwa, G.Pancheri and Y.Srivastava(ed.) Singapore:World Scientific 1988

[25]. R. M. Weiner: Phys. Lett. 232B (1989) 278; 242B (1990) 547

[26]. M. Gyulassy, S. K. Kaufmann, and L. W. Wilson: Phys. Rev. C20 (1979) 2267

[27]. M. Biyajima, O. Miyamura and T.Nakai: Proc. Multiparticle dynamics (Research Institute For Fundamental Physics, Kyoto Univ., 1978) p.139

[28]. M. Biyajima: Phys. Lett. 92B (1980) 193; Prog. Theor. Phys. 66 (1981) 1378

[29]. M. Biyajima, A. Bartel, T. Mizoguchi, N. Suzuki and O. Terazawa: Higher Order Bose-Einstein Correlations in identical Particle Production, (UWThPh-1990-3, SULDP-1990-1) and private communication.

[30]. B. Andersson and W. Hofmann: Phys. Lett. 169B (1986) 364

[31]. M. G. Bowler: Z. Phys. C - Particles and Fields 29 (1985) 617

[32]. Y. Hama and S. Padula: Phys. Rev. D37 (1988) 3237

[33]. G. Lachs: Phys. Rev. B138 (1965) 1012

[34]. L. Mandel and E.Wolf: Phys. Rev. 149 (1966) 1033

[35]. M. Aguilar-Benitez et al., Nucl. Instr. and Meth. 205(1983)79; Nucl. Instr. and Meth. in Phys. Res. A258 (1987)26

[36]. J. L. Bailly et al., Z. Phys. C - Particles and Fields 23 (1984)205

[37]. Aguilar-Benitez et al.: Z. Phys. C - Particles and Fields 31 (1986)441

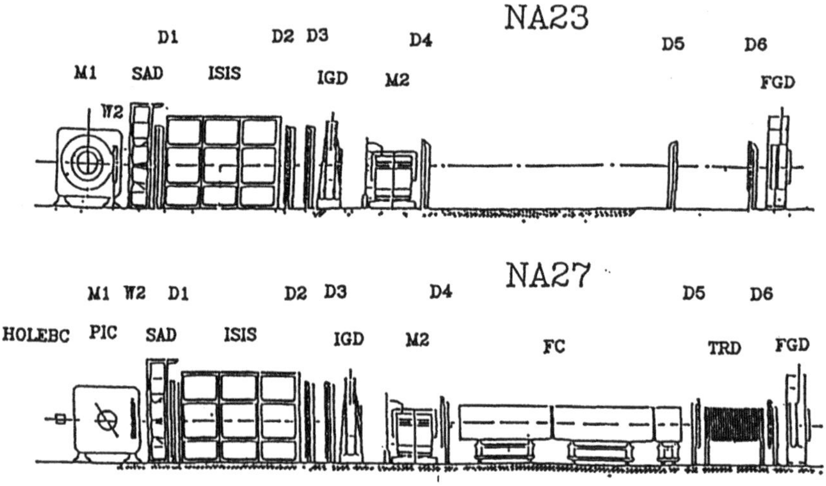


**Fig. 1**

Experimental set-up of the RCBC and the European Hybrid Spectrometer (EHS) of NA23 experiment (upper figure). That of LEBC and EHS of NA27 experiment (lower figure).

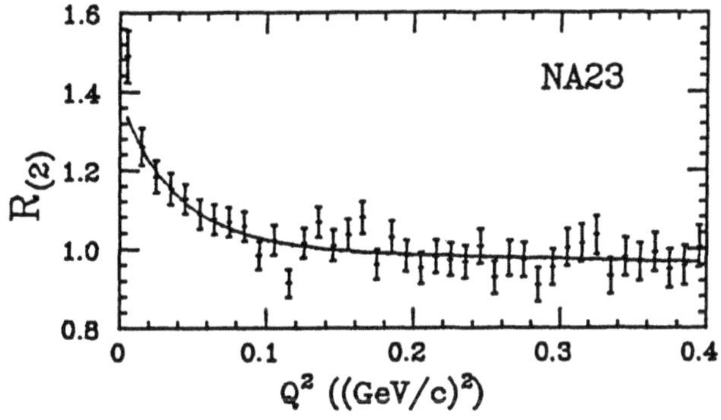


Fig. 2

$R_{(2)}(Q^2)$ data of NA23. The solid curve is a fitted one with the equation (6) and the dotted curve with the equation (7) by dropping the first data point [12]. These two curves are indistinguishable.

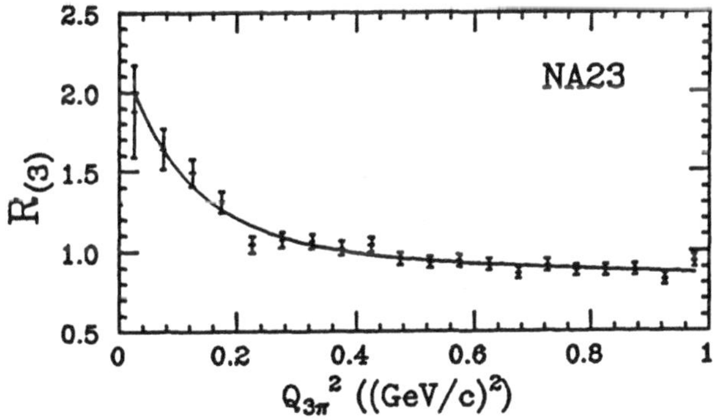


Fig. 3

$R_{(2)}(Q^2_{3\pi})$ for three-particle correlations of NA23 data and a fitted curve with the equation (8).

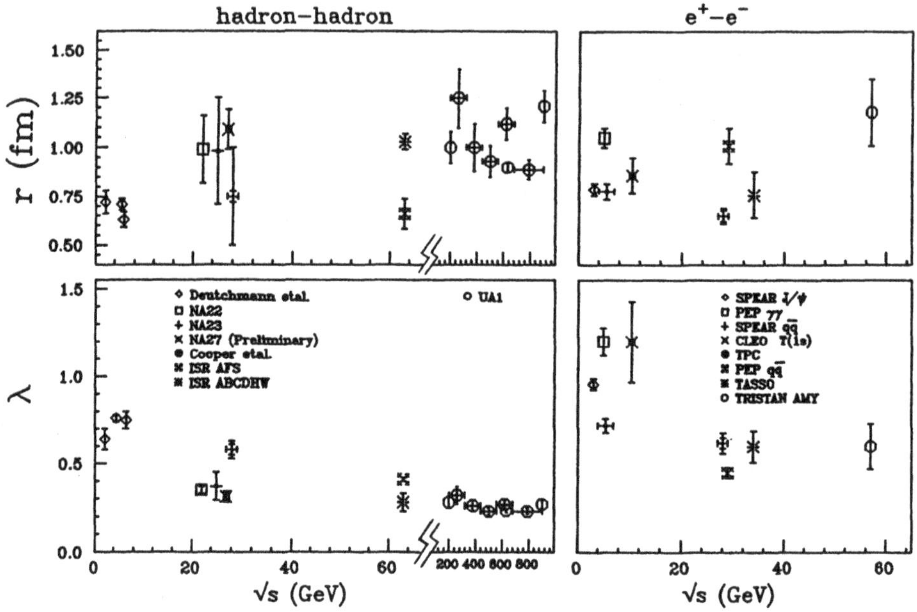


Fig. 4

Summary of experimental results of the source radius (upper figure) and the chaoticity (lower figure) in hadron-hadron and $e^+e^-$ reactions. For NA27 data, the systematic errors due to track mixing procedure are included.

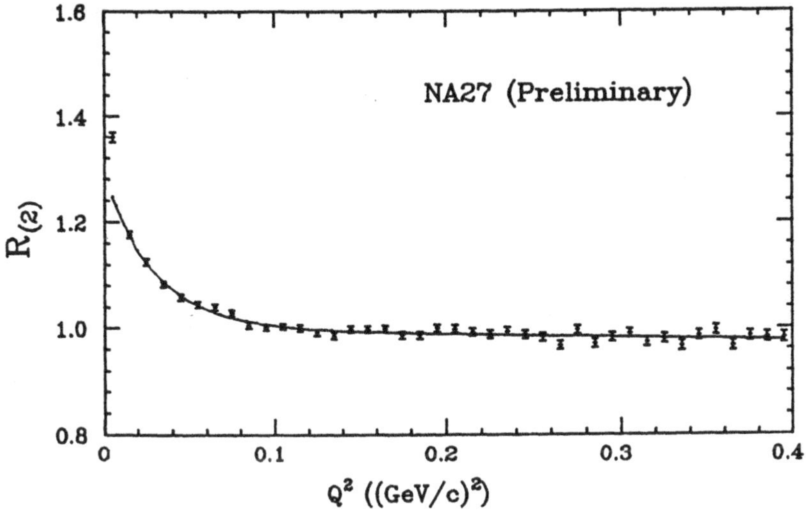


Fig. 5 (a)

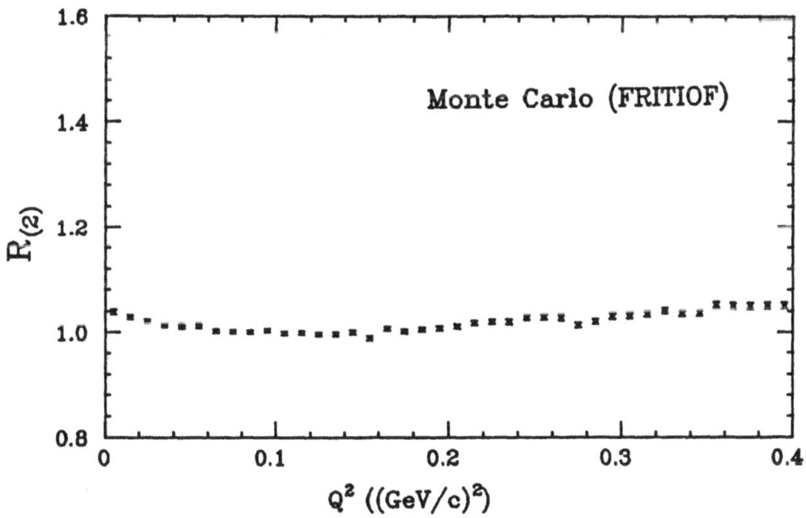


Fig. 5 (b)

(a) $R_{(2)}(Q^2)$ for two-particle correlations of NA27 data. The curve is a result of fit with the equation (6) the dotted curve with the equation (7) by dropping the first data point. These two curves are indistinguishable. (b) $R_{(2)}(Q^2)$ data of the same statistics as (a) generated by FRITIOF Monte Carlo program.

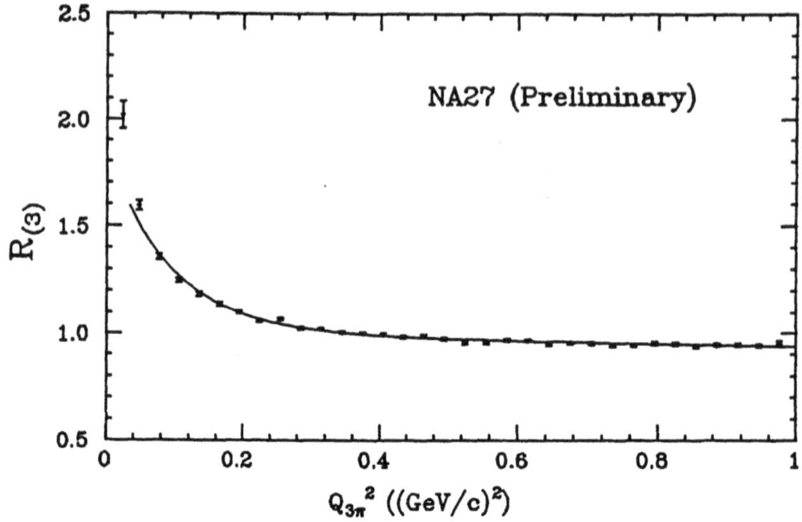


Fig. 6 (a)

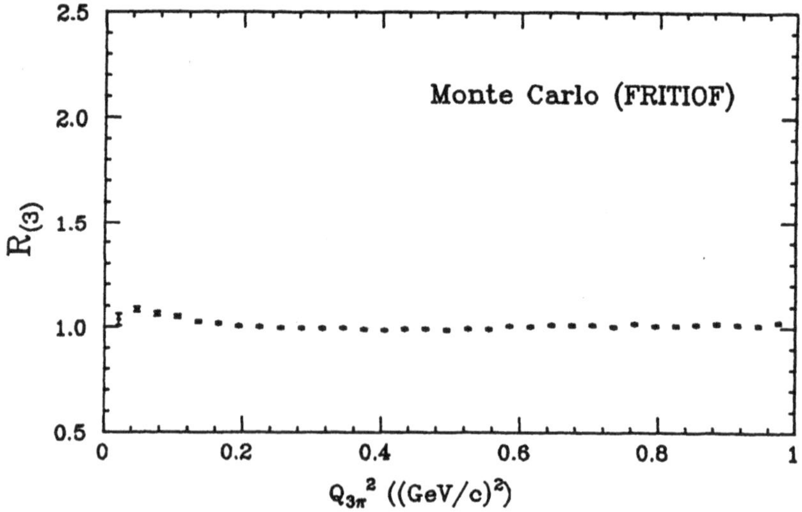


Fig. 6 (b)

(a) $R_{(3)}(Q^2_{3\pi})$ for three-particle correlations of NA27 data. The solid curve is a result of fit with the equation (8) by dropping the first data point ($0. < Q^2_{3\pi} < 0.03$). (b) $R_{(3)}(Q^2_{3\pi})$ data generated by FRITIOF Monte Carlo program with the same statistics as (a).

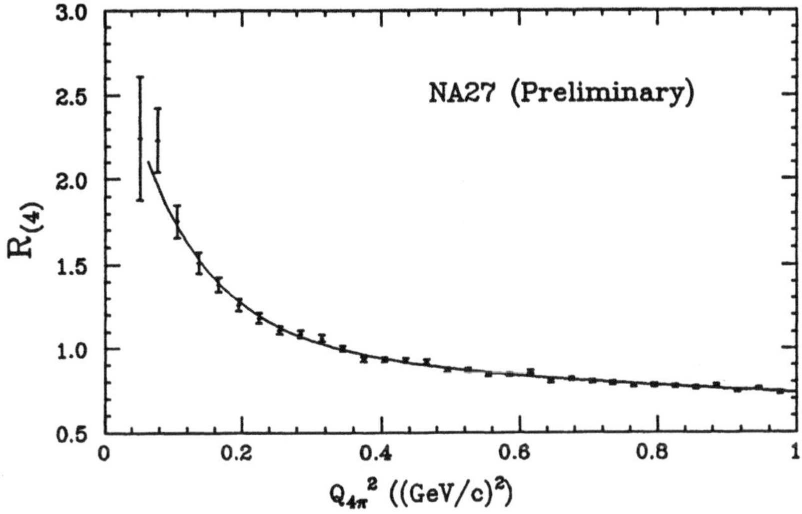


Fig. 7(a)

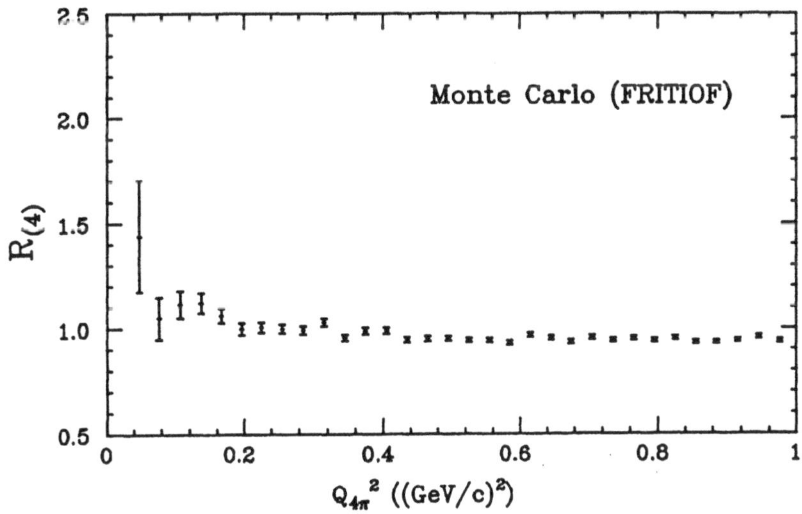


Fig. 7 (b)

(a) $R_{(4)}(Q^2_{4\pi})$ for four-particle correlations of NA27 data. The solid curve is a result of fit with the equation (9) by dropping the first two data points $(0. < Q^2_{4\pi} < 0.06)$. (b) $R_{(4)}(Q^2_{4\pi})$ data generated by FRITIOF Monte Carlo program with the same statistics as (a) .

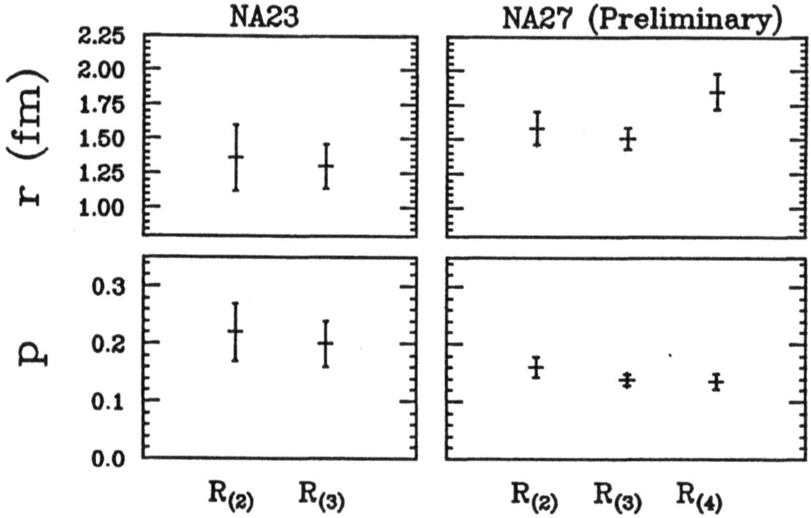


Fig. 8

The summary of the chaoticity $p$'s and the chaotic source size $r$'s with the equations (7), (8) and (9) obtained by data analyses of $R_{(2)}$ for two-particle correlations, $R_{(3)}$ for three-particle correlations and $R_{(4)}$ for four-particle correlations, respectively. For NA27 data, systematic errors due to track mixing procedure are included.

# BOSE-EINSTEIN CORRELATIONS IN THE TARGET FRAGMENTATION REGION IN 200 $A$ GEV $^{16}$O + NUCLEUS COLLISIONS

T. Peitzmann[4], R. Albrecht[1], T. C. Awes[5], M. Bahr[4], P. Beckmann[4,a], F. Berger[4], R. Bock[1], G. Claesson[3], G. Clewing[4], L. Dragon[4], A. Eklund[3], R. L. Ferguson[5], A. Franz[5,b], S. Garpman[3], R. Glasow[4], H. Å. Gustafsson[3], H. H. Gutbrod[1], G. Hölker[4], J. Idh[3], P. Jacobs[2], K. H. Kampert[4], B. W. Kolb[1], P. Kristiansson[3], H. Löhner[4,a], I. Lund[1], F. E. Obenshain[5,b], A. Oskarsson[3], I. Otterlund[3], S. Persson[3], F. Plasil[5], A. M. Poskanzer[2], M. Purschke[4], H. G. Ritter[2], H. Rothgänger[4], S. Saini[5], R. Santo[4], H. R. Schmidt[1], R. Schmidt[4], S. P. Sørensen[5,b], K. Steffens[4], E. Stenlund[3], D. Stüken[4], M. L. Tincknell[5] and G. R. Young[5]

**WA80 Collaboration**

1. *Gesellschaft für Schwerionenforschung, D-6100 Darmstadt, Fed. Rep. of Germany*
2. *Lawrence Berkeley Laboratory, Berkeley, California 94720, USA*
3. *University of Lund, S-22362 Lund, Sweden*
4. *University of Münster, D-4400 Münster, Fed. Rep. of Germany*
5. *Oak Ridge National Laboratory, Oak Ridge, Tennessee 37831, USA*

a. *now at: KVI, University of Groningen, NL-9747 AA Groningen, Netherlands*
b. *University of Tennessee, Knoxville, Tennessee 37996, USA*

**Abstract**

Correlations of positive pions are investigated in the target fragmentation region in 200 $A$ GeV $^{16}$O + nucleus collisions. The pions are measured with the Plastic Ball detector in the WA80 experiment at the CERN SPS. The detailed behaviour of the fitted parameters on rapidity, centrality and the number of spectator nucleons is studied. The data indicate a large source in central O + Au collisions for $y_{lab} \leq 0$. The observations are compatible with a substantial amount of rescattering in the target.

## 1 Introduction

Pion interferometry allows to study the space-time properties of the particle emitting source in high energy collisions [1]. An enhancement in the production of identical pions with similar momentum - known as Bose-Einstein Correlations, the Hanbury-Brown - Twiss effect (HBT) [2] or the Goldhaber - Goldhaber - Lee - Pais effect (GGLP) [3] - has been observed in various experimental studies. A simple theoretical picture of multi-particle production yields the following prediction for the two-particle correlation function:

$$C_2 \equiv \frac{< n^2 >}{< n(n-1) >} \cdot \frac{d^6N/dp_1^3 dp_2^3}{d^3N/dp_1^3 \cdot d^3N/dp_2^3} = 1 \, + \, \lambda \cdot |\tilde{\rho}(p_1^\mu - p_2^\mu)|^2.$$

$\tilde{\rho}$ is the Fourier transform of the distribution of emitters; the most commonly chosen analytic expressions for $\tilde{\rho}$ contain a suitable correlation length $R$. While this parameter seems to measure dynamical properties of the produced strings in $e^+e^-$ or pp collisions, it can provide geometrical information in medium energy heavy ion collisions.

The importance of rescattering of secondaries in the target spectator matter in high energy nuclear reactions has been emphasized in earlier publications [4]. It is therefore of interest to study pion correlations in the target fragmentation region to clarify both the general role of the target spectators and their influence on pion interferometry. In this paper we will present a study of $\pi^+$ correlations in reactions of 200 $A$ GeV $^{16}$O ions with various targets in the target rapidity region ($y_{lab} \leq 1$).

# 2 Data analysis

Data were taken in the WA80 experiment [5] at the CERN SPS during the $^{16}$O run in fall 1986. The pions were identified with the Plastic Ball detector in $30° \leq \theta_{lab} \leq 160°$ using both $\Delta E$-$E$-information and the time delayed signal of the $\pi^+$-decay. Details on the detector design and performance and the identification are given in [6]. Only pions, which are stopped in the detector modules ($E_{kin} \leq 120\,\mathrm{MeV}$), are used. The contamination of misidentified particles was estimated to be of the order of 5 % for all samples. The resolution in relative momentum is $\delta q \approx 10\,\mathrm{MeV/c}$.

Reactions with C, Cu, Ag and Au targets were investigated. Background distributions were obtained by the event-mixing procedure: A pion from one event was combined with a pion from another event with the same apparent $\pi^+$ multiplicity. The correlation functions were corrected for the Coulomb-interaction with the standard Gamow-factor [7]. The data were analyzed both in terms of the absolute value of the four momentum difference: $Q \equiv \sqrt{-(p_1^\mu - p_2^\mu)^2}$ and in terms of the energy- and momentum-difference in the lab-system: $(|\vec{q}|, q_0) \equiv (|\vec{p}_1 - \vec{p}_2|, |E_1 - E_2|)$. For simplicity, gaussian functions have been used both for one- and two-dimensional analysis:

$$C_2(Q) = 1 + \lambda_{inv} \cdot \exp(-\frac{1}{2} R_{inv}^2 Q^2)$$

$$C_2(|\vec{q}|, q_0) = 1 + \lambda_{lab} \cdot \exp(-\frac{1}{2} R_{lab}^2 |\vec{q}|^2 - \frac{1}{2} \tau^2 q_0^2).$$

For the analysis in $Q$ an exponential function:

$$C_2(Q) = 1 + \lambda_{exp} \cdot \exp(-R_{exp} Q)$$

has also been used for comparison. Fits were performed in the region of $Q \leq 400\,\mathrm{MeV}$ by maximizing the logarithm of a likelihood function.

Table 1: Pion Correlation Fit Parameters ($-1 \leq y_{lab} \leq 1$)

| target | Au | Ag | Cu | C |
|---|---|---|---|---|
| $R_{inv}$ (fm) | $2.5^{+0.1}_{-0.1}$ | $3.1^{+0.2}_{-0.2}$ | $3.5^{+0.1}_{-0.1}$ | $4.1^{+0.3}_{-0.3}$ |
| $\lambda_{inv}$ | $0.094^{+0.003}_{-0.003}$ | $0.14^{+0.01}_{-0.01}$ | $0.16^{+0.01}_{-0.01}$ | $0.40^{+0.05}_{-0.02}$ |
| $R_{exp}$ (fm) | $2.5^{+0.1}_{-0.1}$ | $3.3^{+0.4}_{-0.4}$ | $4.2^{+0.3}_{-0.3}$ | $5.7^{+0.6}_{-0.6}$ |
| $\lambda_{exp}$ | $0.162^{+0.006}_{-0.006}$ | $0.22^{+0.02}_{-0.03}$ | $0.29^{+0.03}_{-0.01}$ | $0.92^{+0.14}_{-0.03}$ |
| $R_{lab}$ (fm) | $2.8^{+0.1}_{-0.1}$ | $3.2^{+2.2}_{-0.2}$ | $3.5^{+0.3}_{-0.3}$ | $3.3^{+0.1}_{-0.1}$ |
| $\tau$ (fm/c) | $1.8^{+0.5}_{-0.7}$ | $0.0^{+1.2}$ | $3.4^{+0.8}_{-0.8}$ | $12.1^{+1.4}_{-1.4}$ |
| $\lambda_{lab}$ | $0.116^{+0.003}_{-0.003}$ | $0.16^{+0.01}_{-0.01}$ | $0.20^{+0.10}_{-0.10}$ | $0.60^{+0.04}_{-0.04}$ |

# 3 Results and discussion

Fig. 1 shows correlation functions in dependence of $Q$ for central reactions of $^{16}O$ with different targets in the rapidity region $-1 \leq y_{lab} \leq 1$. A clear enhancement for small values of $Q$ is visible in all cases. A fit with a gaussian parametrization is shown as a solid line. Counterintuitively the radii $R_{inv}$ decrease with increasing target mass, also the intercept $\lambda_{inv}$ goes down for the heavier systems. The correlation functions have also been fitted with an exponential distribution to investigate systematic effects connected to the choice of the parametrization. In addition, the same data sets have been analyzed in $(|\vec{q}|, q_0)$ and fitted with a two-dimensional gaussian. The resulting fit parameters are summarized in table 1.

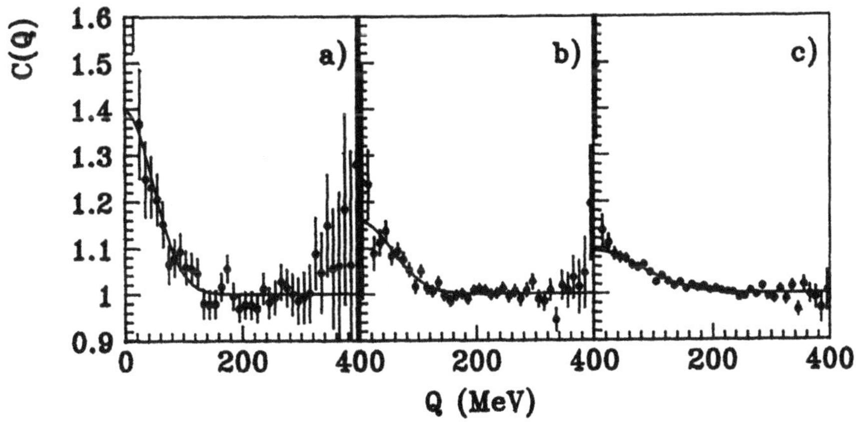


Fig. 1: $\pi^+$ correlation functions in dependence of $Q$ for $-1 \leq y_{lab} \leq 1$. a) O + C, b) O + Cu, c) O + Au.

$R_{inv}$ and $\lambda_{inv}$ ($R_{exp}$ and $\lambda_{exp}$) show a significant decrease with increasing target size. The correlation strength $\lambda_{lab}$ shows a similar decrease for the heavier systems, but the space-component of the correlation length $R_{lab}$ seems to be only weakly dependent on the target mass, the values are close to the charge radius of the projectile: $R_{proj}^{rms} \approx 2.9\,\text{fm}$, while the time parameter $\tau$ exhibits a drastic decrease.

How can the variations in the parameters be understood in our current picture of ultrarelativistic nuclear collisions? A reduction of the value of $\lambda$ may be caused by a variety of mechanisms. Most of these, as e.g. final state interaction, resonance production or coherence, should not show such a drastic variation for different targets.

It is however easy to realize that pions observed in the target rapidity region might undergo very different evolutions. They may be

- directly produced by the participant nucleons or
- rescattered in the target spectator matter.

It has been noted that data from the target fragmentation region indicate rescattering effects. Such rescattering need not affect the pion multiplicities, but it would alter the volume measured by interferometry to the "freeze-out" volume of the excited spectator matter. The rescattered pions might come from very large volumes or late times, and their correlation could thus be hidden at very small values of $Q$ hardly accessible to experiment. In fact, there is an indication of a steep rise in the high statistics O + Au correlation function for $Q \leq 30\,\text{MeV}$ (see fig. 1). These data points do however depend crucially on the Coulomb correction, whereas the overall fit is not very sensitive to this correction.

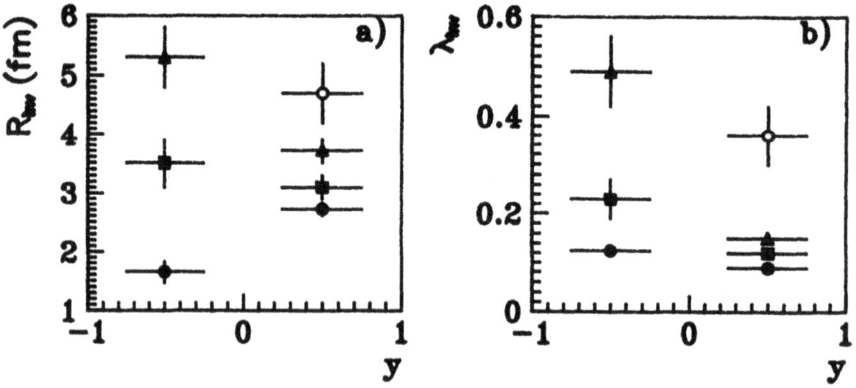


Fig. 2: Fit parameters – a) $R_{inv}$, b) $\lambda_{inv}$ – of the $\pi^+$ correlation as a function of the pion rapidity. The filled circles show O + Au, the squares O + Ag, the triangles O + Cu and the open circles O + C data.

**Rapidity dependence:** Fig. 2 shows the parameters $R_{inv}$ and $\lambda_{inv}$ for the rapidity intervals $-1 \leq y_{lab} \leq 0$ and $0 \leq y_{lab} \leq 1$ for all targets. The variation in the parameters is much stronger for the backward interval, indicating that we observe a physical effect connected to the target and not created by e.g. accidental double hits or other detector effects. The $\lambda_{inv}$-values are systematically smaller for the more forward interval.

**Centrality dependence:** For the system O + Au at 200 $A$ GeV we have studied the centrality dependence. The data were divided into three subsamples selected by the forward energy flow ($\theta_{lab} \leq 0.3°$) measured with the Zero-Degree Calorimeter [8]. We have then used the FRITIOF Monte-Carlo to extract the number of participant nucleons from the forward energy $E_{ZDC}$ event by event. From this we can calculate an effective participant radius $R_{part} \equiv < N_{part} >^{1/3}$ which characterizes the size of the participant volume. We obtained $R_{part}$ = 2.77, 3.34 and 3.82 for the peripheral, medium central and central reactions respectively.

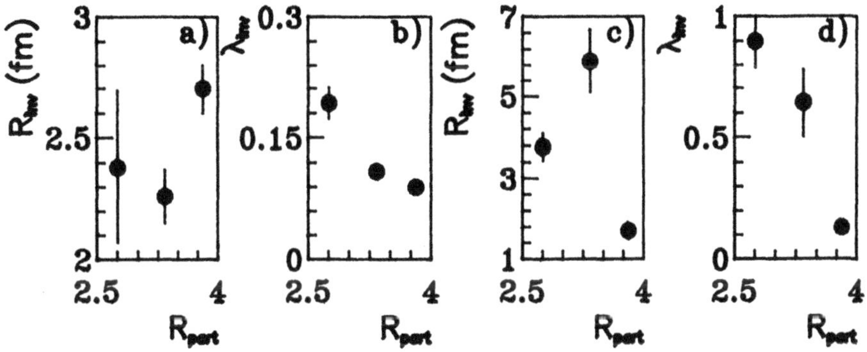


Fig. 3: Fit parameters – a) $R_{inv}$ ($0 \leq y_{lab} \leq 1$) , b) $\lambda_{inv}$ ($0 \leq y_{lab} \leq 1$) , c) $R_{inv}$ ($y_{lab} \leq 0$) , d) $\lambda_{inv}$ ($y_{lab} \leq 0$) – of the $\pi^+$ correlation as a function of the participant radius (see text).

Fig. 3 shows the dependence of $R_{inv}$ and $\lambda_{inv}$ on the participant radius, analyzed seperately in the rapidity intervals $0 \leq y_{lab} \leq 1$ and $y_{lab} \leq 0$. In the more forward region the $R_{inv}$-parameter increases from $R_{inv} \approx 2.3 - 2.4$ fm for the peripheral and medium central reactions to 2.7 fm for the central collisions. The $\lambda_{inv}$-value decreases with centrality - it is small ($\lambda_{inv} < 0.2$) for all cases. In the backward rapidity region the $R_{inv}$-parameter increases from 3.8 fm to 5.9 fm when going from peripheral to medium central collisions and the $\lambda_{inv}$-parameter decreases slightly, but is rather large (0.90 and 0.64 resp.). Both parameters drop to small values for the central reactions – if there is a two-component source, this might be due to the fact that the fit is more sensitive to the smaller component. Indeed, if we divide the data sample again into pions of low and high transverse momentum (fig. 4), we observe a larger component ($R_{inv} = 7.0 \pm 0.8$ fm and $\lambda_{inv} = 0.52 \pm 0.08$) for the lower $p_T$-pions, while only a small source ($R_{inv} = 1.5 \pm 0.2$ fm and $\lambda_{inv} = 0.14 \pm 0.01$) is visible from the high $p_T$-particles.

The parameters are consistent with the assumptions that we observe in the more forward interval only the fraction of pions coming directly from the participant volume

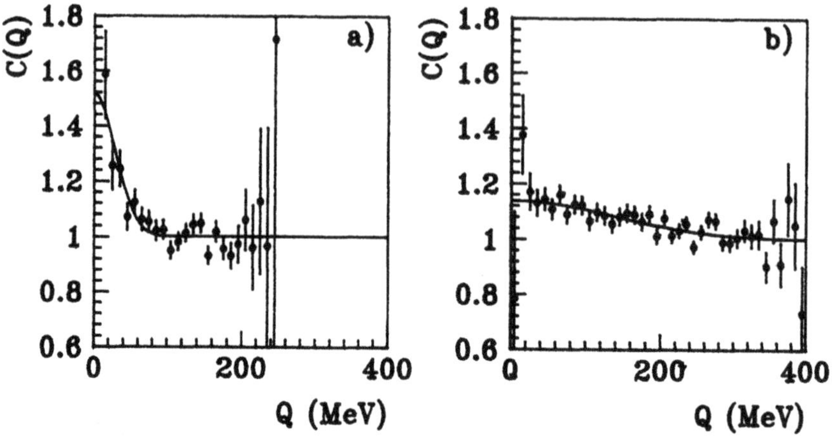


Fig. 4: $\pi^+$ correlation functions in dependence of $Q$ for central O + Au in $-1 \leq y_{lab} \leq 0$. a) $p_T \leq 100\,\mathrm{MeV/c}$, b) $p_T \geq 100\,\mathrm{MeV/c}$

and that in the backward region the full source including spectator matter is seen.

**Spectator dependence:** Finally we can compare systems, which are similar according to the number of participants and the effective center-of-mass rapidity, but which differ by the number of target spectators. We know from analysis with the FRITIOF Monte-Carlo that the data sets given in table 2 are pairwise comparable in this sense. The fit parameters for these systems vary significantly. Both parameters are substantially smaller for the systems with larger number of spectators, when all other relevant parameters are fixed. The presence of spectator matter seems to diminish the "visible" fraction of the pions, and their apparent radius is reduced.

Table 2: Spectator Dependence of Pion Correlation ($-1 \leq y_{lab} \leq 1$)

| system | Cu central | Au medium | C central | Au peripheral |
|---|---|---|---|---|
| $< N_{part} >$ | 39.5 | 37.4 | 19.2 | 21.2 |
| $< y_{cm} >$ | 2.8 | 2.7 | 3.1 | 2.9 |
| $< N_{spect} >$ | 38.4 | 172.9 | 3.2 | 184.3 |
| $R_{inv}$ (fm) | $3.5^{+0.1}_{-0.1}$ | $2.2^{+0.2}_{-0.2}$ | $4.1^{+0.3}_{-0.3}$ | $2.8^{+0.2}_{-0.2}$ |
| $\lambda_{inv}$ | $0.16^{+0.01}_{-0.01}$ | $0.12^{+0.01}_{-0.01}$ | $0.40^{+0.05}_{-0.05}$ | $0.25^{+0.02}_{-0.02}$ |

# 4 Summary and conclusions

We have observed correlations of positive pions in the target fragmentation region for reactions of 200 $A$ GeV $^{16}$O on C, Cu, Ag and Au targets. The interpretation of the enhancement in terms of Bose-Einstein correlations parametrized as a gaussian yields the following results:

1. Applying a simple one component fit gives a decrease in $R_{inv}$ and $\lambda_{inv}$ with increasing target mass. A two-dimensional ($R_{lab}$, $\tau$) analysis indicates a more or less constant $R_{lab}$ and a decrease in the $\tau$-parameter.

2. This decrease in the parameters is obviously connected to the number of spectators. It is consistent with a substantial amount of rescattered pions emerging from a large freeze-out space-time volume of the excited spectator matter in the case of the heavy targets.

3. At backward rapidities in O + Au reactions we observe a large source whose size increases from 3.8 fm for peripheral to 7.0 fm for central collisions. The large component in central collisions is most easily seen with an additional cut in transverse momentum: $p_T \leq 100\,\mathrm{MeV/c}$. The $\lambda_{inv}$-parameters are above 0.5 in all these cases.

4. In the rapidity interval $0 \leq y_{lab} \leq 1$, which gives the dominant contribution to our overall correlation functions, we see in O + Au a smaller source increasing slightly for central collisions. The measured $\lambda_{inv}$-values are smaller than 0.2.

This analysis emphasizes the need to study the dynamics of the so-called "spectator nucleons", which in fact seem to participate strongly in a later stage of the collision. Care should be taken, if one compares pion interferometry in different rapidity regions, when the presence of target spectators may obscur the interpretation.


[1] G.I. Kopylov and M.I. Podgoretskii, Sov. J. Nucl. Phys. 15 219 (1972); 18 336 (1974)

[2] R. Hanbury-Brown and R.Q. Twiss, Nature 178 1046 (1956)

[3] G. Goldhaber et el., Phys. Rev. 120 300 (1960)

[4] R. Albrecht et al., Z. Phys. C - Particles and Fields 45 529-537 (1990)

[5] R. Albrecht et al., Report No. GSI-85-32, Gesellschaft für Schwerionenforschung, D-6100 Darmstadt (1985)

[6] A. Baden et al., Nucl. Instr. Meth. 203 189 (1982)

[7] M. Gyulassy et al., Phys. Rev. C 20 2267 (1979)

[8] G. R. Young et al., Nucl. Instr. Meth. A 279 503-517 (1989)

# INTERFERENCE CORRELATIONS OF IDENTICAL BOSONS WITH SLIGHTLY DIFFERENT MOMENTA IN THE MODEL INCLUDING BOTH ONE-PARTICLE AND MULTIPARTICLE SOURCES

V.L.Lyuboshitz

High Energy Laboratory, Joint Institute for Nuclear Research, Dubna, USSR

ABSTRACT

The effect of Bose statistics on pair correlations of identical pions with slightly different momenta is discussed. The structure of narrow pair correlations is analyzed in the framework of the model in which one or several multiparticle sources are added to a set of independent chaotically located one-particle sources. It is shown that when the distribution of the number of particles from sources is the Poisson distribution, our results are in agreement with the relations obtained in a series of papers on the basis of the coherent state technique. Such correspondence takes place also in the limit of very large multiplicities for any distribution laws. In the case when a n-particle point-like source is in the symmetry centre of the probability distribution of 4-coordinates of m one-particle sources, the dependence of the correlation function of identical bosons on the momentum difference is determined by two groups of space-time parameters in the ratio of $\sqrt{2}$.

1. In the model of independent one-particle sources emitting unpolarized identical particles with spin j, the probability for the creation of two particles with 4-momenta $p_1$ and $p_2$ at the space-time points $x_1 = \{\vec{r}_1, t_1\}$ and $x_2 = \{\vec{r}_2, t_2\}$ is described by the expression

$$W(p_1,p_2) \sim 1 + \frac{(-1)^{2j}}{2j+1}\cos(qx) \quad , \qquad (1)$$

where $q = p_1 - p_2$ , $x = x_1 - x_2$ , $qx = q_0 t - \vec{q}\vec{r}$ .

Eq.1, which characterizes the effect of Bose or Fermi statistics on pair correlations of identical particles, forms the basis of the method for determining the space-time parameters of the region of multiple elementary particle production[1-5]. In the spirit of statistical concepts the function cos(qx) in Eq.1 should be averaged over the space-time distribution of chaotically located sources.

Taking into account only the contribution of Bose or Fermi statistics and neglecting the effect of final state interaction[6], the correlation function of two identical particles in the region of small relative momenta takes a simple form

$$R(q) = 1 + \frac{(-1)^{2j}}{2j+1} \langle \cos [q(x_1 - x_2)] \rangle \quad , \qquad (2)$$

or

$$R(q) = 1 + \frac{(-1)^{2j}}{2j+1} |F(q)|^2 \quad , \qquad (3)$$

where

$$F(q) = \int W(\tilde{x}) \exp(iq\tilde{x}) \, d^4\tilde{x} \quad . \qquad (4)$$

In Eq.4 the function $W(\tilde{x})$ describes the probability distribution of 4-coordinates of one-particle sources and satisfies the normalization condition

$$\int W(\tilde{x}) \, d^4\tilde{x} = 1 \quad .$$

It is evident that $F(0) = 1$, and the function $F(q)$ tends to zero at sufficiently large q. If the distribution $W(\tilde{x})$ has the centre of symmetry, then $\mathrm{Im}F(q) = 0$.

2. We shall consider the correlations of identical bosons, concretely pions. In this case

$$R(q) = 1 + |F(q)|^2 \quad . \qquad (5)$$

Analyzing experimental data, Eq.5 is frequently replaced by a more complicated expression

$$R(q) = 1 + \Lambda |F(q)|^2 \quad , \qquad (6)$$

where the coefficient $\Lambda \neq 1$. As shown in paper [7], there is a number of simple reasons for the appearance of the factor $\Lambda$, for example, the presence of two or several space-time parameters of the process. On the other hand, in many papers the introduction of $\Lambda$ is related to the assumption of pion emission in coherent states and to the existence of a mixture of coherent and chaotic states (see, e.g., [8-15]). The role of coherent states for the description of narrow pair correlations of bosons consists in the fact that there are no correlations due to symmetrization of wave functions between pions in the same coherent state.

But the absence of interference correlations is characteristic of all the situations when several bosons are created in the same quantum state, in particular, for any fixed number of particles as well. In connection with this, it is interesting to investigate the structure of pair correlations of pions with slightly different momenta in the presence of both one-particle and multi-particle sources. The corresponding model has been suggested earlier in our paper [16]. In the framework of such a model it is easy to explain not only the difference of $\Lambda$ from unity but also the appearance of several characteristic dimensions (see also [17]).

3. Following paper [16], let us consider a situation when a set of n identical pions is emitted simultaneously from the same point (or from different points located at distances which are very small in comparison with the wave length $\lambda \sim 1/p$). Then we can speak of a n-particle point-like source. It is essential that in this situation, in contrast to the case of one-partcle sources, the symmetrization effect is absent, and there are no in-

terference correlations. If the generation amplitudes are assumed to be constants, then the probability of pion registration with momenta $p_1$ and $p_2$ does not depend on momentum difference.

We are interested in pair correlations in the inclusive approach when the momenta of some two identical particles are fixed and averaging over the remaining momenta is performed. Then each pair of pions can be considered independently of all the other pairs. Besides, we suggest that the momenta of all the particles are so large that the interference maximum occupies only a very small part of the phase volume and the rare events with three or more particles having nearly equal momenta can be neglected. Under these conditions it is sufficient to restrict ourselves to a simple summation of the two-particle probabilities corresponding to different pairs of pions, taking into account that the production probability is independent of momentum difference for two identical pions from the same multi-particle source, and this probability is proportional to $(1 + \langle\cos[q(x_1-x_2)]\rangle)$ for pions from different sources.

Let us assume that in each interaction one n-particle source and m one-particle sources are excited. We denote the probability of a process, when the momenta of all the pions differ greatly, by $W_o$. How does this probability change when the momenta of any two pions become close to each other ? The answer to this question depends on that which pair is chosen out of the total number of possible pairs $\frac{1}{2}(n + m)(n + m - 1)$. If the momenta of two pions from the n-particle source approach each other, then the probability does not change, but for other combinations the probability change follows the law $(1 + \langle\cos[q(x_1-x_2)]\rangle)$. The number of pairs from the n-particle source is equal to $\frac{1}{2}n(n-1)$, the number of pairs from one-particle sources is $\frac{1}{2}m(m-1)$ and the number of "mixed" pairs is nm. After averaging over space-time distributions of sources, we obtain for the process probability at small relative momenta the following expression:

$$W = \frac{2W_o}{(n+m)(n+m-1)}\left\{\frac{n(n-1)}{2} + \frac{m(m-1)}{2}\left(1 + |F_1(q)|^2\right) + \right.$$

$$\left. + nm\left[1 + \mathrm{Re}(F_1(q)F_2^*(q))\right]\right\} . \qquad (7)$$

Here $F_1(q)$ is the Fourier transform of the function $W_1(\tilde{x})$ describing the probability distribution of 4-coordinates of one-particle sources, and $F_2(q)$ is the Fourier transform of the analogous function $W_2(\tilde{x})$ for the n-particle source (see Eq.4). Thus, narrow pair correlations of identical pions are expressed by the formula

$$R(q) = \frac{W}{W_o} = 1 + \Lambda_1|F_1(q)|^2 + \Lambda_2\, \mathrm{Re}(F_1(q)F_2^*(q)) , \qquad (8)$$

where

$$\Lambda_1 = \frac{m(m-1)}{(n+m)(n+m-1)} , \quad \Lambda_2 = \frac{2nm}{(n+m)(n+m-1)} . \qquad (9)$$

In particular, if the numbers n and m fluctuate in accordance with the Poisson law when passing from one act of interaction to another, then we have

$$\Lambda_1 = \frac{1}{(1+\gamma)^2} , \quad \Lambda_2 = \frac{2\gamma}{(1+\gamma)^2} , \quad \gamma = \bar{n}/\bar{m} . \qquad (10)$$

It should be emphasized that the result Eqs.10 take place for arbitrary multiplicity distributions when the conditions $\bar{n} \gg 1$ and $\bar{m} \gg 1$ are valid (in particular, for fixed numbers $n \gg 1$ and $m \gg 1$ as well).

It is interesting that Eqs.10 are in full accordance with the relations obtained in paper [12/] on the basis of the coherent state approach (see also [13/]). This coincidence is not mere chance: really, for the coherent state the probability distribution over the states with a given number n is the Poisson distribution[18].

4. Taking into account Eqs.10, we can rewrite Eq.8 in the form

$$R(q) = 1 + p^2|F_1(q)|^2 + 2p(1-p)\mathrm{Re}(F_1(q)F_2^*(q)) \quad , \qquad (11)$$

where $p = \bar{m}/(\bar{n}+\bar{m})$ is the "chaoticity" parameter introduced by Weiner[15]. In our model the parameter p describes a relative contribution of chaotically located one-particle sources.

Let us assume that the symmetry centres of the distributions $W_1(\tilde{x})$ and $W_2(\tilde{x})$ coincide and the root-mean-square radii are equal to $r_1$ and $r_2$, respectively. In accordance with the Eq.4, in this case the equalities $\mathrm{Im}F_1(q)=0$ and $\mathrm{Im}F_2(q)=0$ hold. It is easy to see that the root-mean-square radius determining the dependence upon momentum difference q for the second term in Eq.11, which is proportional to $p^2$, is equal to $R_1 = \sqrt{2}\, r_1$. The root-mean-square radius determining the q-dependence of the third term in Eq.11, which is proportional to $2p(1-p)$, equals $R_2 = \sqrt{r_1^2 + r_2^2}$. If $r_2 \ll r_1$, we obtain the relation presented in paper[15]:

$$R_1 = \sqrt{2}\, R_2 \quad . \qquad (12)$$

Let us consider the case when the point-like multi-particle source is in the centre of the spherical-symmetrical Gaussian distribution of independent one-particle sources. Then we have

$$W_1(\vec{\tilde{r}},\tilde{t}) = (2\pi)^{-4} r_o^{-3}\, \tau_o^{-1} \exp(-\vec{\tilde{r}}^2/(2r_o^2) - \tilde{t}^2/(2\tau_o^2)), \qquad (13)$$

$$W_2(\vec{\tilde{r}},\tilde{t}) = \delta^3(\vec{\tilde{r}})\, \delta(\tilde{t}) \qquad ; \qquad (14)$$

$$F_1(q) = \exp(-\tfrac{1}{2}\, \vec{q}^2 r_o^2 - \tfrac{1}{2}\, q_o^2\, \tau_o^2) \; , \quad F_2(q) = 1 \quad . \qquad (15)$$

As a result, if the time parameter $\tau_o = 0$, the correlation function takes the same form as in paper [15/]: (16)

$$R(q) = 1 + p^2 \exp(-\vec{q}^2 r_o^2 - q_o^2 \tau_o^2) + 2p(1-p)\exp(-\tfrac{1}{2}\vec{q}^2 r_o^2 - \tfrac{1}{2}q_o^2\tau_o^2).$$

However, in the framework of our approach the Eqs. 12 and 16 are not of a general character, and they are violated at $r_2 \sim r_1$. These relations are also changed in the case when $r_2 = 0$ but the multiparticle source is located at a definite distance $b = \{\vec{b}, T\}$ from the symmetry centre of the space-time distribution of one-particle sources. Under these conditions we should write instead of Eq.16 the following expression for the correlation function of identical pions: (17)

$$R(q)=1 + p^2\exp(-\vec{q}^2 r_o^2 - q_o^2\tau_o^2) + 2p(1-p)\cos(bq)\exp(-\tfrac{1}{2}\vec{q}^2 r_o^2 - \tfrac{1}{2}q_o^2\tau_o^2).$$

5. Other situations with the participation of multiparticle sources are also possible in the framework of our approach. Some of them have been analyzed in [16/].

Let us consider the general case when we have m one-particle sources with 4-coordinate distribution $W_1(x)$ and N multi-particle sources with 4-coordinate distribution $W_2(x)$, each of them instantaneously emitting n identical pions. We denote the Fourier transforms of these distributions by $F_1(q)$ and $F_2(q)$, respectively. The number of pion pairs when two pions are emitted by the same n-particle source equals $\frac{1}{2}n(n-1)N$. A contribution of these pairs to the generation probability does not change when the pion momenta approach each other. Besides, $\frac{1}{2}m(m-1)$ pairs from one-particle sources make a contribution proportional to $(1+|F_1(q)|^2)$, $\frac{1}{2}n^2N(N-1)$ pairs from different n-particle sources make a contribution proportional to $(1+|F_2(q)|^2)$, and nmN "mixed" pairs, when one of the pions is emitted by the one-particle source and another is emitted by the multiparticle source, make a contribution proportional to $(1+\mathrm{Re}(F_1(q)F_2^*(q))$. In this situation the pair correlations of identical pions with slightly different momenta are described by the expression

$$R(q) = 1 + \Lambda_1|F_1(q)|^2 + \Lambda_2\mathrm{Re}(F_1(q)F_2^*(q)) + \Lambda_3|F_2(q)|^2 \tag{18}$$

with three parameters $\Lambda$ :

$$\Lambda_1 = \overline{m(m-1)}/\beta \quad , \quad \Lambda_2 = 2\bar{n}\bar{m}\bar{N}/\beta \quad , \quad \Lambda_3 = \overline{N(N-1)}(\bar{n})^2/\beta \ , \tag{19}$$

$$\beta = \overline{n(n-1)}\bar{N} + \overline{m(m-1)} + \overline{N(N-1)}\bar{n}^2 + 2\bar{n}\bar{m}\bar{N} \quad . \tag{20}$$

When the multiplicities n and m are distributed in accordance with the Poisson law (and also for any distributions with $\bar{n}\gg 1$, $\bar{m}\gg 1$), we have: $\Lambda_1 = 1/\overline{(1+\gamma N)^2}$ ,

$$\Lambda_2 = 2\gamma\bar{N}/\overline{(1+\gamma N)^2}, \qquad \Lambda_3 = \overline{N(N-1)}\gamma^2/\overline{(1+\gamma N)^2} \quad , \tag{21}$$

where $\gamma = \bar{n}/\bar{m}$ . An analogous result has been obtained in paper [14] by means of the coherent state technique.

If N=1, then $\Lambda_3=0$, and the Eqs.8 and 9 follow directly from Eqs.18-20. If one-particle sources are absent (i.e., m=0), then $\Lambda_1 = \Lambda_2 = 0$, and we come to the formula

$$R(q) = 1 + \Lambda|F_2(q)|^2 \tag{22}$$

with one parameter $\Lambda<1$:

$$\Lambda = (\bar{n})^2\overline{N(N-1)}/[\overline{n(n-1)}\bar{N} + \overline{N(N-1)}(\bar{n})^2 \quad . \tag{23}$$

In the case of the Poisson distribution Eq.22 gives

$$\Lambda = 1 - \bar{N}/\overline{N^2} \quad . \tag{24}$$

This expression coincides with the relation obtained in paper [12] by means of the coherent state approach.

Analogously, the three-particle correlations can be considered in the framework of our model.

6. Thus, the properties of correlations of bo-

sons with nearly equal momenta, which are usually explained by means of a special role of coherent states, can be interpreted in the framework of the simple model including both multi-particle and one-particle sources. It should be emphasized that our results are related to arbitrary multiplicity distributions, in particular, to reactions with a fixed number of particles. In this sense they are of a more general character.

The coincidence of our concrete formulae with the relations obtained in terms of coherent states corresponds to the particular case of the Poisson distribution of bosons from sources (or to the limit of very large multiplicities for any distributions).

It should be noted once again that, from our point of view, the essence of the problem is that there are no interference correlations of the pions which are in the same quantum state. The concrete interpretation of coherent states as superpositions over the number of bosons is of no importance since correlations contain no contributions from the interference of states with a different number of particles. Besides, in the case of charged pions the coherent states are the superpositions of states with different electric charges. In the framework of modern conceptions the existence of such superpositions is prohibited by the well-known rule of superselection following from the exact conservation of electric charge[19/]. This circumstance prevents a literal application of the coherent state technique for the description of pair correlations of $\pi^+$ or $\pi^-$-mesons with slightly different momenta. However, such an application is not needed since the superposition properties of coherent states are not really used. At the same time multi-particle sources were in fact first introduced just due to the conception of coherent states.

The author is grateful to R.Lednicky and M.I.Podgoretsky for valuable discussions.

References

1. Goldhaber G., Goldhaber S., Lee W. and Pais A., Phys.Rev. 120, 300 (1960).
2. Grishin V.G., Kopylov G.I. and Podgoretsky M.I., Yad. Fiz. 13, 1116 (1971) /Sov.J.Nucl.Phys. 13, 638 (1971)/.
3. Kopylov G.I. and Podgoretsky M.I., Yad.Fiz. 18, 656 (1973) /Sov.J.Nucl.Phys. 18, 336 (1973)/.
4. Kopylov G.I., Phys.Lett. B50, 472 (1974).
5. Cocconi G., Phys.Lett. B49, 459 (1974).
6. Lednicky R. and Lyuboshitz V.L., Yad.Fiz. 35, 1316 (1982) /Sov.J.Nucl.Phys. 35, 770 (1982)/.
7. Lednicky R. and Podgoretsky M.I., Yad.Fiz. 30, 837 (1979) /Sov.J.Nucl.Phys. 30, 432 (1980)/.
8. Fowler G.N. and Weiner R.M., Phys.Lett. B70, 201 (1977).
9. Fowler G.N. and Weiner R.M., Phys.Rev. D17, 3118 (1978).
10. Bartnic E.A. and Rzazewski K., Phys.Rev. D18, 4308 (1978).
11. Fowler G.N., Stelte N. and Weiner R.M., Nucl.Phys. A319, 349 (1979).
12. Gyulassy M, Kaufman S.K. and Wilson L.W., Phys.Rev. C20, 2267 (1979).
13. Biyajima M., Phys.Lett. B92, 193 (1980).
14. Biyajima M., Progr.Theor.Phys. 66, 1378 (1981).
15. Weiner R.M., Phys.Lett. B232, 278 (1989).
16. Lednicky R., Lyuboshitz V.L. and Podgoretsky M.I., Yad. Fiz. 38, 261 (1983) /Sov.J.Nucl.Phys. 38, 153 (1983).
17. Paul H., Rev.Mod.Phys. 58, 209 (1986).
18. Glauber R.J., Phys.Rev.Lett. 10, 84 (1963).
19. Streater R.F. and Wightman A.S., "PCT, Spin, Statistics and All That" (Benjamin, New York, 1964).

# Some different analysis of the AFS-data on interferometry

B. Lörstad
Department of Particle Physics, Lund University
S-22362 Lund, Sweden

**Abstract**

A comparison of the interferometry effect in hadronic jets and in general *pp* collisions shows striking similarities. The correlation between charged pions of the same charge is shown to depend only on Q, the absolute value ot the four-momentum transfer between the pions. The correlation function in Q is the same in these samples of very different selection, it is more peaked than a single Gaussian and can be parametrized by an exponential function or as a sum of two Gaussians.

A string model can accomodate these features and explain the similarities between these samples. A relativistic hydrodynamical model can be tuned to describe the data from general *pp* collisions, whereas a simple quark-gluon plasma model is not that successful. A static approach of pure quantum-mechanical considerations with partial coherence is shown not to connect the Bose-Einstein effect and the multiplicity distributions with consistent parameters. It is argued that interferometry is becoming an important tool in the study of dynamical properties at high energy.

## 1 Introduction

In this talk I will discuss the data on interferometry collected by the Axial Field Spectrometer collaboration (AFS) at the CERN-ISR. These data have been the subject of different analyses using different physical models and I will summarize the main results. In the next section I will present the apparatus pointing out the reasons why these data are well suited for a detailed analysis of the Bose-Einstein effect. Samples with two different triggers have been analysed, a jet sample and a minimum bias sample. The jet sample was triggered on events with two separated hadronic jets and represents a large sample of hadronic jets with energy of about 10 GeV in each jet. These jets are similar to those produced in $e^+e^-$ annihilations but constitute larger statistics. Further details are found in ref.[1]. The minimum bias sample represents the general $pp$ event, but only in the central region of phase space as only a rapidity region of -.8 to .8 is covered by the apparatus. The comparison of the Bose-Einstein effect in these data sets shows, section 3, an identical behaviour, which is parametrized by a dependence of the correlation function on only the absolute value of Q, the four-momentum transfer between the two identical pions[2]. The Q-dependence is more peaked than the Gaussian function usually assumed. This result has a large impact on the value of the correlation function near small Q values, the so-called $\lambda$-parameter. There is a rich literature on the interpretation of small $\lambda$ values as evidence for coherence production now to be reconsidered[3].

The interferometry effect has been used to measure the shape of the pion emitting region[2,4], section 4. These measurements are explained by Lorentz contraction effects going from the $\pi\pi$ system to the total centre of mass system, where the shape measurement are done. The findings result in a consistent picture of the Bose-Einstein effect as mainly depending on Q. The dependence on Q can be parametrized as the sum of two Gaussians or as an exponential. Such a form can be given by a string model, discussed in detail by M.Bowler in his contribution to this meeting.

This string model considers only the space-time evolution of the hadronization processs without explicit geometrical considerations. This emphasizes that it is certainly insufficient to take into account only the geometrical aspects when interpreting the interferometry effect. The dynamics of the space-time evolution will influence the two-particle correlation function and different models can thus be tested against data. A general model of the interaction applicable to high energy hadronic interactions is the Bjorken inside-outside cascade model[5], which was used by Kolehmainen and Gyulassy in a calculation of the two-pion correlation[6]. The basic requirements of the model are fulfilled by the minimum bias sample: the distribution in rapidity is uniform in the central region, the charged pions are identified, and the transverse-momentum distribution is fairly well reproduced at low $p_T$ by the pseudothermal model used. This model can be reasonably adjusted to the data[7], section 5.

Another approach is suggested by R.Weiner[8] and based on quantum statistical considerations only. In this model the particle field is assumed to be a superposition of one chaotic and one coherent part. Weiner points out that the empirical relation with an ad hoc chaoticity or incoherence factor $\lambda$ which is usually introduced to describe the experimental data together with a Gaussian is in contradiction with quantum statistics and he gives a formula with two coupled Gaussians derived from this theory. This has a further interest as this formulation connects the Bose-Einstein correlation to the multiplicity distributions, permitting a new consistency check, section 6.

Quark-gluon plasma models have been developed for the study of heavy-ion interactions at very high energy[9,10]. Such a model has been applied to this minimum bias sample. The $p_T$ and $y$ distributions are reproduced to very high $p_T$ but a detailed correlation analysis shows disagreements with interferometry data, section 7.

## 2 Data

The data were taken with the Axial Field Spectrometer at the CERN Intersecting Storage Rings. Most data were obtained from proton-proton collisions at $\sqrt{s} = 63$ GeV.

The AFS has been described in detail elsewhere[11]. The interaction region is surrounded by an inner hodoscope, a cylindrical drift chamber, and a uranium/copper/scintillator calorimeter with full azimuthal coverage.

Charged particle tracking was provided by the cylindrical drift chamber, operated in an axially symmetric magnetic field. The chamber was azimuthally segmented into independent $4^0$ sectors. The active volume extended over $2\pi$ in azimuth, except for two $16^0$ wedges in the vertical plane, used for mechanical support. Coordinates in the transverse plane were provided by drift-time measurements with a spatial resolution of $230\mu m$, and those along the beam direction by charge division with 1.5 cm resolution. The pulse-height measurements also provide dE/dx information which is used for particle identification.

Quality cuts on the reconstructed charged-particle tracks in the drift chamber include a minimum of measurements being one more than half of the number of readout wires per sector. This will ensure that the measurements due to the passage of one particle are not divided into two separate track candidates. When two particles pass through the same $4^0$ sector of the drift chamber, the charge division information can not be used even when the drift-time measurements can resolve the tracks. With the stringent requirement in order to be sure not to treat one particle track as two fictitious tracks, pairs of the same charge will sometimes be lost when the azimuthal separation is lower than $4^0$. Instead of trying to correct for this inefficiency we require at least $4^0$ azimuthal separation of tracks of equal charge. The loss of particle pairs for the study of the Bose-Einstein effect is marginal as the mean opening azimuthal angle is quite large, around $25^0$ for the minimum bias sample. The gain is that we can trust the measurements down to the smallest Q-values, as they constitute a safe sample of independently measured particle pairs.

A full Monte Carlo simulation of the drift chamber response shows that the acceptance for charged particles having $p_T > 200$ MeV/c is very close to 100% in the central region, $-0.7 < y < 0.7$. The error in the absolute value of the four-momentum transfer Q is less than 15 MeV/c. The dE/dx information from the tracks is used to separate out most of the protons and kaons from the pions. Kaons could be identified in the momentum range 0.3 to 0.7 GeV/c, and protons in the momentum range 0.4 to 1.1 GeV/c.

## 3 Evidence of non-Gaussian Q-dependence

The phenomenon of Bose-Einstein correlations is usually studied in terms of the ratio $R$ of the joint probability for a pair of identical particles to be emitted, $P(p_1, p_2)$ to the product $P(p_1) \cdot P(p_2)$ of the single-particle probabilities, where $p_1, p_2$ are the four-momenta of the two particles. For fully chaotic sources this ratio is given by

$$R = \frac{P(p_1,p_2)}{P(p_1)\cdot P(p_2)} = \frac{P(p_1,p_2)}{P_0(p_1,p_2)} = 1 + \rho^2(Q), \tag{1}$$

where $Q = p_1 - p_2$ is the four-momentum difference between the identical bosons, and $\rho(Q)$ is the Fourier transform of the space-time source distribution, normalized to unity as $Q \to 0$. $P_0(p_1,p_2)$ is a reference distribution which in the ideal case incorporates all the two-particle correlations except the Bose-Einstein correlation, i.e. correlations due to energy-momentum conservation and other correlations due to event structure. In this study we have minimized these correlations by selecting i)for the jet sample only single jets and ii) for the minimum bias sample only the central part of the events which does not show any special structure. We have restricted the transverse momentum of the particles to be in the range $0.2 < p_T < 2.0$ GeV/c. The higher cut-off quarantees that particle correlations are not influenced by phase-space limitations, the lower cut-off quarantees a good acceptance of nearly 100%. A Monte Carlo study has proven the effectiveness of the reference distribution calculation. This ratio R is identified with the two-particle correlation function, C(Q).

A source density distribution fixed in space-time and described by Gaussians separable in space and time is mostly assumed in hadronic interactions

$$\rho(x) = const.\exp(-\frac{x_T^2 r_T^2}{2} - \frac{z^2 r_z^2}{2} - \frac{t^2\tau^2}{2}), \tag{2}$$

where $r_T(r_z)$ is the transverse (longitudinal) radius and $\tau$ is the spread in production times. The corresponding correlation function is

$$C(Q_T, Q_z, Q_0) = 1 + \lambda\exp(-Q_T^2 r_T^2 - Q_z^2 r_z^2 - Q_0^2\tau^2) \tag{3}$$

You may remark that this formalism is not Lorentz covariant.One often omits the "time" factor regarding C as a function of the transverse and longitudinal components of the pair momentum difference only. The factor $\lambda$, which can take values between 0 and 1, is the so-called incoherence or chaoticity factor which takes into account that the interference, for various reasons, may not be complete.

At high energy hadronic jets are produced. Their mechanism of production is described phenemenologically by string models[12]. Hadronic jets have a very different structure longitudinal to the jet axis as compared to transversal to the jet- axis. It is tempting to try to visulize the size of the jet by interferometrical methods. Strings are described by a Lorentz covariant formulation, why jet-data are fitted to a pair of variables invariant to boosts along the string direction, $Q_T^2$ and $Q_L^2$. $Q_T^2$ is the square of the four-momentum difference transverse to the jet-axis, whereas $Q_L^2$ is given by

$$Q_L^2 = Q_z^2 - Q_0^2 \tag{4}$$

where $Q_z$ is the longitudinal momentum difference and $Q_0$ the energy difference between the two particles. $Q_L^2$ is the variable conjugated to $t^2 - z^2$, which describes the source density in a one dimensional string model and is thus related to the longitudinal development of the jet.

The data from the study of hadronic jets[2] are divided into a two dimensional array in $Q_T^2$ and $Q_L^2$ and the correlation function is fitted to the form

$$C = N(1 + \lambda \exp(-r_T^2 Q_T^2 - r_L^2 Q_L^2)), \quad (5)$$

where N is a normalization factor. A good fit is obtained with the parameters $r_T \approx r_L$. In Fig.1 some slices in $Q_T^2$ and $Q_L^2$ are shown with the fit superimposed upon the data. The data given in the figure have been normalized so that N=1. We remark that the correlation function follows the Gaussian form even for negative $Q_L^2$-values, providing evidence that this combination, $Q_z^2 - Q_0^2$, is the relevant combination to describe the data.

Thus the correlation function shows the same dependence on $Q_T^2$ as on $Q_L^2$, consequently, it is well parametrized as a Gaussian in the single variable $Q^2 = Q_T^2 + Q_L^2$. Similar results have been obtained from jet studies in $e^+e^-$ annihilations[13].

Also the general *pp* events have been analysed in a similar way, i.e. the interesting axis is here the beam axis instead of the jet axis. Fitting Eq.(5) using the boost-invariant variables $Q_T^2$ and $Q_L^2$ it is found also in these events that $r_T \approx r_L$[2,14].

In order to expand the interesting region of small momentum transfers we will display the correlation function as a function of Q rather than the most often used $Q^2$. Figure 2 a shows the two-particle and reference distributions as a function of Q, normalized to the tail of the two-particle distribution. Figure 2 b displays the ratio R of these distribution and it clearly shows that the correlation is more peaked than a simple Gaussian (dashed line). It can be parametrized as a sum of two Gaussians (full line) or with an exponential function.

Figures 3 a and 3 b show that the correlation function in minimum bias data is very similar to the one in hadronic jets even though the two-particle and reference distributions are quite different.

Earliar interferometry studies have not had enough statistics to show this peaked structure[3], why most results are given using a single Gaussian to extract the parameters $\lambda$ and $r_G$:

$$R = 1 - \lambda \exp[-(r_G \cdot Q)^2] \quad (6)$$

Another used formulation, in fact corresponding to a similar shape of the measured experimental correlation function, is

$$R = 1 - \lambda \frac{\left(\frac{2J_1(r_B \cdot q_T)}{(r_B \cdot q_T)}\right)^2}{(1 - (Q_0 \cdot \tau)^2)} \quad (7)$$

Eq.(7) corresponds to pion sources uniformly distributed over the surface of a sphere with radius $r_B$; in some models $\tau$ represents the lifetime of the sources, in others it is ascribed to the depth of the emission layer[15]. Here $q_T$ is the component of $\vec{p}_1 - \vec{p}_2$ transverse to $\vec{p}_1 + \vec{p}_2$, $Q_0$ is the energy difference between the two identical bosons, and $J_1$ is a Bessel function of first order.

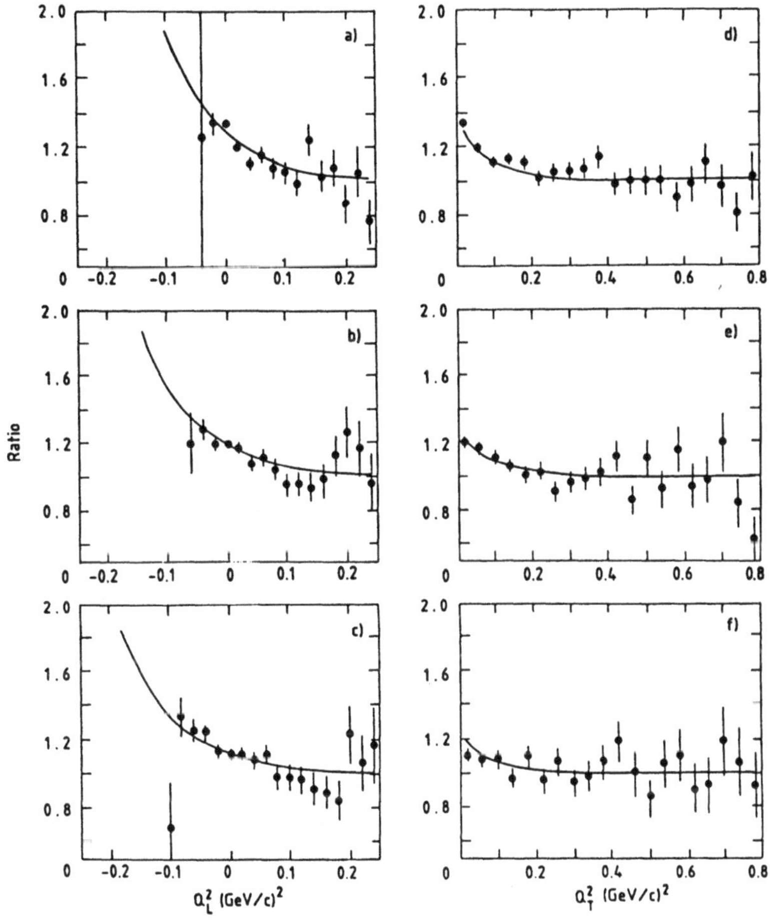


Figure 1: **a-f** Representation of a part of the fit in $(Q_T^2, Q_L^2)$. **a-c** show tha data versus $Q_L^2$ in slices of $Q_T^2$: **a** $0.00 < Q_T^2 \leq 0.04(GeV/c)^2$, **b** $0.04 < Q_T^2 \leq 0.08(GeV/c)^2$, **c** $0.08 < Q_T^2 \leq 0.12(GeV/c)^2$. **d-f** show the data versus $Q_T^2$ in slices of $Q_L^2$: **d** $-0.01 < Q_L^2 \leq 0.01(GeV/c)^2$, **e** $0.01 < Q_L^2 \leq 0.03(GeV/c)^2$, **f** $-0.03 < Q_L^2 \leq -0.01(GeV/c)^2$

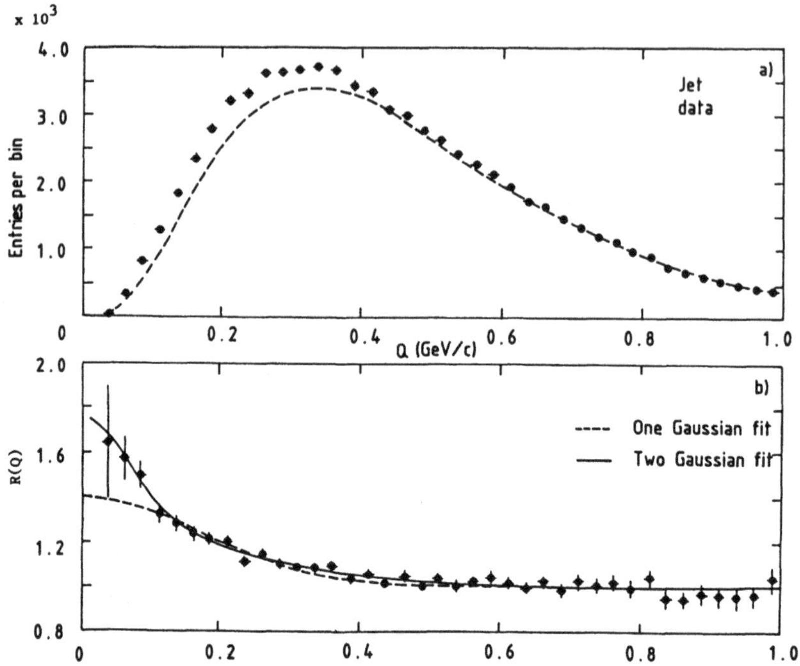


Figure 2: **a** The two-particle and the reference distributions for particle combinations within a hadronic jet as a function of Q, the four-momentum transfer. **b** The correlation function R(Q), i.e. the ratio of the two-particle distribution to the reference distribution. The full line shows the fit with the sum of two Gaussians, whereas the dashed line is the fit to one Gaussian.

Eq.(7) has mostly been used to investigate the structure of the interaction region in hadron-hadron collisions, whilst Eq.(6) is preferred in the analysis of the jet events in $e^+e^-$ collisions.

Both formulae turn out to be nearly indistinguishable in shape, and both fit the measured correlations quite well, formula(7) giving a radius parameter $r_B$ approximately twice as large as $r_G$ of Eq.(6), as can be seen from the exponential approximation of Eq.(7),see also[16].

$$R \approx 1 - \lambda \exp(-r_B^2/4 \cdot q_T^2 - \tau^2 Q_0^2) \tag{8}$$

The determination of the parameter $\lambda$, the incoherence or chaoticity parameter, is subject to large correction factors and systematic errors, whereas the determination of the radius parameter is not that much influenced. Many effects could affect the value of $\lambda$, such as the creation of pions through the decay of resonances, final state interactions and the prescence of several different pion sources[3]. Most sources of experimental difficulties, like particle misidentification and problems with close

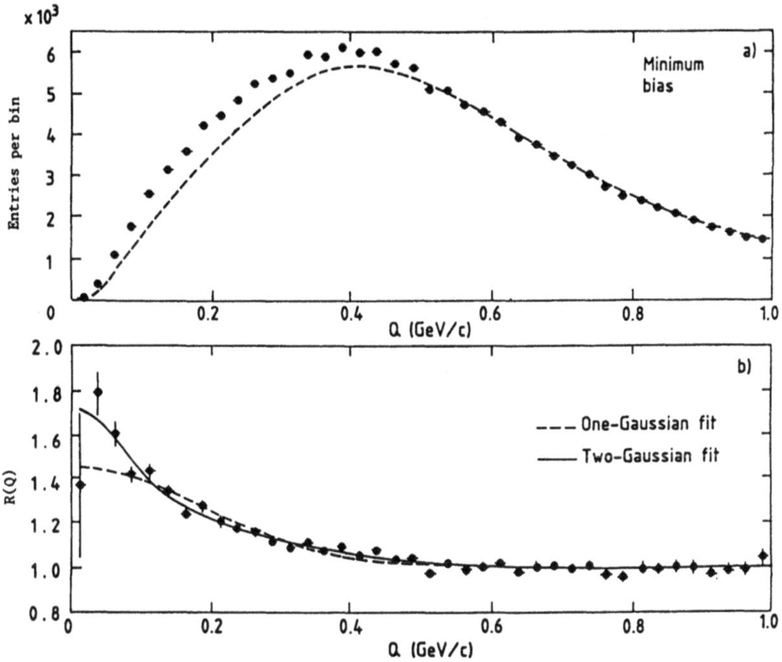


Figure 3: **a** The two-particle and the reference distributions for minimum-bias-triggered data as a function of Q, the four-momentum transfer. **b** The correlation function R(Q), i.e. the ratio of the two-particle distribution to the reference distribution. The full line shows the fit with the sum of two Gaussians, whereas the dashed line is the fit to one Gaussian.

tracks, tend to dilute the enhancement why measured $\lambda$-values are to be considered as lower limits.

Furthermore we have here evidence for another important point: the shape of the

correlation function must be measured with high accuracy at low Q-values as the supposed Gaussian shape do not fit the data. The factor $\lambda$, parametrizing the strength of the correlation, has been extracted by single Gaussian fits with typical values in the region $0.4 - 0.6$. The more peaked exponential form gives larger values of $\lambda$. Generally we can conclude that even if many experiments report values of $\lambda < 1$, no clear evidence for coherence has been observed.

The peaked structure parametrized with a second Gaussian could be due to the resonance effects[17]. It is well known that many of the pions in the jets are produced via resonances. Short-lived resonances, such as the $\rho$-resonance, will have decay lengths of only a few fermi, and their decay pions will show interference with directly produced pions as well as with those produced via other short-lived resonances. This might explain the observed structure of the two-particle correlation.

Another interpretation is possible using a string model. String models have been very successful in interpreting hadronic jets. The total source size is rather large: of the order of 10 fm for a 10 GeV jet which is much larger than the effect seen by interferometry experiments. The reason is the strong correlation between particle production points and particle momentum in the string model. Hadrons with approximately the same momentum are in fact produced only in a region of invariant size of about 1 fermi, which is in fact what is measured by interferometry. Andersson and Hofmann[18] have made an estimate of interference effects for particles produced in a jet within the Lund string model. Their approximate form ressembles an exponential function and hence the measured correlation function, see figure 4. In this calculation the influence of resonances and final state interactions was not taken into account. M. Bowler[19] has analysed these effects in great detail as they could have a large impact on the measured correlation. It turns out that the resonance effect is rather small in the model and that final state interactions tend to cancel, even though single terms may be quite important. For a discussion of the comparison of this model to different data sets, see Bowlers contribution to this meeting.

The string model gives an approximate dependence on Q, the absolute value of four-momentum transfer, and a correlation function for charged pions corresponding to what we measure both within hadronic jets and in the general $pp$ event. It is astonishing to see such a clear string effect in the minimum bias events which have a rather low multiplicity density, dn/dy $\approx$ 1.5. The reason could be that the probability to get two particles of the same mometum is so low that particles produced from the same string are selected. Only at higher particle densities effects due to a more complicated string structure might be seen[3,20].

To distinguish the interpretation of the peaked structure as due to an old-fashioned resonance effect or due to a string effect more experimental information is needed. Studies of $\pi^0\pi^0$ combinations or $K^{\pm}K^{\pm}$ might be able to clarify the interpretation. The resonance influence on $K^{\pm}K^{\pm}$-correlations is quite different from that on $\pi^{\pm}\pi^{\pm}$-combinations, whereas the string model gives something similar to or even narrower than what is seen for $\pi^{\pm}\pi^{\pm}$-combinations. $K^{\pm}K^{\pm}$ correlations as reported[21] have similar characteristics to $\pi^{\pm}\pi^{\pm}$ correlations but the statistical uncertainties are too large to permit a conclusion. The LEP experiments accumulating large samples of events with identified hadronic jets might be able to perform a conclusive study.

The $\pi^0\pi^0$ combination can be neigbours on a string why a wider correlation ought to result. Such a widening as compared to $\pi^{\pm}\pi^{\pm}$ correlations should be a signature of a string effect. However the $\pi^0\pi^0$ correlation due to the Bose-Einstein effect is experimentally difficult to isolate due to the large influence on the correlation function from the $\eta$ decay into $\pi^0\pi^0\pi^0$.

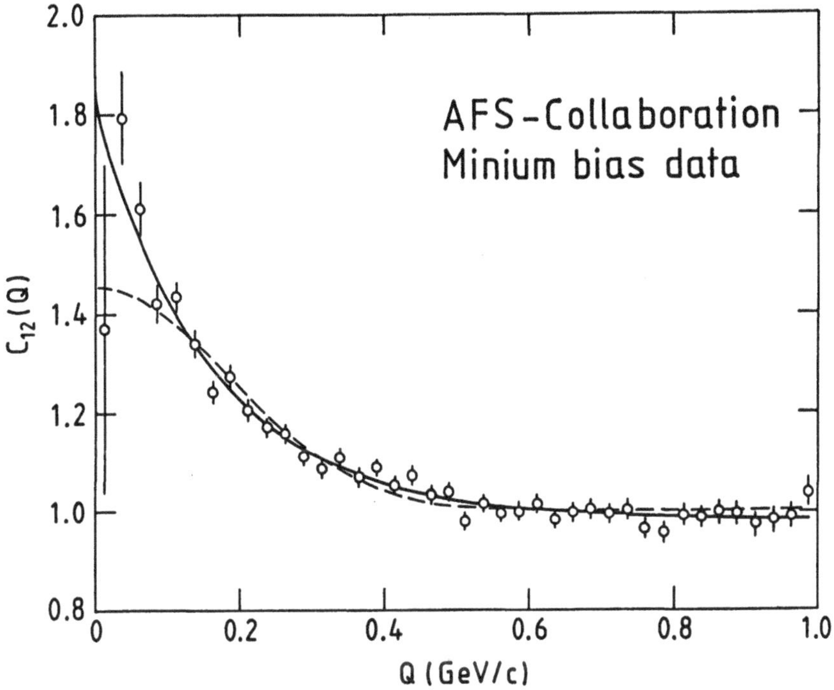


Figure 4: The correlation function of two pions displayed as a funtion of Q. The full line is the result of a fit to an exponential function, the dashed line to a Gaussian

## 4 Shape measurements of the pion emission region

The simple dependence on the absolute value of Q of the Bose-Einstein effect in these data can be interpreted as evidence that the pion emission is spherical in the $\pi\pi$ centre-of-mass system. The shapes in other systems as the total centre-of-mass system should then be given by Lorentz boosts from the $\pi\pi$ system. We will see that this is the case, at least qualitatively.

The shape of the emission region in space can be determined by a two-dimensional fit to the function

$$R = 1 + \lambda \exp(-r_T^2 Q_T^2 - r_z^2 Q_z^2) \tag{9}$$

where z-axis is a significant axis and T stands for transversal to this axis. This formula is not in contradiction to Eq.5, as you naively might think. The radius $r_z$ is in fact connected to the Lorentz invariant radius $r_L$ via a $\gamma$-factor, between the $\pi\pi$-system and the new system: $r_z = r_L/\gamma$.

In the new system the line of flight of the $\pi\pi$-system is given by

$$\vec{n} = \frac{\vec{p_1} + \vec{p_2}}{\mid \vec{p_1} + \vec{p_2} \mid} \tag{10}$$

The component of Q parallel to $\vec{n}$, $Q_{\parallel}$ , and $Q^0$ are related:

$$Q_{\parallel} = (\vec{p_1} - \vec{p_2}) \cdot \vec{n} = \frac{Q_0}{\beta} \tag{11}$$

where $\beta$ is the velocity of the $\pi\pi$-systems:

$$\beta = \frac{\mid \vec{p_1} + \vec{p_2} \mid}{E_1 + E_2} \tag{12}$$

This gives immediately that the $Q_L$-component, invariant to boosts along $\vec{n}$, will be given as

$$Q_L^2 = Q_{\parallel}^2 - Q_0^2 = \frac{Q_{\parallel}^2}{\gamma^2} \tag{13}$$

From this formula the relation $r_z = \frac{r_L}{\gamma}$ follows directly when Eqs.(9) and (5) are used for the fitting.

Qualitatively this effect is found: in the hadronic jet system the $\pi\pi$-system moves longitudinally with the jet-axis and it is found that $r_{jet-axis} = 0.43 \pm 0.06$ fm< $r_T = 0.70 \pm 0.04$ fm$^2$. In the minimum bias sample, restricted to the central region $y \leq 0.8$, the $\pi\pi$-system moves mainly transversal to the beam-axis and $r_{beam-axis} = 1.3 \pm 0.1$ fm is found to be larger than $r_T = 0.72 \pm 0.04$ fm, ie an prolate shape[4]. In another pp experiment[22] at somewhat lower energy an oblate shape is found. This can in fact be consilient considering that most $\pi\pi$-systems in these events go forward thus contracting the longitudinal radius.

Quantitatively we can calculate mean $\gamma$-factors of the samples to check the relation $r_z = \frac{r_L}{\gamma}$. Using Gaussian fits the consistency is not very great, however with exponential functions instead of Gaussians for the extraction of the radii the values are consistent[23]. This result emphasizes that exponential functions are preferable: they fit the data better with fewer parameter which are consistent in different centre of mass systems.

## 5 The cascade model

In the cascade model there is a correspondence between the production site in space-time of a particle and its momentum, in particular longitudinal rapidiy, as in the string model. However the strict production scheme of the string model is changed to a hydrodynamical development, suitable for a description of collisions between larger and more complicated systems such as nucleus-nucleus collisions. The requirements of the model are however fulfilled by the minimum bias events: fairly uniform distribution in rapidity, identified pions and a transverse-momentum distribution reproduced at low $p_T$ by the pseudo-thermal model used[6]. The first step in the analysis is the extraction of the parameter T from the invariant particle distribution:

$$P_1(p) = C \cdot K_0(\frac{m_T}{T}) \tag{14}$$

where C is a normalization constant, $m_T$ the transverse mass of the pion, and $K_0$ a modified Bessel function. The fit was performed in the interval $0.22 \leq m_T \leq 1.00 GeV/c^2$, where the lower limit is imposed to protect from non-uniform detector acceptance and the upper excludes the familiar change of slope of the distribution at high energy. The result is $T = 206.3 \pm 0.1$ MeV.

The correlation equation is a function of 4 variables:$m_{T1}, m_{T2}, q_T$, and $\Delta y$.

$$R = 1 + \lambda \frac{| G_{12} |^2}{G_{11} G_{22}} \tag{15}$$

where

$$\begin{aligned} G_{12} &= C \cdot K_0(\sqrt{u}) \exp(-\frac{q_T^2 R_T^2}{4}), \\ u &= 2 m_{T1} m_{T2} (\frac{1}{4T^2} + \tau^2) \cosh(\Delta y) + (m_{T1}^2 + m_{T2}^2)(\frac{1}{4T^2} - \tau^2) + \\ & i(\frac{\tau_0}{T})(m_{T1}^2 - m_{T2}^2) \end{aligned} \tag{16}$$

In performing such a 4-dimensional fit, the number of bins and number of entries had to be carefully balanced against each others as the statistics is at the limit to allow such a complicated fit. The fit was based on the principle of maximum likelihood applied to Poisson distributions. Inserting the value of T already achieved the following results were obtained:

$R_T = 0.87 \pm 0.07$ fm, $\tau_0 = 1.49 \pm 0.21$ fm/c, and $\lambda = 0.60 \pm 0.07$.

The value of the radius parameter is compatible with the previous results using other parametrizations, the proper time of freeze out $\tau_0$ is of the order 1 fm/c as expected in these reactions, whereas $\lambda$ is significantly higher than what found in the single Gaussian fits. The fitted correlation function is projected to the variable Q, earlier shown to be the main variable. In Fig. 5 the solid curve shows this projection superimposed on the data points. We see that this model gives a correlation function considerably more peaked than a single Gaussian in correspondence with the data.

## 6 Static approach with partial coherence

In the quantum statistical model[8] the resulting particle field is considered to be made up of one coherent and one chaotic part. This gives a resulting chaoticity $p$, defined through $p = \frac{<n_{chaotic}>}{<n>}$, where $< n_{chaotic} >$ is the mean number of particles arising from the chaotic source, and $< n >$ the mean total number of particles. With Gaussian forms for both constituent fields, the two-particle correlation function reads:

$$R = 1 + 2p(1-p)\exp(-(\Delta y)^2 r^2) + p^2 \exp(-2(\Delta y)^2 r^2) \tag{17}$$

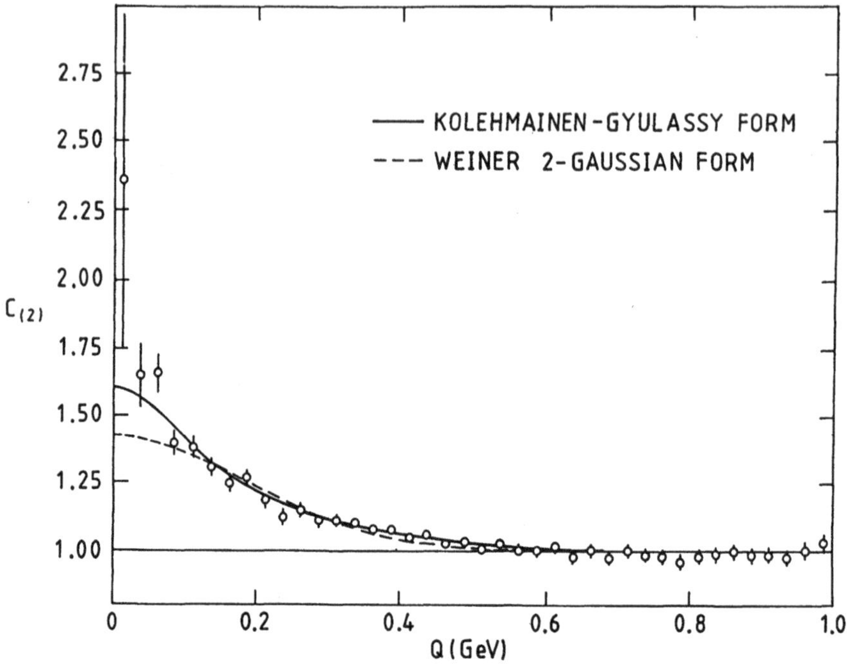


Figure 5: Data points shows the correlation function vs Q. The superimposed solid curve is calculated from the KG model using parameter values obtained from the fit described in the text. The dashed curve represents the fit with Weiner's form, Eq.(17).

where y is a "stationary" but not uniquely determined variable characterizing the particle field, and r the conjugate correlation quantity. For a nearly uniform rapidity distribution, one expects a single value of the chaoticity and Eq. (17) to be valid. The appropriate variable to describe the Bose-Einstein correlations in these data is Q, why we apply the Eq.(17) with $\Delta y = Q$ even though Q is a complicated quantity in the quantum-statistical model mixing the stationary and non-stationary variables $y$ and $p_T$. The following results are arrived at:

$r = 0.72 \pm 0.03$ fm, $p = 0.24^{+0.03}_{-0.02}$.

The characteristic peak at low Q values present in the data is, however, not recovered, see Fig. 5. Furthermore the $p$ value is significantly higher than the predicted value of $p \leq 0.1$ extracted from a study of multiplicity distributions in this energy range[8].

Analysing the data in terms of $\Delta y$, the difference in rapidity, reveals a single Gaussian structure. A fit with Eg.(17) yields: $r = 1.59 \pm 0.10, p = 0.10 \pm 0.01$. This result is insensitive to the $p_T$ interval. We note the agreement with the $p$ value extracted from the multiplicity distributions. However, this rapidity correlation

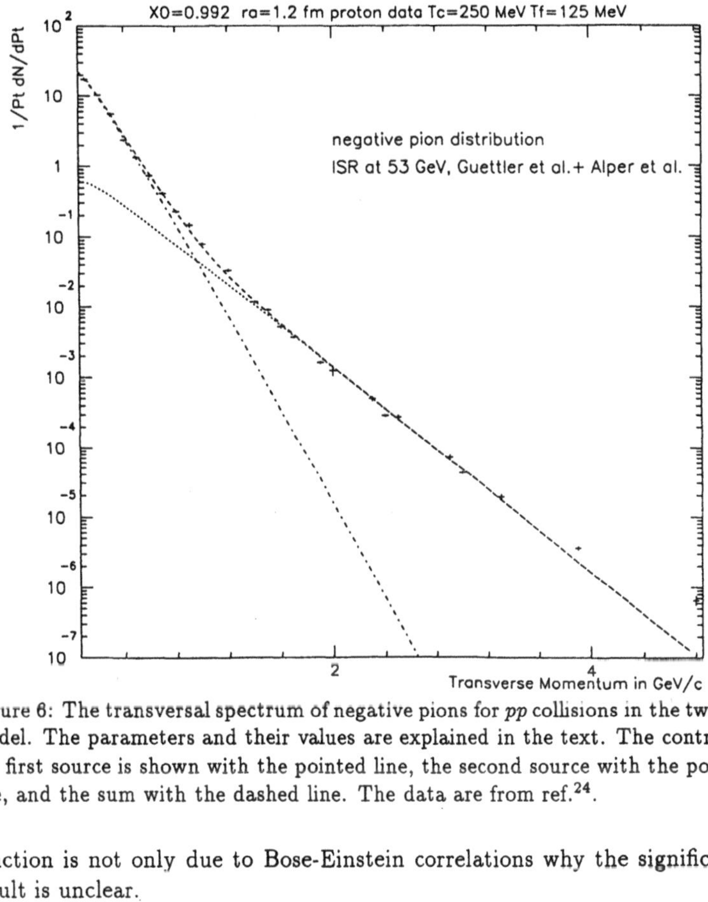


Figure 6: The transversal spectrum of negative pions for $pp$ collisions in the two-component model. The parameters and their values are explained in the text. The contribution from the first source is shown with the pointed line, the second source with the pointed-dashed line, and the sum with the dashed line. The data are from ref.[24].

function is not only due to Bose-Einstein correlations why the significance of this result is unclear.

## 7 The quark-gluon plasma model

Another model with roots in the Bjorken model is the model by Sinyukov et al.[10] discussed in more detail at this meeting by Yu.M.Sinyukov. Two sources of particle production are considered. In the first source there is a mixed phase between quark-gluon plasma and hadrons. When all plasma is transformed to hadrons the second source, of pure hadron gas, cools down to freezeout and the rest of the pions is released. Only pion production is considered. This hydrodynamical model is also intended for collisions between large nuclei, the aim of this analysis is to see if and when the model can not describe the $pp$ data. The single particle distrbutions in $p_T$ and $y$ can be fairly well described with this model. The Fig. 6 shows that the $p_T$-distribution is reproduced over a large region in $p_T$, from 0.1 to around 3.5

GeV/c with suitable parameters: $\chi_0$=.992, the proportion of pion gas in the first stage,$r_a$ = 1.2 fm the projectile-target overlap radius,$T_c$ = 250 MeV the critical temperature and $T_f$ = 125 MeV being the final temperature of pion freezeout. The parameters of the model are fixed by the adjustment to the single particle distributions and some interferometry predictions can be tested. The transverse radius is in agreement with the data,$R_T \leq 1$ fm, but the prediction of $R_z$ is far from what found in data: $R_z \approx 5$ fm to be compared to $R_z = 1.3 \pm 0.1$ fm[2,4].

# 8 Summary

A comparison of the interferometry effect in hadronic jets and general *pp* collisions shows striking similarities. The correlation between pions of same charge is shown to depend only on Q, the absolute value of the four-momentumtransfer between the pions. The correlation function in Q is the same in these samples of very different selection, it is more peaked than a single Gaussian and can be parametrized by an exponential function or as a sum of two Gaussians.

A string model can accomodate these features which makes the similarities significant, however data on other particle combinations, $K^{\pm}K^{\pm}$ and $\pi^0\pi^0$, are demanded to clarify if this is really a string effect or due to a more oldfashioned resonance effect. The relativistic hydrodynamical model of Kolehmainen and Gyulassy can also describe these data. A static model of pure quantum-mechanical considerations is not in accordance with the peaked structure of the correlation function nor with multiplicity constraints. A simple quark-gluon plasma model can reproduce the single particle spectra of *pp*-collisions but interferometry predictions are not fulfilled. This emphasizes that the Bose-Einstein effect at high energies is sensitive to the dynamical features of the interactions not only geometrical sizes and its study constitutes an important tool to clarify dynamical properties.

I would like to express my gratitude to Professor R.Weiner for the invitation to the CAMP Workshop in Marburg where different aspects of the interferometry analysis were extensively discussed.

# References

[1] T.Åkesson et al., *Z.Phys.C -Particles and Fields* **30**,27 (1986)

[2] T.Åkesson et al., *Z.Phys. C - Particles and Fields* **36**, 517 (1987)

[3] Recent reviews are:

B. Lörstad, *Int. Journal Mod. Phys.* **A4**, 2861 (1989) W. Zajc, *in Hadronic Multiparticle Production, P. Carruthers ed., World Scientific Publ. 1988* G. Goldhaber in *Proc. First Int. Workshop on Local Equilibrium in Strong Interaction Physics*, eds. D.K. Scott and R.M. Weiner, (World Publishing Co. 1985)

[4] T.Åkesson et al., *Phys. Lett.* **187B**, 420 (1987)

[5] J.D.Bjorken ,*Phys. Rev.* **D27**,140 (1983)

[6] K.Kolehmainen,M.Gyulassy *Phys. Lett.* **203B**,320 (1988)

[7] K. Kulka and B. Lörstad, *Z. Phys. C - Particles and Fields* **45**, 581 (1990)

[8] R.M. Weiner *Phys. Lett.* **232B**, 278 (1989) R.M. Weiner: Coherent States in High Energy Physics revisited. Invited talk at the meeting Hadronic Matter in Reactions(LESIP III),Tucson, Arizona (1988)

[9] G.F. Bertsch, *Nucl. Phys.* **A498**, 173c (1989)

[10] Yu.M. Sinyukov, V.A. Averchenkov and B. Lörstad, preprint *Lund* LUNFD6(NFFL-7055)1990

[11] H.Gordon et al., *Nucl. Instrum. Methods* **193**,303 (1982) O.Botner et al., *Nucl. Instrum. Methods* **196**,315 (1982)

[12] B. Andersson et al., *Phys. Rep.* **97**,31 (1983)

[13] M. Althoff et al., *Z.Phys.C - Particles and Fields* **30**,355 1986)

[14] J.L. Bailly et al. *Z. Phys. C - Particles and Fields* **40**, 215 (1988)

[15] G. Kopylov and M.I. Podgoretsky,*Sov. J. Nucl. Phys.* **18**, 336 (1974) and **19**,215 (1974); G. Kopylov, *Phys. Lett.* **50b**,472 (1974); G. Cocconi, *Phys. Lett* **49B**, 459 (1974)

[16] C. Albajar et al., *Phys. Lett.* **226B**, 410 (1989)

[17] P.Grassberger, *Nucl. Phys. B* **120**,231 (1977)

[18] B. Andersson, W. Hofmann, *Phys. Lett.***169B** , 364 1986)

[19] M.G. Bowler, *Phys. Lett.* **180B**,299 (1986); *Z. Phys. C - Particles and Fields* **39** ,81 (1988) and *Z. Phys. C - Particles and Fields* **46** , 305 (1990)

[20] T.Åkesson et al., *Phys. Lett.* **129B**, 269 (1983)

[21] T.Åkesson et al., *Phys. Lett.* **155B**, 128 (1985)

[22] C. De Marzo et al. *Z. Phys. C - Particles and Fields* **33**, 363 (1984)

[23] B. Lörstad in *Proc. XVIII Symp. on Multiparticle Dynamics*, Tashkent, USSR, 1987, eds. I. Dremin and K. Gulamov (World Scientific Publishing Co., 1988) p. 111.

[24] B.Alper et al., *Nucl. Phys. B* **100**,237 (1975) K.Guettler et al., *Phys. Lett.* **64B**,309 (1976)

# Importance of Charge Correlations for Multiparticle Production at High Energies

C. C. Shih

Department of Physics, University of Tennessee, Knoxville, TN 37996-1200

**Abstract**

The role of charge correlations in multihadron distributions at high energy is investigated. A generalized quantum statistical approach with a partially coherent stochastic field ensemble requires at least three parameters: correlation length, chaoticity and charge coupling strength.

# 1 Introduction

One of the most debated issues in multihadron production processes at high energy has been the topic of correlations and fluctuations of the multiplicity distributions $P_n$ as a function of a restricted phase space region[1]. The experimentally observed normalized factorial moments

$$f_q = \sum_n n(n-1)...(n-q+1)P_n/(\sum_n P_n)^q$$

as a function of diminishing rapidity window $\delta y$ were interpreted first by Bialas and Peschanski[2] as an indication of "intermittency". At that time most dynamical models seriously underestimated this broadening of the multiplicity distribution $P_n$, and it was natural to conjecture that this phenomenon is due to a previously unappreciated property of high energy collisions such as "intermittency", "self similarity" or fractal dimensions[1].

Shortly after the study of Bialas and Peschanski[2], some of us advocated the possibility that this increase of fluctuations of $P_n$ is a reflection of the underlying quantum statistical (QS) field correlations with a given correlation length[4, 5]. Work in this direction has recently been continued by Carruthers and Sarcevic[6] as well as Capella, Fialkowski and Krzywicki[7]. These authors attempted to relate the observed two particle correlations functions to the bulk of "intermittency" data without relying heavily on detailed models.

Statistical approaches to the above issue have so far been based on several alternative ideas: (1) an effective partial coherent quantum statistical (QS) formulation inspired by quantum optics[4]; (2) a linked pair formulation, analogous to two-particle correlations inspired by galaxy distributions[6, 7]; or (3) a Gaussian statistical field ensemble analogy[8]. To study the similarities and differences of these models all of which have a finite correlation length, we present here a generalization of a QS formulation of particle production using the language of density matrix and coherent states which is well established in elementary particle physics[9]. Our formulation focuses on rapidity correlations among different charges treated by field quantities. The reasons are: (1) Most statistical models have failed to emphasize the importance of charge correlations and do not provide a proper distinction between the $n_-$,$n_+$ and $n_{ch}$ distributions[4, 6, 8]; earlier QS analyses did not study correlations in rapidity[9]. (2) There is presently some confusion as to whether the normalized factorial moments are better described by a classical or a fully chaotic Gaussian statistical ensemble[8]. (3) Correlations between opposite charges are stronger than the correlations between like charges over a wide range of rapidity $y$[10]. Yet, most experimental data on multiplicity distributions do not consider charges. In order to have a good understanding of the $n_{ch}$ multiplicity distributions, the role of charge correlation has to be clarified. (4) Most of the dynamical models do not provide an adequate prescription for the two particle correlation functions and the related "Intermittency" behavior. Furthermore, additional theoretical assumptions are required to address the multihadron distribution of the positive, negative and the all-charge correlations[11]. An analysis along this direction may facilitate a better understanding of the underlying dynamics.

## 2 Basic Formulation

An easy measure of the strength of charge correlations in multiplicity distributions can be obtained through the cross correlation $r(+,-)$ given by

$$r(+,-) = <(n_+ n_-) - <n_+><n_->> /(var(n_+)var(n_-)), \tag{1}$$

where $var(n_\pm) = \sqrt{<n_\pm^2 - <n_\pm>^2>}$. Clearly, $0 \leq r(+,-) \leq 1$. It is related to the moments of the charge $n_{ch}$ and $n_\pm$ multiplicity distributions through

$$2 <n_+ n_-> = <n_{ch}^2> - <n_+^2> - <n_-^2> .$$

Thus,

$$r(+,-) = (var(n_{ch})^2 - var(n_+)^2 - var(n_-)^2)/(2var(n_+)var(n_-)). \tag{2}$$

If the various multiplicity distributions are further represented by negative binomials, we get

$$var(n_\lambda)^2 = <n_\lambda> + \frac{<n_\lambda>^2}{k},$$

so that

$$r(+,-) = \frac{<n_{ch}>^2}{2<n_+><n_+>}\left(\frac{1}{kch} - \frac{<n_+>^2}{<n_{ch}>^2 k_+} - \frac{<n_->^2}{<n_{ch}>^2 k_-}\right)/[(\frac{1}{<n_+>}+\frac{1}{k_+})(\frac{1}{<n_->}+\frac{1}{k_-}]^{\frac{1}{2}}$$

$$\approx 2\left(\frac{1}{kch} - \frac{1}{4k_+} - \frac{1}{4k_-}\right)/[(\frac{1}{<n_+>}+\frac{1}{k_+})(\frac{1}{<n_->}+\frac{1}{k_-})]^{\frac{1}{2}}. \quad (3)$$

Notice that the bound $r(+,-) \geq 0$ is translated into the requirement

$$\frac{1}{k_{ch}} \geq \frac{1}{4k_+} + \frac{1}{4k_-} .$$

For example, at NA22 energy[12] the negative binomial fits to the $n_{ch}$ and $n_{\pm}$ $p\pi^+$ multiplicity data for the central region $|y| < 0.25$ give $<n_{ch}>= .94, \frac{1}{k_{ch}} = 0.36, <n_+>= 0.51, \frac{1}{k_+} = 0.18, <n_->= 0.44, \frac{1}{k_-} = 0.17$. The corresponding value of $r(+,-)$ is 0.2. Thus even in this region, the correlation between the positive and the negative charge remains rather strong and not negligible.

Let us briefly review the relevant properties of the particle production formulation in the quantum statistical language. We begin with the hadronic inclusive distribution with momentum $\vec{k}$ given formally by[13]

$$P_{1;\lambda}(\vec{k}) = \frac{1}{\sigma_\lambda}\frac{d^3\sigma_\lambda}{d^3\vec{k}} = Tr(\rho a^+_\lambda(\vec{k})a_\lambda(\vec{k})). \quad (4)$$

Here $\rho$ is the density matrix and $a_\lambda(\vec{k})$ are the destruction operator of a particle with a definite charge $\lambda = +$ or $-$. Similarly the two particle inclusive distribution of momenta $\vec{k_1}$ and $\vec{k_2}$ is given by

$$P_{2;\lambda_1,\lambda_2}(\vec{k}_1,\vec{k}_2) = \frac{1}{\sigma_{\lambda_1,\lambda_2}}\frac{d^6\sigma_{\lambda_1,\lambda_2}}{d^3\vec{k_1}d^3\vec{k_2}} = Tr(\rho a^+_{\lambda_1}(\vec{k}_1)a^+_{\lambda_2}(\vec{k}_2)a_{\lambda_2}(\vec{k}_2)a_{\lambda_1}(\vec{k}_1)). \quad (5)$$

Here $\lambda_1$ ( $\lambda_2$ ) is the charge of the particle with momentum $\vec{k}_1$ ( $\vec{k}_2$). For identical particles, these quantities are the central quantities used in the Bose-Einstein correlation analyses[14]. The general $q$-th order hadronic inclusive distribution is also given formally by[9]

$$P_{\lambda_1,..\lambda_q}(\vec{k}_1,..,\vec{k}_q) = \frac{1}{\sigma_{\lambda_1,\lambda_2,..,\lambda_q}}\frac{d^{6q}\sigma_{\lambda_1,\lambda_2,..,\lambda_q}}{d^3\vec{k_1}d^3\vec{k_2}..d^3\vec{k_q}} = Tr(\rho a^+_{\lambda_1}(\vec{k}_1)..a^+_{\lambda_q}(\vec{k}_q)a_{\lambda_q}(\vec{k}_q)..a_{\lambda_1}(\vec{k}_1)). \quad (6)$$

The above inclusive distribution is related to the normalized factorial moments. For example, by integrating over the phase space of $\vec{k}_1, \vec{k}_2, ..., \vec{k}_m$ and summing over all intermediate states of $\vec{k}$ in terms of the field operators $a_\lambda(\vec{k}) = \int d^3x e^{i\vec{k}\vec{x}} \Psi_\lambda(\vec{x})$, we get

$$F_{q,-} = < n_-(n_- - 1)..(n_- - q + 1) > = Tr(\rho \Psi_-^+(x_1)...\Psi_-^+(x_q)\Psi_-(x_q)...\Psi_-(x_1)). \quad (7)$$

Similar expressions can be found for other charge combinations.

In general, field correlation functions of various orders are needed. The dynamics of the production processes undergo a major simplification if the higher order correlation functions can be reduced to second order correlations through an appropriate two-point reduction formulation. This can be achieved whenever the the underlying fields are of Gaussian stochastic nature. If global charge conservation does not play a major role in the central rapidity region[15], we can then consider the two-particle correlation function as a matrix in the charge space of the $n_\pm$ distributions with the correlations between the charges constructed through the density matrix $\rho$ only. Consider the two-particle correlation

$$\Gamma(\vec{x}_1, \vec{x}_2; \lambda_1; \lambda_2) = Tr(\rho \Psi_{\lambda_1}^+(\vec{x}_1)\Psi_{\lambda_2}(\vec{x}_2)) = \Gamma(1,2)\,. \quad (8)$$

We break the underlying field into two parts,

$$\Psi_\lambda(y, \vec{p}_T) = \Psi_{\lambda,coh}(y, \vec{p}_T) + \Psi_{\lambda,chao}(y, \vec{p}_T) \quad (9)$$

$$\Psi_{\lambda,coh}(y, \vec{p}_T) = \sum a'_{m,\lambda,0} \Psi_\lambda^{(m)}(y, \vec{p}_T). \quad (10)$$

Here the base of the expansion $\Psi_\lambda^{(m)}(y, \vec{p}_T)$ are the eigenstates of the two-particle correlation function $\Gamma(1,2)$. While the coherent field $\Psi_{\lambda,coh}(y, \vec{p}_T)$ is a constant with respect to the ensemble average, the chaotic component $\Psi_{\lambda,chao}(y, \vec{p}_T)$ is a Gaussian stochastic field with zero mean $< \Psi_{\lambda,chao}(y, \vec{p}_T) >= 0$. Notice that in the P-representation the operator field $\Psi_\lambda$ can be replaced by the corresponding c-number field $\psi_\lambda(\vec{x})$ through the coherent fields; the specific assumption of Gaussian stochasticity then implies that the many-particle field correlation functions of the chaotic field $\Psi_{\lambda,chao}(y, \vec{p}_T)$ can factor into the products of the two-particle field correlation functions[16]. This connects our QS formulation to traditional quantum optics[17]. For example, we have

$$\mu_q = (q-1)!B_q + q!\overline{B_q}, \tag{11}$$

$$B_q \sim \prod_{j=1}^{q} \int dy_j \int dp_{T,j}^2 \Gamma(1,2)...\Gamma(q-1,q)\Gamma(q,1), \tag{12}$$

$$\overline{B}_q \sim \prod_{j=1}^{q} \int dy_j \int dp_{T,j}^2 \Gamma(1,2)...\Gamma(q-1,q)\psi_{coh}(y_1, \vec{p}_{T,1})\psi_{coh}^{+}(y_q, \vec{p}_{T,q}). \tag{13}$$

Here all the charge indices are suppressed for convenience. At this point we further assume that the $\vec{p}_T$ averaging has been performed [18] and concentrate only on the rapidity and charge correlations. The richness in structure of the rapidity distribution, and the eventual simplicity of the QS model can then be simply demonstrated. We notice that the overlapping integrals of the chaotic field possess structures [17] similar to the linked-pair structures proposed by Carruthers and Sarcevic[6, 7]. There are, however, two types of diagrams. Eq.(12) are formally linked-pair diagrams [8], while eq.(13) are linked-ring[6, 7]. We next assume that the 2-field correlation functions $\Gamma(1,2)$ can be approximated by a Lorentzian spectrum[4, 16, 19]:

$$\Gamma(y_1, y_2; -, -) = \Gamma(y_1, y_2; +, +) - \frac{1}{A}\Gamma(y_1, y_2; +,\ ) = \Gamma_0 exp(\frac{|y_1 - y_2|}{\xi}), \tag{14}$$

$$p_\lambda = Tr(\rho\Psi_{\lambda,chao}(y)^\dagger \Psi_{\lambda,chao}(y))]/[Tr(\rho\Psi_\lambda(y)^\dagger \Psi_\lambda(y))] = p_0, \quad for\ all\ \lambda \tag{15}$$

reflecting that the underlying statistics is "stationary" in $y$ for the central rapidity region[18]. Here the constant $A$ measures the relative strength of coupling between the opposite charge fields as compared to the like charge fields. For example, the normalized charge factorial cumulants $f_{2,ch} = F_{2,ch}/F_{2,ch}^q$ for the region $|y| < \delta_y/2$ are given by

$$\mu_{2,ch} = (1-p_0)p_0(A+1)\overline{B}_2 + \frac{1}{2}p_0^2(A^2+1)B_2 \tag{16}$$

$$f_{2,ch} = 1 + \mu_{2,ch} \tag{17}$$

where the $B_j$ and $\overline{B_j}$ are known analytically[4, 16] up the the 8-th order. For $B_2$, we have

$$B_2(\beta) = \overline{B}_2(2\beta) = (e^{-2\beta} + 2\beta - 1)/(2\beta^2) \qquad \beta = |\delta_y|/\xi \ . \tag{18}$$

Similarly, $\mu_{q,-}$ and $f_{q,-}$ are obtained by setting $A = 1$ in the corresponding expressions above,i.e.,

$$f_{q,-} = f_{q,ch} \mid_{A=1} . \tag{19}$$

This is equivalent to the effective one-component model used earlier in reference 4. The above parameter $A$ can be replaced by a directly measurable parameter $C$, the ratio of the probability density between the unlike and like charge at $y = 0$, i.e.,

$$\begin{aligned} C &= <|\psi_{chao,+}(0)|^2|\psi_{chao,-}(0)|^2> /[<|\psi_{chao,-}(0)|^2|\psi_{chao,-}(0)|^2>], \\ &= [(1-p_0)p_0A + p_0^2A^2]/[(1-p_0)p_0 + p_0^2]. \end{aligned} \tag{20}$$

The factorial moments depend on three parameters: (1) $C$, the relative strength of the unlike-charge cross-coupling compared to the like-charge coupling. (2) $p_0$, the over all chaoticity parameter. (3) $\xi$, the correlation length of the chaotic fields.

# 3 Conclusion

The following observations emerge from the above analysis:

(1) A quantum statistical formulation of the multiplicity distributions can be formulated to describe the $n_{ch}$, and $n_-$ normalized factorial moments for diminishing symmetrical rapidity windows. At least three control parameters ($C$, $p_0$, and $\xi$ ) are required at each energy. They can be fairly uniquely determined at the lower NA22 energy, but are not completely determined by the UA5 $n_{ch}$ multiplicity data alone. At both energies a substantial amount of coherence may be present[21]. Because of the finite correlation length, all the normalized factorial moments show saturation at small rapidity windows.

(2) Given the strong correlation between the positive and the negative charges, the charged multiplicity fluctuation is not dominated by a pure Bose-Einstein effect between identical particles[4, 14]. Serious modification is needed to improve the earlier formulations by using a straightforward quantum optical analogy[4] to describe the $n_{ch}$ and $n_-$ data simultaneously.

(3) Since there is a significant difference between the $n_{ch}$ and $n_-$ distributions, it is important to maintain the proper charge assignment at each energy. A modification of the linked-pair analysis at the NA22 energy is therefore needed to confirmed the energy independence of the parameters in this representation [6].

(4) It is more desirable to retain a partial coherent formulation instead of assuming a fully chaotic Gaussian ensemble as was discussed in reference 5. A more detailed analysis indicated the possibility of ambiguities in the parameter space, which can only be specified more realistically, when other characteristics of the many particle system are specified[16].

Many important factors have been ignored in the present analysis. The most important of all is the " nonstationary " effect in rapidity $y$, which may be nonzero even in the central rapidity region. Other factors due to transverse $\vec{p}_T$ and leading particle clustering may also play some role. When all these factors are considered, we believe, that charge correlations will still remain to be indispensable. Detailed analyses are currently in progress which do include the above factors as well as other quantities such as forward-backward correlations, "multi-bin" analyses and a "bin-average" often used in a typical "intermittency" analysis.

Acknowledgments. This research was supported in part by the National Science Foundation, the Tennessee Science Alliance and Deutsche Forschungsgemeinschaft. We thanks Drs. P. Carruthers, R. M. Weiner, E. Friedlander, C. Y. Wong and M. Plumer for extensive discussions. We also wish to thank the University of Marburg for its hospitality, where the draft was prepared.

## References

[1] P. Carruthers, J.Modern Phys. A4 (1989) 5587.

[2] A Bialas and R. Peschanski, Phys. Lett. 207 (1988) 59.

[3] C. Chiu and R. Hwa University of Oregon and University of Texas preprint 1990.

[4] P. Carruthers, R. Friedlander, C. C. Shih and R.M. Weiner, Phys. Lett. 222B,(1989) 487;

[5] M. Gyulassy, LBL preprint 26831 (1989).

[6] P. Carruthers and I.Sarcevic Phys. Rev Letts 63 (1989) 1562.

[7] A. Capella, K. Fialkowski and A. Krzywicki,Phys. Lett. 230B (1989) 149.

[8] E. A. De Wolf, Acta Phys. Pol. B21 (1990) 611.

[9] Coherent States, J.Klauder and B. Skagerstam editors, World Scientific (1985).

[10] N. N. Biswas, et al., Phys. Rev. Lett. 35, (1975) 1059. The underlying two-particle correlation function is roughly consistent with what was measured at slightly lower energy.

[11] G. Gustafson, University of Lund preprint, LU TP 90-5 (1990).

[12] M. Adamus et.al., EHS/NA22 Collaboration Z. Phys. 37 (1988) 215; I.V. Ajinenko et al., EHS/NA22 Collaboration, preprint (1990).

[13] D. Horn and R. Silver Annals of Phys. 66, (1971) 509; J. C. Botke, D. J. Scalapino and R. L. Sugar Phy. Rev. D9 (1974) 813.

[14] M Gyulassy, S.K. Kauffmann and L. W Wilson, Phys. Rev.C 2267, 1967; R. M. Weiner, Phys. Lett. B232 (1989) 278.

[15] C.C.Shih and X.C. He, Phys. Rev. D38 (1988) 3397.

[16] C. C. Shih, Phys. Rev. A40 (1989) 1456; also to be published in Phys. Rev. D.

[17] B. Saleh, "Photoelectron Statistics" Springer-Verlag, (1978); A.K.Jaiswal and C.L.Mehta, Phy. Rev. A2,(1970) 168.

[18] R.Friedlander, X.C.He, C.C.Shih, and R.M.Weiner, submitted to Phys. Rev. D; C. C.Shih, Proceeding of the "International Workshop of Intermittency", Santa Fe, NM, March 1990.

[19] A similar analysis can be done for the positive charge multiplicity distribution. A more detailed analysis of the positive charge multiplicity distributions requires a knowledge of the leading particle effects.

[20] R.E. Ansorge et al., Z Phys. C37 (1988) 191.

[21] The coherence observed here be may modified by the effects of $\vec{p}_T$[18].

# BOSE-EINSTEIN CORRELATIONS IN PION PRODUCTION AT TRISTAN

R.C. WALKER

*National Laboratory for High Energy Physics (KEK)*
*Tsukuba, Ibaraki 305, Japan*

*University of Rochester*
*Rochester, NY 14627, USA*

ABSTRACT

**Measurements of Bose-Einstein correlations in the production of like-signed pion pairs have been made with the AMY detector at TRISTAN. An event sample of 3800 hadronic events with $\langle\sqrt{s}\rangle$=57.2 GeV was used. These measurements indicate a strength of the enhancement of $\lambda = 0.60 \pm 0.13 \pm 0.08$, with a typical correlation size of $R_0 = 1.18 \pm 0.17 \pm 0.10$ fm. This value of $\lambda$ is consistent with measurements at lower energies; however, $R_0$ is somewhat larger than lower energy results have indicated.**

## 1. Introduction

Descriptions of the theoretical basis for Bose-Einstein correlations in the production of identical bosons (pions) can be found in a number of sources;[1–3] only a brief review will be given here. Given a chaotic (thermal) source of pions, and requiring that the pion wave function be symmetric under the exchange of identical bosons, the correlation function between two like-signed pions is related to the Fourier transform of the pion source distribution, $\hat{\rho}(k_1 - k_2)$, by:

$$C(k_1, k_2) = 1 + |\hat{\rho}(k_1 - k_2)|^2 \tag{1}$$

where $k_1$ and $k_2$ are the pion momentum. In order to extract this correlation function from data, it is customary to compare the distribution of like-signed pion pairs to unlike-signed pairs. This is due to the fact that the unlike-signed pair distribution contains much of the same physics as the like-signed distribution (*i.e.* phase space, momentum and angular distributions), but does not contain Bose-Einstein correlations. Differences exist, however, due to resonance decays and other phenomenon.

The ratio of the distributions, $R \equiv N_{like}/N_{unlike}$, is frequently parameterized in terms of a Gaussian function of the four-momentum difference, $Q$:

$$R(Q^2) = 1 + \lambda e^{-R_0^2 Q^2} \tag{2}$$

$$Q^2(k_1, k_2) \equiv -(k_1 - k_2)^2 = M_{12}^2 - 4M_\pi^2 \tag{3}$$

where $M_{12}$ is the invariant mass of the two pion system. The parameters $\lambda$ and $R_0$ are related to the strength of the enhancement and the size of the source, respectively, and are determined by fits to the data. Although this expression is only empirical, it has been shown to describe $e^+e^-$ annihilation data well over a wide range of energies[4-7].

## 2. Experimental Data

### *2.1. AMY Detector*

The AMY detector contains cylindrical tracking and shower detectors located inside a 3 T solenoidal magnet, and has been well documented elsewhere.[8] Charged particles were detected using a 4-layer cylindrical array of tube-type drift cells followed by a 40-layer cylindrical drift chamber (CDC) with 25 axial and 15 small-angle stereo layers, covering the polar angle region $|\cos\theta| < 0.87$. From Bhabha scattering events, a resolution of $\delta p_T/p_T = 0.7\% p_T(\mathrm{GeV})$ was determined. A cylindrical electromagnetic calorimeter, consisting of 20 layers of interspersed lead and resistive plastic proportional tubes, is situated radially outside of the tracking chamber, covering the angular region $|\cos\theta| < 0.73$. Muons were detected using four layers of drift cells and one layer of plastic scintillators, situated outside of the iron return yoke and covering the angular region $|\cos\theta| < 0.74$.

"Good" tracks in the CDC were determined by requiring a minimum number of wire hits, as well as vertex and polar-angle cuts. Multiple tracks caused by "curlers" (*i.e.* spiralling low momentum particles that reenter the CDC, resulting in two or more reconstructed tracks) were eliminated. The sample of events used were defined by the standard AMY hadronic event selection criteria.[9] These criteria eliminated background events from beam-wall, beam-gas, $\tau^+\tau^-$, two-photon, and radiative Bhabha events; as well as guaranteeing no large energy leakage into the beam pipe. Backgrounds from these processes were estimated to be $< 2\%$.

Hadronic particle identification was not possible using the AMY detector. However, it was possible to eliminate tracks caused by electrons or muons using data from the Shower Counter and Muon Counter. It was required that there be at least five hadronic tracks, with at least two positively and two negatively charged, for the event to be included in this analysis.

*2.2. Calculation of Correlation Function*

For each event, all possible pairs of charged tracks were formed and $Q^2$, the negative of the difference of the four-momenta squared, was calculated. Two histograms of the number of particle pairs were formed; one of like-signed pairs and one of unlike-signed pairs. These distributions are shown in Fig. 1. The ratio of these two distributions was the correlation function, after making corrections for differences not related to the Bose symmetrization requirement of the like-signed pions. Such corrections are described below.

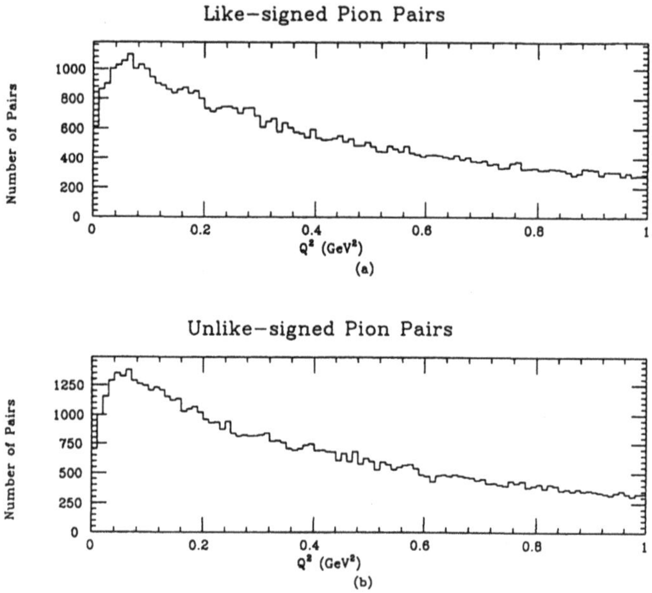


**Figure 1 – The distributions of (a) like-signed particle pairs and (b) unlike-signed particle pairs, for the AMY hadronic data sample, versus $Q^2$.**

Differences between the like- and unlike-signed samples due to resonance decays, normalization, and detector effects were corrected using Monte-Carlo simulation techniques. This simulation was based on Lund 6.3[10] for 5-flavor parton production and the Lund string fragmentation model[11] for the hadronization process. It should be noted that the Monte Carlo does not contain Bose-Einstein correlations. From this simulated data, the $Q^2$ distributions of like- and unlike-signed pairs were calculated using the same cuts as for the real data. The measured correlation function was then divided by the ratio of these distributions.

Like-signed pairs of non-identical particles ($\pi$,$K$,p) do not contain Bose-Einstein correlations. The proportion of like-signed pairs that were identical pions was determined using the Monte-Carlo simulation, and was fit to a polynomial, $f_\pi(Q^2)$. According to the Lund 6.3 Monte-Carlo calculation, approximately 68% of all like-signed pairs were $\pi\pi$ pairs, with only a slight variation in $Q^2$.

Like-signed pions exert a mutually repulsive electromagnetic force, while unlike-signed pions feel an attractive force. This Coulomb interaction alters the distribution $R(Q^2)$ by an amount[12]:

$$R(Q^2) \rightarrow R(Q^2)\exp(2\pi\alpha M_\pi/Q) \tag{4}$$

where $2\pi\alpha M_\pi \approx 6.4$ MeV. This correction is about 9.5% at a $Q^2$ value of 0.005 $(\mathrm{GeV/c})^2$.

## 3. Results

The ratio of like- to unlike-signed pairs as a function of $Q^2$, after applying all corrections, is shown in Fig. 2. The spectrum was parameterized by:

$$R(Q^2) = N_0(1 + f_\pi(Q^2)\lambda e^{-R_0^2 Q^2})(1 + \gamma Q^2) \tag{5}$$

where $N_0$ is a normalization constant to take into account the different number of like- and unlike-signed pairs, $\lambda$ and $R_0$ are the parameters of the Bose-Einstein correlation function, the term involving $\gamma$ is used to take into account long-range correlations that exist (*e.g.* charge and energy conservation), and the function $f_\pi(Q^2) \approx 0.68$ was discussed previously. A fit to this spectrum yields parameters of: $N_0 = 1.021 \pm 0.014$, $\lambda = 0.603 \pm 0.126$, $R_0 = 1.182 \pm 0.170$ fm, $\gamma = -0.041 \pm 0.025$ $(\mathrm{GeV/c})^{-2}$, with $\chi^2/\mathrm{dof} = 92.5/96$; this is also shown in Fig. 2. The fact that $N_0$ and $\gamma$ are approximately consistant with 1.0 and 0.0, respectively, indicates a good reliability of the Monte-Carlo corrections (without such corrections, values of $N_0 = 0.765$ and $\gamma = 0.148$ $(\mathrm{GeV/c})^{-2}$ were obtained).

A full study of the systematic uncertainties involved in this study, particularly those from the Monte-Carlo corrections, has not yet been carried out. An attempt has been made, however, to estimate some of the following sources of uncertainty:

**Curlers.** In order to eliminate multiple "curler" tracks that were caused by a single low-momentum particle following a helix through the CDC, some like-signed pion pairs at small relative angles may have been eliminated. This effect is limited to only the first $Q^2$ bin of the pair distributions, however, and good agreement was found

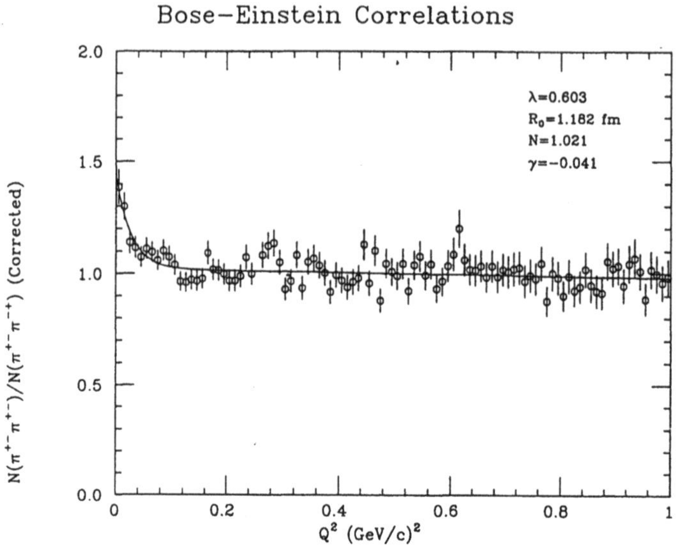


**Figure 2** – **Ratio of like- to unlike-signed particle pairs from the AMY data, after making all corrections.**

between data and Monte-Carlo simulations. A 3% uncertainty has been assigned to this first bin to account for any errors, leading to a systematic uncertainty of $\pm 0.015$ in $\lambda$ and $\pm 0.025$ fm in $R_0$.

**Resonance effects.** A $\pm 30\%$ uncertainty in the $\pi^+\pi^-$ distribution resulting from $\eta$ and $\eta'$ decays has been included. This results in roughly a $\pm 2.5\%$ uncertainty in the ratio of like- to unlike-signed pairs in the region $Q^2 < 0.06$ (GeV/c)$^2$. Based on this, an uncertainty in $\lambda$ of $\pm 0.037$ and a $\pm 0.066$ fm uncertainity in $R_0$ has been included.

**Monte-Carlo statistics.** A systematic uncertainty equal to 0.42 times the statistical uncertainty was assigned to each of the fit parameters to account for the statistical precision of the Monte-Carlo (22,000 events).

**Detector resolution.** The effect of the resolution of the particle tracking was determined by simulating the Bose-Einstein enhancement, with parameters $\lambda = 0.60$ and $R_0 = 1.18$ fm, in the Monte Carlo. After simulating detector resolution effects and fitting the resulting correlation function, extracted values of $\lambda$ and $R_0$ were 0.56 and 1.18 fm, respectively. A systematic error of 0.04 was thus assigned to $\lambda$.

Further studies of the systematic uncertainties is currently underway.

## 4. Comparison to Previous Measurements

The results for the fit parameter $\lambda$ versus center-of-mass energy are shown in Figure 3(a) for this experiment, as well as various lower energy $e^+e^-$ experiments.[4–7] Also shown are the thresholds for charm and bottom production. The results show $\lambda$ near a maximum value of 1.0 at charm-production threshold. The value decreases above the charm threshold, and is further decreased as the bottom threshold is crossed. This reduction is thought to be caused by the fact that a signficant number of pions are produced not as a result of the inital hadronization, but rather as the decay products from the heavy quarks. Since the quarks travel a significant distance (compared to $R_0 \approx 1$ fm) before decaying, these pions are effectively "uncorrelated" within the experimental resolution of the detectors. However, this phenomenon is not well understood.[3,7] The results from AMY are consistent with other results above bottom threshold.

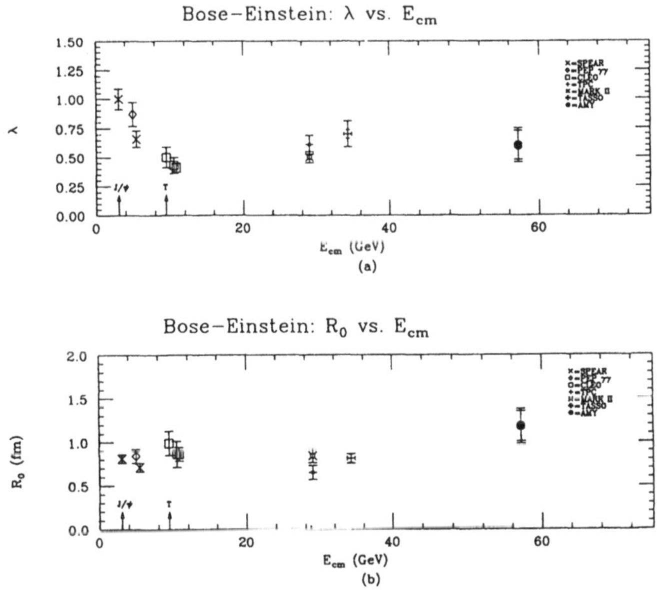


Figure 3 – (a) The Bose-Einstein parameter $\lambda$, shown versus center-of-mass energy, for a variety of experiments. (b) The Bose-Einstein parameter $R_0$.

In Figure 3(b) the results for $R_0$ are shown. In this case, the typical correlation distance of $R_0 \approx 0.8$ fm is energy independent below the AMY energy range. The results from AMY, however, indicate a possible increase in $R_0$ to above 1 fm; however, the discrepency is only 2.2 standard deviations. Increased statistics, which will be achieved over the next three years of TRISTAN running, along with improved tracking with the AMY 1.5 detector, should make it possible to determine whether this discrepency is real, or only a statistical fluctuation.

## 5. Conclusion

Bose-Einstein correlations in $\pi\pi$ production have been measured in $e^+e^-$ annihilations with the AMY detector at TRISTAN, using a hadronic event sample of 3800 events with $\langle\sqrt{s}\rangle$=57.2 GeV. Results show correlations with a strength of $\lambda = 0.60 \pm 0.13 \pm 0.08$, in agreement with lower energy results. The typical correlation size was determined to be $R_0 = 1.18 \pm 0.17 \pm 0.10$ fm, which is slightly larger than lower energy results have indicated. An increased data sample presently being taken with the AMY 1.5 detector at TRISTAN at $\sqrt{s} = 58$ GeV should help clarify if any changes in hadronic production in the TRISTAN energy range are being observed.

## References


1. G. Goldhaber, *et al.*, *Phys. Rev.* **120**, 300 (1960).
2. A. Giovannini and G. Veneziano, *Nucl. Phys.* **B130**, 61 (1977).
3. I. Juricic, Ph.D. thesis, Lawrence Berkeley Laboratory, (1987).
4. TPC Collaboration, H. Aihara, *et al.*, *Phys. Rev.* **D31**, 996 (1985).
5. CLEO Collaboration, P. Avery, *et al.*, *Phys. Rev.* **D32**, 2294 (1985).
6. TASSO Collaboration, M. Althoff, *et al.*, *Z. Phys.* **C30**, 355 (1986).
7. Mark II Collaboration, I. Juricic, *et al.*, *Phys. Rev.* **D39**, 1 (1989).
8. AMY Collaboration, T. Kumita, *et al.*, submitted to *Phys. Rev.*
9. T. Mori, Ph.D. Thesis, University of Rochester, UR-1104, (1988).
10. T. Sjostrand, *Com. Phys. Comm.* **43**, 367 (1987).
11. B. Andersson, *et al.*, *Phys. Rep.* **97**, 33 (1983).
12. M. Gyulassy, *et al.*, *Phys. Rev.* **C20**, 2267 (1979).

# BOSE-EINSTEIN CORRELATIONS AS A SIGNATURE OF MIXED PHASE AND SOFT PION PHENOMENA IN A+A COLLISIONS

Yu.M.SINYUKOV, V.A.AVERCHENKOV
Institute for Theoretical Physics, SU-252130 Kiev, USSR

B.LORSTAD
Department of Particle Physics, Lund University, S-22362 Lund, Sweden

ABSTRACT

We have calculated the pion transversal momentum spectrum and the pion correlator in outward, sideward and longitidinal directions for S+S collisions at 200 GeV/c and we reproduce preliminary experimental data. We used a two-temperature model that allows us to explain the abundance of pions at small and high $p_T$. The interferometrical analysis at different $p_T$ allows the measurement of the lifetime and other parameters for the two sources to describe the evolution of the system formed in A+A collisions.

## 1. INTRODUCTION

There is still no evidence that a QCD plasma is formed in ultra-relativistic nuclear collisions but several puzzling observations have been made in the study of these collisions. One of them is an increase in the production of soft pions. It seems that there existed two pion sources $N_1$ and $N_2$ with temperatures $T_1$= 150 - 180 MeV and $T_2$= 50 - 80 MeV, respectively, radiating pions from hydrodynamical Bjorken tubes [1] . It has been shown [2] that such behaviour can be explained if the initial state of matter is in a mixed quark-gluon-hadron phase. The high temperature of the pion spectrum is due to the direct radiation of pions from a time-like hypersurface formed by a rarefaction shock wave. The deflagration process ceases when the internal part of the mixed phase converts into hadron phase due to longitudinal expansion. The unburnt part of the matter becomes the second low-temperature source. In fact, a similar picture was advanced recently by Shuryak [3] , suggesting a mechanism that leads to a long lifetime of a thermalized hadron system, forming the low-temperature spectrum for the pions emitted. The idea is that the pion dispersion curve in the thermalized matter is modified making it difficult for the pions to leave the system.

In the framework of the above-mentioned picture, see [2,3] , it is possible to describe the pion transversal momentum spectrum as we shall show for $^{32}S$ + S collisions at 200 GeV/A. The main point we shall discuss how pion interferometry data can be used to further test the idea of two different sources for pion production in A+A collisions.

## 2. THE MODEL

Our approach was recently described in ref. [2]. We assume that the intermediate system, which is formed after the A+A collision, is thermalized at the initial proper time $\tau_0$ which is about 1 fm. At this time the matter is in a mixed phase with pion entropy density $S_\pi$ and pion entropy concentration $\chi_0$ . The initial system occupies a volume with transversal radius $R_T(\tau_0)$, being about $R_{nuclear}$ , and has the transversal velocity $v_T(\tau_0) = 0$. It expands in a longitudinal direction in boost-invariant manner [4] . The rapidity boundaries in cms are $\pm$ Y.

During the time $\tau_0$ to $\tau_h$ the first source acts. Its mechanism is a direct conversion of mixed phase with temperature $T_c$ into a free pion gas with a temperature $T_1$ near the rarefaction shock front. According to the results of ref. [2] the maximum entropy prduction $S_\pi/S_m \approx 1.1$ is realised at $T_1 = T_c/\sqrt[4]{2}$. The shock front moves inside faster than in standard deflagration models [5] because of a pressure decrease near the shock front, which occurs if free pions radiate directly from a time-like hypersurface formed by a rarefaction shock front [2] . The equation for the velocity of the front and its solution is described in ref.[2].

The deflagration process stops at the proper time

$$\tau_h = \frac{1-\chi_o(1-r)}{r}\tau_o \quad \text{where } r = \frac{g_\pi}{g_{qg}} \approx 0.07\,[6] \qquad (1)$$

The unburnt hadron part of the matter is the second source of pions and is difficult to describe in detail. It can undergo transversal expansion and its radius can increase from the initial $R_h = R_{shock}(\tau = \tau_h)$ to $R_f$, when the system decays. It can be rather homogenous in the transversal direction because of an increase in viscosity due to an increase in the mean free path of pions. Shuryak,s suggestion that the pion system has a long lifetime because the pions can not leave the system before it becomes cool enough [3] is most interesting. In any case we shall define the lifetime of the system as a free parameter and try to connect it with pion interferometry data. We connect $\tau_f$ and $R_f$ in a simple way. In the expanding pion gas at temperatures $T_\pi \lesssim m$ ,the general number of pions does not change very much[7]. The latter condition gives the equation we are interested in

$$n_\pi(T_c)R_h^2\tau_h = n_\pi(T_f)R_f^2\tau_f \qquad (2)$$

where $n_\pi(T)$ is the pion density at the temperature T .

We hope that we include at least qualitatively the main points of the picture, namely, the existence of two sources with two different effective temperatures and lifetimes, i.e. the deflagration source and the unburnt hadronic part of the matter.

The general formula describing the spectra of one and two identical pions which radiate from arbitrary (timelike or spacelike) hypersurfaces $\sigma$ was obtained in refs [2,8].

These can be expressed in terms of the auxiliary function

$$J(p_1,p_2) = \int_\sigma d\sigma_\mu \left[ p_1^\mu \theta(\lambda_{p_1})\theta(2-\lambda_{p_1}) + 2u^\mu p_1\cdot u_\pi \theta(\lambda_{p_1}-2)\right] \exp(i(p_1\cdot x - p_2\cdot x))\varphi\left(\frac{p_1\cdot u_\pi}{T_\pi}\right) \quad (3)$$

where the integration is performed on the hypersurface $\sigma$ defined by the normal 4-vector $n^\mu(x)$. $u_\pi^\mu(x)$ is the collective velocity of the pion gas that is emitted from $\sigma$ near point x, $\varphi$ is the Bose-Einstein heat factor for pions, $\theta$ is the Heaviside function and

$$\lambda_p \equiv \lambda_p(x) = \frac{p\cdot n(x)}{u_\pi(x)\cdot p \; u_\pi(x)\cdot n(x)} \quad (4)$$

The one-particle spectrum is expressed through $J(p,p)$:

$$p^0 \frac{d^3N}{d^3p} = J(p,p) \quad (5)$$

The Bose-Einstein correlator is

$$R(p_1,p_2) = \frac{\mathrm{Re}\,(J(p_1,p_2)J(p_2,p_1))}{J(p_1,p_1)J(p_2,p_2)} \quad (6)$$

The first deflagration source $J_1$ radiates from the time-like hypersurface (see [2] for details) and has the temperature $T_1 = T_c/\sqrt[4]{2}$ and collective velocity in longitudinal and transversal directions. The second source radiates from the space-like hypersurface [2] and has the temperature $T_2 = T_f$. In the transversal pion spectrum the both sources contribute,

$$\frac{d^3N}{d^2p_T dy} = J(p,p) = J_1(p,p) + J_2(p,p) \quad (7)$$

The Bose-Einstein correlations are described by eq. (6) with $J(p_1,p_2) = J_1(p_1,p_2) + J_2(p_1,p_2)$. The longitudinal, sideward and outward momentum differences take form

$$q_{long} = p_{1L} - p_{2L}\,;\; q_{out} = \frac{\vec{q}_T\cdot\vec{K}_T}{K_T}\,;\; q_{side} = \left|\vec{q}_T - \frac{\vec{K}_T}{K_T} q_{out}\right|$$
$$\vec{K}_T = (\vec{p}_{1T} + \vec{p}_{2T})/2 \qquad \vec{q}_T = \vec{p}_{1T} - \vec{p}_{2T} \quad (8)$$

## 3. THE RESULTS FOR S + S COLLISIONS AND DISCUSSION

Firstly we fixed the parameters of the model from the preliminary data on single-pion spectra for symmetrical central S+S collisions [1,9]. The main parameters are the critical temperature $T_c$, the decay temperature for a hadron fluid $T_f$, and the initial entropy concentration $\chi_0$ of the hadron phase in the mixed phase. Our model reproduces the experimental data for the transversal pion spectrum using $T_c$ = 200 MeV, $T_f$ = 80 MeV and $\chi_0$ = 0.75, Fig.1.

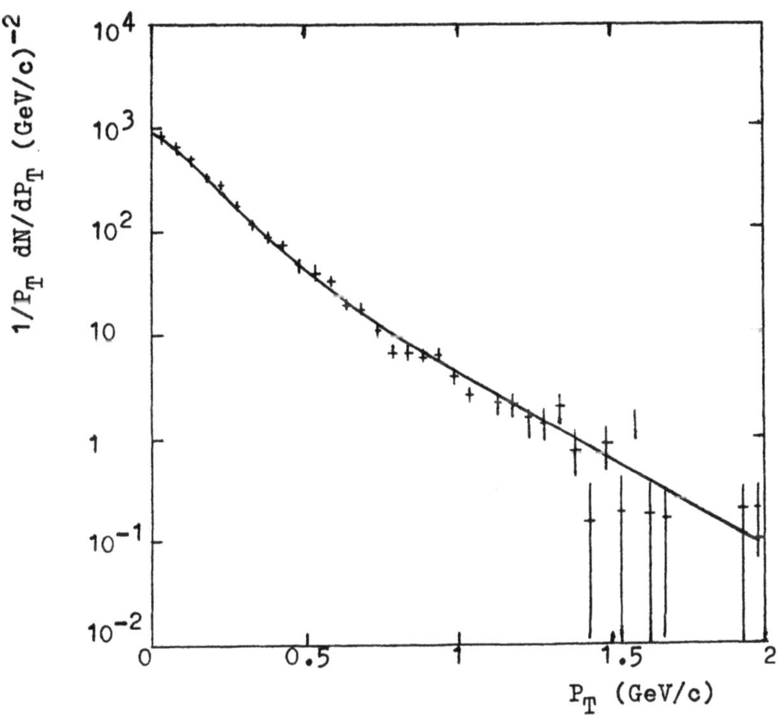


Fig.1. The transversal spectrum of negative pions for S+S collisions at 200 GeV/c in two-component model. The preliminary data are from [1].

The values for the initial proper time $\tau_0$ = 1.15 fm. The mixed phase exists from proper time $\tau_0$ until $\tau_h$= 5 fm. The second source, having at $\tau_h$ an initial radius $R_h$= 3fm, is a hadron fluid which exists from $\tau = \tau_h$ and at $\tau = \tau_f$ radiates a cool pion gas with temperature $T_f$= 80 MeV.Our results on one-particle spectra do not depend on the time $\tau_f$ according to eq. 2. The value $\tau_f$ may reflect very specific properties of the hadron fluid.

In order to measure the time $\tau_f$ we propose to use the interferometrical analysis as the interferometrical data are very sensitive to the value of $\tau_f$. In Fig. 2 we demonstrate the behaviour of the "pure" out-correlator R($q_{out}$, $q_{side}$=0, $q_{long}$=0; $K_T$) at two values $\tau_f$=15 fm/c and $\tau_f$ = 30 fm/c in different regions of transversal momentum $K_T$.

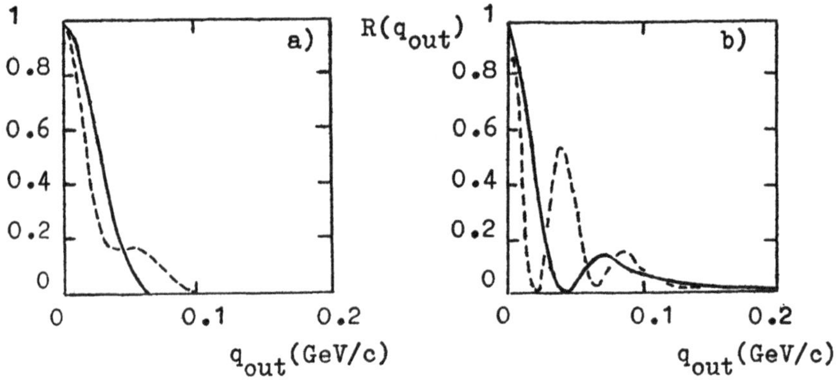


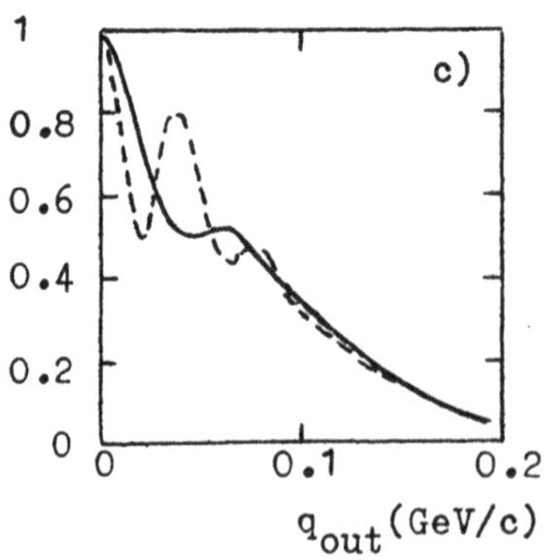


Fig.2

Out-correlator for S+S collisions as a function of $q_{out}$. The parameters of the model are the same as in Fig.1.Solid lines correspond to a lifetime of the system of $\tau_f$ = 15 fm/c dashed lines to $\tau_f$= 30 fm/c. a) $K_T$ = 100 MeV/c, b)$K_T$ = 350 MeV/c, c) $K_T$ = 600 MeV/c

In the region $K_T$ = 0.1 GeV/c (Fig. 2a) the second source dominates. This results in a fast fall of the correlator.

So we can measure in this soft pion region mostly effects from the low-temperature source. In the region $K_T$=0.35 GeV (Fig.2b), where the intensities of the two sources are aproximately equal one can see strong interference between two sources because they are separated in time. The half-period of the oscillation $\Delta q_0$ allows the determination of the time $\tau_f$ directly: $\tau_f \approx \pi/\Delta q_0$. In the region $K_T$ = 0.6 GeV/c the first source dominates over the second source. After the fast fall of the correlator in the region $0 < q_{out} < \Delta q_0 = \pi/\tau_f$ we can see the tail of the broad correlator, that corresponds to the first source itself.

Not side- neither long-correlator show oscillatory behaviour. We obtaine the Lorentz contraction of the longitudinal interferometrical size becouse of the hydrodynamical expansion just as it was considered in detail in [8].

4. SUMMARY

We have shown that the transversal pion spectrum in A+A collisions at CERN-SPS energies can be explained by a model with two sources. We find that the interferometrical analysis is most sensitive to important moments of the evolution of the matter formed after collision. Most interesting is the behaviour of the out-correlator for different regions in $p_T$. At small $p_T$, the correlator "feels" mostly the second low-temperature source, at large $p_T$ it "feels" partly the first mixed phase source and partly the interference effect between the two sources. At average $p_T$ the out-correlator shows strong interference effects between two sources. It is then possible to measure the time $\tau_f$ of the hadron fluid existence. So the idea of a longlived pion fluid can be directly tested by pion interferometry.

# REFERENCES


1. Strobele H. et al. (NA35 collaboration), Z.Phys. C38, 89 (1989).
2. Sinyukov Yu.M., Z.Phys. C43, 401 (1989).
3. Shuryak E.V., CERN-TH-5386/89 and Phys.Lett. B231, 175 (1989).
4. Gorenstein M.I., Sinyukov Yu.M., Zhdanov V.I., Phys. Lett. B71, 199 (1977);
   Bjorken J., Phys.Rev. D27, 140 (1983).
5. Friman B., Kajantie K., Ruuskanen P.V., Nucl.Phys. B266, 468 (1986).
6. Ruuskanen P.V., Acta Phys.Pol. B18, 551 (1987).
7. Gersdorff H., McLerran L., Kataja M., Ruuskanen P.V., Phys.Rev. D34, 794 (1986).
8. Sinyukov Yu.M., In "Proceedings of Quark Matter 88", Nucl.Phys. A498, 151c (1989).
9. Harris J.W. (NA35 collaboration), In "Proceedings of Quark Matter 88", Nucl.Phys. A498, 133c (1989).

# On the interpretation of the GGLP formula

T. Csörgő, J. Zimányi
Central Research Institute for Physics,
H-1525 Budapest 114. P.O.B. 49., Hungary

**Abstract**

The GGLP effect is caused by the Bose–Einstein statistics of identical particles. The authors point out that the "radius" parameter $R$ of the $C(Q_I)$ form of the correlation function can be related to the width of the proper time distribution in case of very strongly correlated spacetime and momentum space.

## 1 The puzzle of variables of the correlation function

In elementary particle reactions the functional form of the two pion correlation function,

$$C(\vec{p}_1, \vec{p}_2) = \frac{< n_\pi >^2}{< n_\pi(n_\pi - 1) >} \frac{< N(\vec{p}_1, \vec{p}_2) >}{< N(\vec{p}_1) >< N(\vec{p}_2) >}, \tag{1}$$

was studied with high statistics experiments. No significant dependence of $C$ was found on any other combinations of the two pion momenta except a dependence on the Lorentz invariant difference $Q_I = M^2 - 4m^2$, where $M^2$ is the invariant mass of the pair. The two pion correlation function depends on $Q_I$ only. This observation holds in $e^+ + e^-$ annihilation performed in various circumstances, as well as for lepton–hadron and hadron–hadron reactions, for reviews on data and models see refs.[2,3]. The puzzle is that the measured correlation functions are consistent with the exclusive $Q_I$ dependence, because even the Kolehmainen–Gyulassy form [4] suggests a factorizable $Q_T$ dependence. So do the other forms, which are based on more static models. The Andersson-Hofmann model [5], which takes into account the Bose–Einstein symmetrization for the pion pairs emitted from a string and fits the $e^+ + e^-$ data surprisingly [6] successfully, doesn't suggests an explicit fitting form.

## 2 Strongly correlated spacetime and momentum space – a possible solution to the puzzle

Let's suppose that particles are emitted after a point-like interaction of the target and projectile in case of an ultrarelativistic collision. The rapidity of a particle is $\eta = \frac{1}{2}\ln(\frac{E+p_z}{E-p_z})$. The momentum distribution of the emitted pions is given by $g(\vec{p})$, normalized to unity. Because of the approximate boost invariance along the collision axis the particle world lines can be continued back to the collision point, at least on the average. The only parameter determining the spacetime evolution is the proper time $\tau = \sqrt{t^2 - x^2 - y^2 - z^2}$. If the overwhelming majority of the collisions are soft, the particles have much less momenta in the transverse direction than along the $z$ axis on the average. If the radius of the primary interaction region is reasonably smaller than the average radius of pion production in the transverse direction, one ends up with an approximately radial expansion in the transverse direction, because of axial symmetry. This type of correlations between $x^\mu$ and $p^\mu$ are reflected in the limit of infinite bombarding energy by the expressions $\tau = \sqrt{t^2 - z^2}$, $\frac{\vec{r}}{t} = \frac{\vec{p}}{E} = \vec{v}$, $\gamma = \cosh(\eta)$, $E = m_T\gamma$. Let's denote the probability distribution of pion production in proper time by $\frac{1}{N}\frac{dn}{d\tau} = H(\tau)$, normalized to unity. Then the very strongly correlated phasespace distribution of pion production points is given as:

$$f(t,\vec{r};\vec{p}) = g(\vec{p}) \int_0^\infty d\tau H(\tau)\delta^{(4)}(x^\mu - \tau\frac{p^\mu}{m_T}) \qquad (2)$$

We call attention to the fact that this formula containes the same correlations between $(t, z)$ and $(E, p_z)$ as the ideal inside-outside distribution of [4]. However we take the correlations to the extremes supposing them valid in transverse direction too. (At this point we do not restrict ourselves to any specific production mechanism. Still we point out that eq. (2) seems to be reasonable for cases when some of the pions are produced from resonance decays. If a pion comes from an energetic resonance emitted from the center, the $x^\mu$ of pion production will be proportional to the momenta of the resonance $p^\mu$, and this momentum will be about the pion momentum (corrections are due to the width of the resonance)). Also, we do *not* specify the momentum dependence and the proper time distribution of the pion production points.

In plane wave approximation the Yano-Koonin formula gives the following two pion multiplicity distribution:

$$P(\vec{p}_1;\vec{p}_2) = \int d^4x \int d^4x' f(x;\vec{p}_1)f(x;\vec{p}_2)\{1 + \cos[(x - x')\cdot(p_1 - p_2)]\} \qquad (3)$$

Substituting the eq. (2) distribution for extremely correlated phasespace production points we gain:

$$C(Q_I; m_{T,1}, m_{T,2}) = 1 + Re[\tilde{H}(\frac{Q_I^2}{2m_{T,1}})\tilde{H}(\frac{Q_I^2}{2m_{T,2}})]. \qquad (4)$$

where

$$\tilde{H}(x) = \int_0^{+\infty} d\tau H(\tau)e^{ix\tau} = \int_{-\infty}^{+\infty} d\tau\Theta(\tau)H(\tau)e^{ix\tau} \qquad (5)$$

denotes the Fourier transformed distribution of pion production points in proper time. One has to observe the following features of (4) two pion correlation function: i, The correlation function measures the proper time distribution of the pion production. ii, The correlation function depends dominantly on the invariant momentum difference of the pions, because the transverse mass of the pions has an exponential distribution. After averageing over the $m_T$ -s, they can be replaced by their minimum value $m_{T,0}$:

$$< C(Q_I) >_{m_T} \approx C(Q_I; m_{T,0}, m_{T,0}). \qquad (6)$$

The model predicts: the larger the $m_{T,0}$ transverse mass cut the wider the correlation function and the smaller the *effective* source size. As an example, let's consider the Lorentzian proper time distribution: $H(\tau) \propto \frac{1}{(\tau-\tau_0)^2+\sigma_\tau^2}$ In this case we get a correlation function which oscillates within a Gaussian shape with shortening periods, as given by

$$< C(Q_I) >_{m_T} = 1 + \cos(\frac{\tau_0 Q_I^2}{m_{T,0}})\exp(-\frac{\sigma_\tau Q_I^2}{m_{T,0}}). \qquad (7)$$

This form starts as a Gaussian but falls a bit more rapidly than that, because of the cosine factor, and goes slightly below 1, before the Gaussian factor damps the oscillations under detection errors, if $\tau_0$ and $\sigma_\tau$ are about the same. However this form is undistinquishable from the standard Gaussian shape if $\tau_0 << \sigma_\tau$. In this case the the Gaussian size, $R$ can be related with the width of the Lorentzian as $R^2 = \sigma_\tau/m_{t,0}$. For elementary particle interactions the typical source size is $R \approx 0.8$

fm, the transverse mass cut can be estimated by 300 MeV, thus the width of the Lorentzian $\sigma_\tau \approx 1.$ fm will not be very much different from $R$.

In ref.[7] the pion production was described as a convolution of resonance production and decay,

$$H(\tau) = \int_0^\infty dx R(x) D(\tau - x) \tag{8}$$

where $R(x)dx$ is the probability of producing a resonance in the interval $dx$ around the proper time $x$ and $D(\tau - x)d\tau$ gives the probability that the resonance emits a pion in $d\tau$ interval around $\tau$, under the condition the resonance was produced at $x$. For this type of proper time distributions the correlation function factorizes because of the faltung theorem:

$$C_d(Q_I) = 1 + Re[\tilde{R}^2(\frac{Q_I^2}{2m_{T,0}})\tilde{D}^2(\frac{Q_I^2}{2m_{T,0}})] \tag{9}$$

Now we leave the infinite number of possible examples and comment on the meaning of the intercept or "chaocity" parameter $\lambda$ of the GGLP formula $C(Q_I) = 1 + \lambda \exp(-Q_I^2 R^2)$, ref. [1]. In the literature and at this workshop the effects of partially coherent fields on the two pion correlation function were reinvestigated, refs. [8]. This formalism can be extended to our case, too. Note, that the intercept parameters are very sensitive to an *a priori* assumption on the shape of the correlation function. Representative examples are the recent NA35, E802 data [9] and the results of SPACER calculations [7]. The different possible fitting forms (e.g. Gaussian or exponential) result in a 100% uncertainity in the value of the intercept. That's why the authors think that the presently reported $\lambda < 1$ values are not conclusive whether the source had a coherent component or not. Clearly, further increase in the statistics and a settling of the problem of the different possible functional forms would be a very welcomed and a very hard job for our experimental collegaues.

As a summary, the authors pointed out that the $R^2$ parameter of the GGLP formula can be interpreted as a rate of the width of a Lorentzian proper time distribution over the transverse mass cutoff $\sigma_\tau / m_{T,0}$, if the spacetime and momentum space variables of pion production are extremely strongly correlated. This latter assumption gives an explanation of the experimantal puzzle that the two pion correlation depends on $Q_I$ only.

# 3 Acknowledgements

The authors would like to express their thanks to professor N. L. Balazs and to the Department of Physics, State University New York at Stony Brook, for kind hospitality. One of us (T. C.) thanks to professor R. Weiner for funding his participation at the workshop and making this contribution possible. This work was supported by the NSF - Hungarian Academy of Sciences grant NSFINT 2044B.

# 4 References

1. G. Goldhaber at al, *Phys. Rev.* **120** (1960) 300.
2. B. Lörstadt, *Int. J. Mod. Phys.* **A12** (1989) 2861-2896.
3. P. Mättig, *Phys. Rep.* **177** (1989) 141.
4. K. Kolehmainen, M. Gyulassy, *Phys. Lett.* **B180** (1986) 203.
5. B. Andersson, W. Hofmann, *Phys. Lett.* **B169** (1986) 364.
6. M. G. Bowler, *Phys. Lett.* **B231** (1989) 203.
7. T. Csörgő, J. Zimányi, J. Bondorf, H. Heiselberg, S. Pratt, preprint KFKI-1989 56/A, to appear in *Phys. Lett. B*
8. R. M. Weiner, *Phys. Lett.* **B232** (1989) 278; J. Andreev, R. M. Weiner, *this volume.*
9. Contributions of AFS, NA35 and E802 collaborations, *this volume*

# MEASURING THE TWO PARTICLE CORRELATION LENGTH IN PERIPHERAL HEAVY ION COLLISIONS

**F.S. Navarra** and **M.C. Nemes**

Instituto de Física, Universidade de São Paulo
C.P. 20516, 01498 São Paulo, S.P., Brazil

**ABSTRACT**

We show that from the experimentally measured differential cross section for two particle emission in peripheral collisions it is possible to estimate the two particle correlation length inside the nuclear matter in a model independent way.

The measurement of two particle correlations in high energy reactions is often used to estimate the size (and the lifetime) of the emitting source[1], which is an essential ingredient in the determination of the energy density and hence of the conditions for the creation of quark–gluon plasma (QGP).

Even if one forgets all the limitations of the estimates imposed by final state interactions and the formation and decay of ressonances, the interpretation of those measurements is not clear. As was pointed out by Weiner and Andreev[2] on can never separate the radius of the source from some correlation length.

In this note we propose the use of coincidence experiments as a tool for determining the two particle correlation length inside the nuclear matter.

Most of the interferometric studies performed so far adress to central collisions where QGP might be formed and where at hight energies (~ GeV) one expects to achieve clear separation between the target, projectile and the intermediate fireball (like for example in the participant–spectator model) which is responsible for the bulk of particle production. This fireball is what is called source, whose radius interferometry is supposed to provide.

In peripheral collisions one has almost no particle production, no intermediate fireball formation (and hence no source sizes to measure) and no QGP. Still one can perform particle correlation measurements with the emitted nucleons and find out the two particle correlation length ($r_c$) inside the nuclear matter. There is no possible confusion with the source radius which in this case is not defined. While the size usually measured by interferometry is associated with a later ("freeze–out") stage of the collision, the length $r_c$ provides us with some information about the initial stage of nuclear matter in peripheral collisions.

The correlation length is a very interesting quantity in itself since it could help in discriminating between different nuclear reaction mechanisms. If it is found to be large (compared to the nucleon size) this would be an evidence against models which assume that reactions are a sum of two–body collisions (e.g. independent cascading, nucleon–nucleon superposition, etc.) and would reveal a coherent type interaction. The physics behind large correlation lengths is very exciting and involves quark clusters or field theoretical descriptions of the nuclear matter.

In what follows we develop a simple and model–independent description of two nucleon emission in which the only parameter is the two particle correlation length $r_c$ which can then be estimated by comparing our predictions for the differential cross section with the experimentally measured one.

We make use of the formalism advanced some time ago by Feshbach and Zabek[3] and later by Bertulani et al[4]. In this version we take the Coulomb interaction into account. One can find all technical details in our previous paper[5].

To a first order perturbation approximation the amplitude for two nucleon emission in relativistic peripheral collisions (see figure 1) can be written as

$$a_{fi} = \frac{1}{i\hbar}\int dt\, e^{i\omega t}\int d^3r_1 \int d^3r_2\, \Psi_f^*(\vec{r}_1,\vec{r}_2)\,[V_s(r_1,t) + V_s(r_2,t) +$$

$$+ V_c(r_1,t) + V_c(r_2,t)]\,\Psi_i(\vec{r}_1,\vec{r}_2) \qquad (1)$$

where $V_c(r_i,t)$ and $V_s(r_i,t)$ represent the Coulomb and strong potential respectively, acting between the projectile and the $i^{th}$ nucleon and $\hbar\omega = E_f - \varepsilon_i = \epsilon_1 + \epsilon_2$ + binding energy of the pair (= 16 MeV). $\epsilon_1$ and $\epsilon_2$ are the kinetic energies of nucleons 1 and 2. Since we are going to use standard expressions for the potentials, the most important physics will rely on the choice of the nucleonic pair wave function:

$$\Psi_f(\vec{r}_1,\vec{r}_2) = e^{i\vec{k}_1\vec{k}_1} \cdot e^{i\vec{k}_2\vec{k}_2} \qquad (2)$$

where $\vec{k}_1$ and $\vec{k}_2$ are the wave vectors of the emitted nucleons.

The initial wave function is given by

$$\Psi_i(\vec{r}_1,\vec{r}_2) = N \exp\left\{-\frac{r_1^{\,2}}{2\alpha_T^{\,2}}\right\} \cdot \exp\left\{-\frac{r_2^{\,2}}{2\alpha_T^{\,2}}\right\} \times \left[1 - \exp\left\{\frac{|\vec{r}_1 - \vec{r}_2|^2}{r_C^{\,2}}\right\}\right] \qquad (3)$$

where N is the appropriate normalization constant and $\alpha_T$ the Gaussian width related to the radius and diffuseness of the target density $\alpha_T = 2\sqrt{aR_T}$. $r_c$ is the two particle correlation length. It is the only free parameter in our calculations and hence can be estimated from experimental data. In order to simplify the calculations the Gaussian approximation for the strong potential has been apodted[4,5]. Accordingly the potential created by the projectile centered at the position (X,Y,Z) with $b = \sqrt{X^2+Y^2}$ and $Z = vt$

is given by

$$V_s(\vec{r}_i,t) = \gamma V_g \exp\left\{-\frac{(X - x_i)^2}{\alpha_p^2}\right\} \exp\left\{-\frac{(Y - y_i)^2}{\alpha_p^2}\right\} \exp\left\{-\frac{(Vt - z_i)^2}{\alpha_p^2}\right\} \quad (4)$$

where $\gamma = (1-v^2/c^2)^{-1/2}$ is the standard relativistic factor introduced here to take into account the Lorentz contraction of the nucleonic density of the projectile which generates the potential. The strength $V_g$ and the range parameter $\alpha_p$ are

$$V_g = \frac{V_0}{2} \exp\left\{\frac{R_p}{2a}\right\} \qquad \alpha_p = 2\sqrt{aR_p} \quad (5)$$

where $R_p$ is the radius of the projectile.

The Coulomb potential reads

$$V_c(\vec{r}_i,t) = \frac{Z\ e\ \gamma}{\sqrt{(x_i-x)^2 + (y_i-Y)^2 + \gamma^2(z_i-vt)^2}} \quad . \quad (6)$$

The differential cross section is obtained just integrating (1) over the coordinates and over the impact parameter (see details in ref. 5):

$$(2\pi)^6 \frac{d\sigma}{d^3\ k_1\ d^3\ k_2} = 2\pi \int_{R_p+R_T}^{\infty} db\ b\,|a_{fi}(b)|^2 \quad . \quad (7)$$

The correlation function can be written as

$$C(k_1,k_2) = \left[ \frac{d\sigma}{d^3k_1 d^3k_2} \right] / \left[ \frac{1}{\sigma} \frac{d\sigma}{d^3k_1} \frac{d\sigma}{d^3k_2} \right] . \tag{8}$$

Although data are available (sse for example the contributions of J.P. Alard and D. Ardouin to this meeting) a straightforward comparison of (7) and (8) with them is not meaningfull because:

i) the data refer only to central collisions ($b < R_p+R_T$) and our calculations to peripheral ($b > R_p+R_T$) ones.

ii) Assuming that final state interactions between the emitted nucleons is much weaker (i.e. negligeable) in large than in small relative momentum emission, an accurate estimate of $r_c$ requires the measurement of back–to–back (or nearly) emission. Until now data refers mainly to small relative momentum correlations.

The above mentioned data can be used for our purposes if the triggering conditions are changed and if special attention is drawm to back–to–back emission.

We evaluate now expression (7) as a function of kinetic energies of the outcoming nucleons $\epsilon_1,\epsilon_2$ which are emitted at $\theta_1 = 0^o$ and $\theta_2 = 180^o$. The values of the parameters appearing in the above formulas are $a = 0.65$ fm , $V_0 = 50$ MeV and $m = 938$ MeV. None of them is properly a free parameter. $V_0$ and $a$ are chosen in such a way that (4) reproduces the Woods–Saxon potential. In figure 2 we present the contour plots of the double differential cross section as a function of $\epsilon_1$ and $\epsilon_2$ for different values of $r_c$ , namely: 0.3 , 0.7 , 2.0 , 3.0 fm . One can see that the cross section is very sensitive to $r_c$ both in the shape and in the position of the maximum and hence is an adequate tool for estimating the two particle correlation length.

## ACKNOWLEDGEMENTS

This work was supported by FAPESP. One of us (F.S.N.) would like to aknowledge all the organizers of the CAMP meeting for their hospitality and R.M. Weiner, I. Andreev, J.P. Alard, P. Dupieux and O. Ardouin for very stimulating discussions.

## REFERENCES

1. See for example the contribution of G. Goldhaber and W. Zajc to these proceedings.
2. I.V. Andreev and R.M. Weiner, "*Soft pion coherence, coherence length and radius in Bose–Einstein correlations*", Marburg University preprint, 1990.
3. H. Feshbach and M. Zabek, *Ann. Phys.* **107** (1977) 110.
4. C.A. Bertulani et al., *Mod. Phys. Lett.* **14** (1989) 1315.
5. M.E. Bracco, F.S. Navarra and M.C. Nemes, "*Coulomb versus strong interaction in correlated nucleon emission*", São Paulo University preprint, 1990.

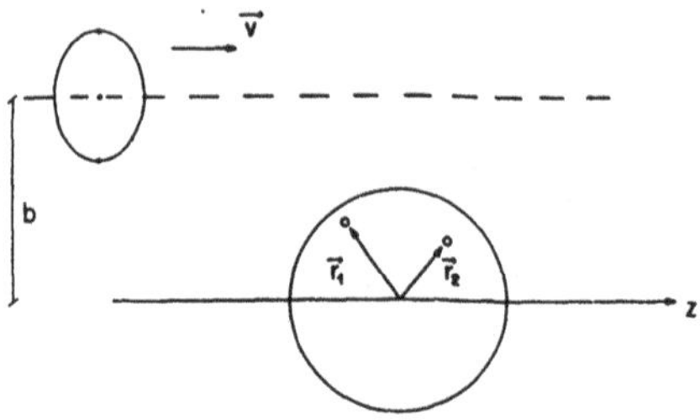


Figure 1. Relativistic collision between a Lorentz contracted nucleus and a nucleon pair in the laboratory frame.

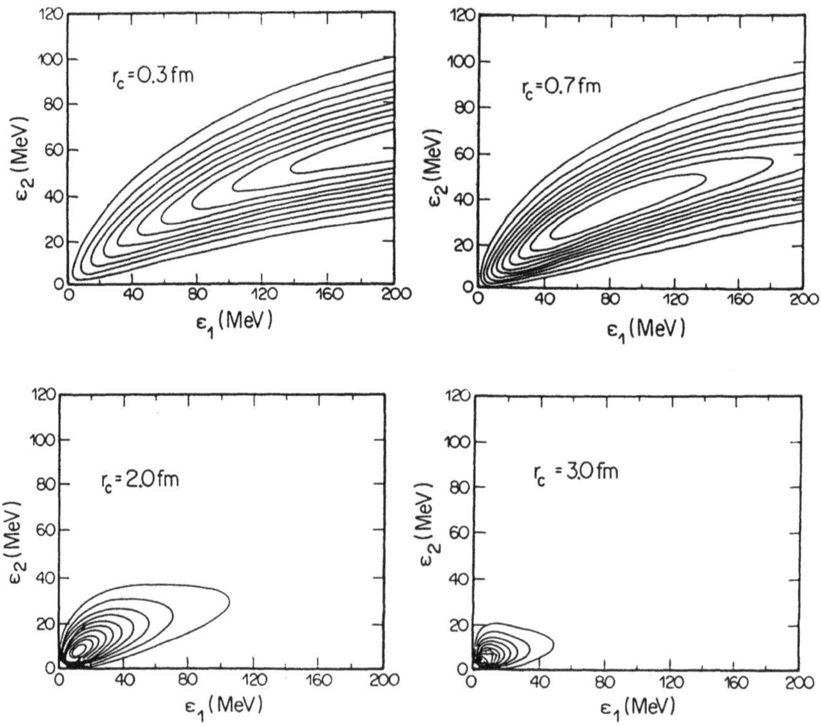


Figure 2. Differential cross section eq.(7) as a function of the nucleon kinetic energies $\epsilon_1, \epsilon_2$ for the $^{40}Ca+^{40}Ca$ system at $E_{lab} = 14.5$ GeV/u in the case of back–to–back ($\theta_1 = 0^o$ and $\theta_2 = 180^o$) emission. We present the cross section contour plots for four different values of the two particle correlation length: $r_c = 0.3$ , 0.7 , 2.0 and 3.0 fm.

## II. Sessions on Fluctuations

Chairmen: *D. Strottman, W. Hofmann, G. Zinovjev, and V. Grishin*

# Multiple Particle Production Processes in the "Light" of Quantum Optics

by

Erwin M. Friedlander

Lawrence Berkeley Laboratory, University of California, Berkeley CA 94611, USA

## 1. A bit of history

Is there an obsession with cascade processes ?

Ever since the observation [1] that high-energy "nuclear active" cosmic-ray particles — today we would say hadrons — create bunches of penetrating particles upon hitting targets, a controversy has raged about whether these secondaries are created in a "single act" (a sort of "mini-Big Bang", probably first termed by Heisenberg [2] as "Multiple Production") or whether — in analogy to the succession of ("particle-poor") bremsstrahlung-pair production processes in electromagnetic cascades — many hadrons are just the result of an intra-nuclear cascade, yielding one meson in every step (this used to be termed "Plural Production" [3]).

I cannot escape the impression that: a) the latter kind of model appeals naturally as a consequence of an innate bio-(or should one say anthropo- ?) morphism in our way of thinking (multiplication by cell division, rarity of twins and multiplets. . .), and that b) in one guise or another it has tenaciously survived to this day, also for hadron-hadron collisions, via multi-peripheral models to the modern parton shower approach.

Do "true" multiple processes occur in nature ?

The above remarks are not meant to imply that the cascade-type of approach to multiparticle production processes is the wrong one, but rather as a reminder of the fact that — in spite of its many successes — it is not the only one. Indeed, from the very beginning of theoretical consideration of multiparticle production [4] [5],the possibility of many particles arising from a single "hot" system ("fireball" ?) has been explored, with many fruitful results, not the least of which are the $s^{1/4}$ dependence of the mean produced particle multiplicity and the "thermal" shape of the $p_T$ spectra.

An important consequence of the thermodynamical-hydrodynamical models (hereafter THM's) is that particle emission is treated in analogy to black-body radiation, implying for the (mostly bosonic!) secondaries a set of specific Quantum-Statistical (hereafter QS) properties, very similar to those observed in quantum optics.

From here on I shall try to review a number of implications and applications of this QS analogy in the study of multiplicity distributions of the produced secondaries (MD's).

I will touch only in passing another very important topic of this class, viz. the Bose-Einstein two-particle correlations; it will be amply covered by other speakers at this meeting and in particular by Professor Gerson Goldhaber.

## 2. Facts and Fits

Approximate Negative Binomial Behavior of MD's

For a long time people have paid relatively little attention to the shape of the multiplicity distributions; they rather concentrated on the mean multiplicity <n>. It should have been clear from the very beginning, however, that P(n) contained a lot of "dynamical" information. Thus e.g., if the THM's are indeed the adequate framework of true multiparticle production, the QS analogy implies that for a single emitting source the P(n) should have the single-cell Bose-Einstein form

$$P_{B.E.}(n)=\frac{1}{1+m}\left(\frac{m}{1+m}\right)^n \qquad (1)$$

where m=<n>; for the general case of k cells in phase space the k-fold convolution of such distributions is the well-known Planck-Polya distribution

$$P_{NB}(n|m,k)=\frac{(n+k-1)!}{n!\,(k-1)!}\left(\frac{k}{m+k}\right)^k\left(\frac{m}{m+k}\right)^n \qquad (2)$$

known in mathematical terms as the negative binomial (NB).

If, on the other hand, particles are independently emitted from the fireball (Eq.(2) implies a lot of correlations !; see below) then n would be distributed according to the Poisson law (PD)

$$P_{Poiss}(n|m)=\frac{e^{-m}m^n}{n!} \qquad (3)$$

It should be noted here that both Eqs. (2) and (3) could be arrived at starting from very different viewpoints and models; thus e.g. the PD could be simply obtained as the limit of the NB if the number of cells increases indefinitely; conversely, an intrinsically Poisson distributed multiplicity n for which the expectation value m is a random variable obeying a gamma distribution (e.g. due to inelasticity fluctuations, to be discussed in Appendix A), n will end up by following an overall NB distribution.

The earliest reliable observations of P(n) were obtained in bubble chamber experiments at the CERN PS at $\sqrt{s}$ = 9 GeV and indicated a Poissonian shape. Later observations at higher energies ( [6] and [7] ) showed more and more marked departures from the PD, as can be seen from Fig. 1 (full lines=PD).

A good measure for the departure of an MD from the PD are the factorial cumulants fq (also known as Mueller functions), to be considered in more detail below, which should be zero for the PD in all orders q>1. The best known of these, $f_2$ is given by:

$$f_2=\langle n^2\rangle-\langle n\rangle^2-\langle n\rangle \qquad (4)$$

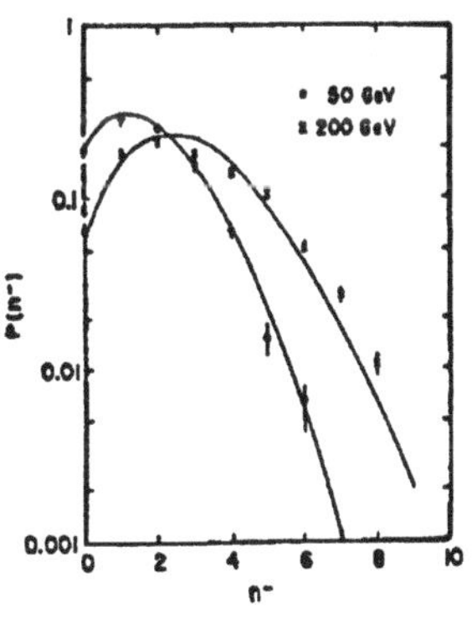


Fig. 1

Experimental values of $f_2$ obtained with hydrogen bubble chambers up to the highest energies covered in early FNAL experiments are plotted in Fig. 2 against <n>.

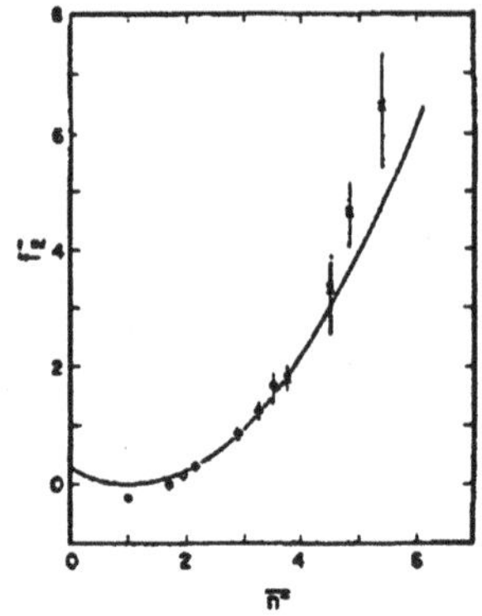

Fig. 2 ↑

It is obvious that with increasing energy, $f_2$ increases to large positive values, i.e. the distributions become broader than the PD. When strong departures from Feynmann scaling and hence from its asymptotic consequence known as KNO scaling were established at considerably higher cms energies, the NB was suddenly "re-discovered" [8] and it was shown that it provides indeed a reasonable description of the MD's over a large range of s-values. That this description is far from perfect is shown in Fig. 3 obtained when more accurate measurements at the CERN SPPS [9] noticed systematic departures from an NB.

Fig. 3
→

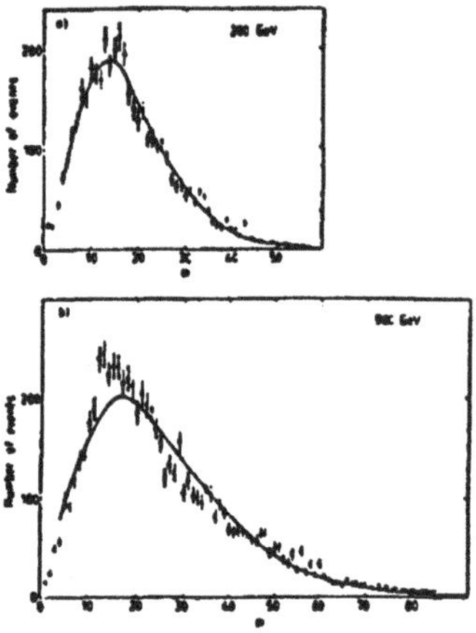

Centered Rapidity Windows

If the MD of "complete" events, in which produced particles from all regions of phase space were counted, have been an object of study for a relatively long time, it took quite a while to recognize [10] the fact that significant "dynamical" information could be revealed if restricted regions of phase space — more precisely, in rapidity — were treated in each event as a "sub-event" and then the MD of such sub-events were investigated. This concept was also re-discovered at the SPPS [11], and then exploited by many experimental groups. Most large-scale investigations concerned rapidity windows centered about zero cms rapidity (usually called "cuts" or "symmetrical windows" going from -Yc to Yc). The main facts established in this field (e.g., [12]) are the good fit yielded by the NB in practically all experiments and a systematic lowering of the width parameter k ( i.e. a widening of the MD) with decreasing window size Yc. (Fig. 4, below).

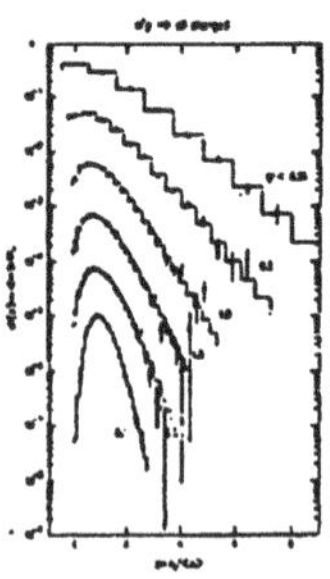

Fig. 4

Shifted Rapidity Windows

The behavior of sub-events collected from windows shifted away from cms rapidity ("shifts" or "asymmetrical windows") ([10], [13] ) reveals that along with the widening of the MD due to decreasing window size (as seen already in the cuts), the width of the MD decreases as the center of the accepted window shifts away from y=0 (ref. 10 and 13). The quality of NB-fits also deteriorates.

Forward-Backward Correlations

A special case of an asymmetrical window is one starting at y=0 and extending up to some $Y_c$; if Yc extends to infinity (or rather to its allowed kinematical limit) a whole hemisphere in rapidity, containing nF particles is covered; in studying correlations with the number $n_B$ of particles observed in the mirror reflection of the accepted interval, one then speaks of "Forward-Backward" correlations. These would be non-existent (i.e. the slope of the $<n_F>=a+bn_B$ regression line would be zero) if the particles were Poisson distributed and the sampling done by opening one hemisphere only would be simply binomial. In fact, significant slopes b are observed, which furthermore increase with $\sqrt{s}$.

## Chaotic and Coherent Sources

Bose - Einstein Correlations: The Unmistakable Signal

The fact that any source emitting hadrons (mainly bosons) should obey the laws of QS is obvious; *how* this occurs in detail is a much more complex problem, which has not been completely explored to this day. The unmistakable signal that QS is at work is the observation of strong (like-sign) two- pion correlations [14] with the characteristic signature of Bose-Einstein (BE) statistics, namely the strong enhancement of the correlation function at low 4-momentum differences [15][16]. Along with this signal, all experiments [17] have also shown that the correlation never reaches its maximum allowed degree (a value of 2 at zero intercept); the most natural interpretation (though by far not the only one!) [18] for this effect is that the source is not 100% chaotic, as would be expected from a purely thermalized system, but that a certain degree of coherence is present in the boson radiation. Details (both theoretical and experimental) of these investigations will fill at least one of the sessions of our workshop. Except for one (essential) connection, to be mentioned further on in my talk, I shall from now on leave this subject to my other colleagues and concentrate on some consequences of the "co-existence" of chaoticity and coherence in the source, for the multiplicity distribution of the emitted particles.

Pure and Partially Coherent Sources

The simplest assumption about the origin of the created secondaries is that they are emitted from a single source; if the field has a chaotic ($\pi_{ch}$) and a coherent ($\pi_c$) component[19]:

$$\pi(x)= \pi_c(x)+ \pi_{ch}(x) \tag{5}$$

such that

$$<n_{ch}>=<| \pi^2_{ch}|> , \quad <n_c>=<| \pi^2_c|> \tag{6}$$

the mean number of particles emitted will be:

$$<n>=<| \pi^2|>=< n_c>+< n_{ch}> \tag{7}$$

and one may define a coefficient of chaoticity ( for short "chaoticity") p:

$$p = \frac{<n_{ch}>}{<n>} , \quad [0 \leq p \leq 1] \tag{8}$$

We are interested in the shape (or , lacking that, at least some numerical characteristics) of the resulting multiplicity distributions (MD's).

The case p=0 corresponds to purely coherent sources. For a single source the predicted MD coincides, oddly, with the prediction for a set of independently emitted particles , i.e. with the Poisson distribution ( PD, Eq.3). As was already noted [20]), a superposition of many coherent sources leads to a generalization of the PD known as the hyper-Poisson (HP) distribution,

$$P_{HP}(n|\alpha,\beta)=\frac{\beta!\,\alpha^{n}}{{}_1F_1(1;\,\beta+1;\,\alpha)\;(\beta+n)!} \tag{9}$$

where ${}_1F_1$ is the confluent hypergeometric function. The HP obviously reduces to the PD if $\beta\rightarrow 0$ . An example of how well the HP describes MD' s, which are definitely wider than the PD ($f_2>0$ ), is shown in Fig.5 [21]. It would be tempting to assume that all reactions are intrinsically coherent, and to assign all the widening of MD's with increasing $\sqrt{s}$ to the change of the additional parameter $\beta$. However, at very large $\sqrt{s}$ the fits deteriorate [21] and it becomes obvious that some chaoticity has to be present, as was already indicated by the BE correlations.

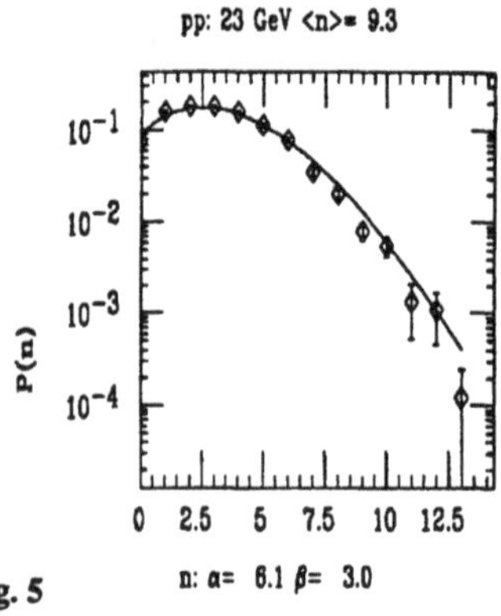


Fig. 5

The general case of a superposition of coherent and chaotic fields (described by Eqs.(5) - (8) ) is easier to consider if the variable x is identified with a kinematic variable which – in analogy to time in optics – has the property of stationarity. Such a variable, in the case of multiparticle production, appears to be ( at least for the bulk of the created secondaries) the longitudinal rapidity y. (Its conjugate is then the boost variable.) The stationarity is revealed in the two-particle correlation

$$<\pi_{ch}(y)^{+}\,\pi_{ch}(y')> \,\sim\, e^{-\frac{|y-y'|}{\xi}} \tag{10}$$

for the chaotic component , which is a function of $|y-y'|$ only. The quantity $\xi$ is the rapidity correlation length of the chaotic component (and appears as such in the BE correlations, when measured in terms of y -). If $\xi\rightarrow\infty$ (as one would expect, e.g. in the vicinity of a phase transition) then P(n) is known to obey the so-called Partially Coherent Laser Distribution (PCLD, also known as Glauber-Lachs.) :

$$P_{PCLD}(n\mid m,p,k)=\left(\frac{k}{k+mp}\right)^{k}\left(\frac{mp}{k+mp}\right)^{n}\exp\left(-\frac{mk(1-p)}{k+mp}\right)\;L_n^{k-1}\left(-\frac{k^2(1-p)}{p(k+mp)}\right) \tag{11}$$

the $L_n^{k-1}$ are generalized Laguerre polynomials, k is the number of "cells" in phase space, and m the mean multiplicity. It is easy to see that the PCLD has as limiting cases the PD if $p\rightarrow 0$ and the NB if $p\rightarrow 1$ . It appears to give [22] a qualitative description of the widening of MD's with $\sqrt{s}$, provided p increases with $\sqrt{s}$ by about an order of magnitude in the energy range covered by the CERN ISR and SPPS colliders. This is illustrated in Fig.6, which shows the ratios $R_q$ between the computed and the observed normalized moments

$$C_q \equiv \frac{\nu_q}{\nu_1^q}, \quad \nu_q=<n^q> \tag{12}$$

along with the $\sqrt{s}$ – dependence of the estimated p–values.

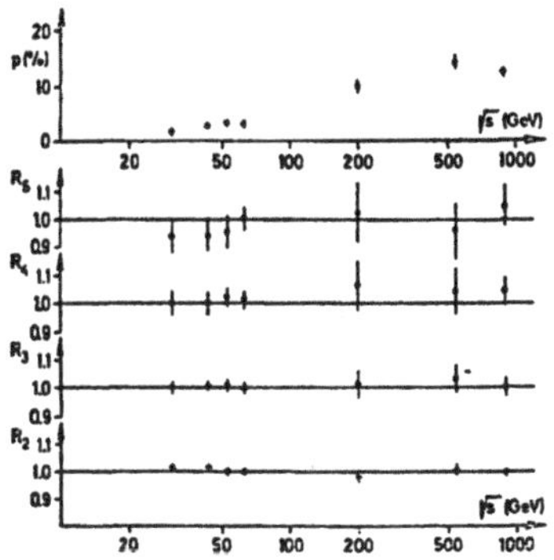


Fig. 6

Forward cone in p-A collisions: Self-Induced Transparency?

The relative width of the PCLD can be measured by its normalized factorial moments of order q,

$$\Phi_q \equiv \frac{F_q}{F_1^q}\,, \quad F_q = \langle n(n-1)\ldots(n-q+1)\rangle \qquad (13)$$

which are given by [22] :

$$\Phi_q = q!\left(\frac{p}{q}\right)^q L_q^{k-1}\left(-k\,\frac{1-p}{p}\right) \;;\ \text{PD:}\ \Phi_q = 1\;;\ \text{NB:}\ \Phi_q = \frac{(q+k-1)!}{k^q(k-1)!} \qquad (14)$$

they thus decrease when $p \to 0$, i.e. with increasing coherence of the source. One is tempted to connect the narrowing of the MD's for "shifted" rapidity windows ( as the center of the "shift" moves away from $y_{cms}$ towards the kinematic limit $Y_c$ ) with a decrease of the chaoticity p.

On the other hand, it is known [23] that in p-A collisions the particle density near $Y_c$ is insensitive to the target thickness (measured by $A^{1/3}$); in other words, the target appears to be transparent to the fastest (high − y ) bosons. This brings to mind an effect well known in Quantum Optics [24], namely the so-called self-induced transparency when a coherent beam traverses a medium with at least one meta-stable state. Fig.7 [10] compares for proton-Emulsion collisions at 200 GeV four Y windows (covering four quarters of the kinematically accepted y range ( going from $-Y_c$ to $Y_c$ ):

a) with respect to "nuclear transparency", as evidenced by the variation (or lack thereof) of the particle density in the given window with the multiplicity $N_h$ of slow target fragments (a good measure of target thickness) and

b) the shape of the multiplicity distribution in the same window. The straight lines are one-cell BE MD's while the curves are the best fitting PD. It is apparent that as one approaches the projectile fragmentation region, "coherence" (low p ) and "transparency" are well correlated. Although each of these effects, taken separately, is susceptible to a few other interpretations , their simultaneous occurrence is very suggestive of the quantum-optical analogy.

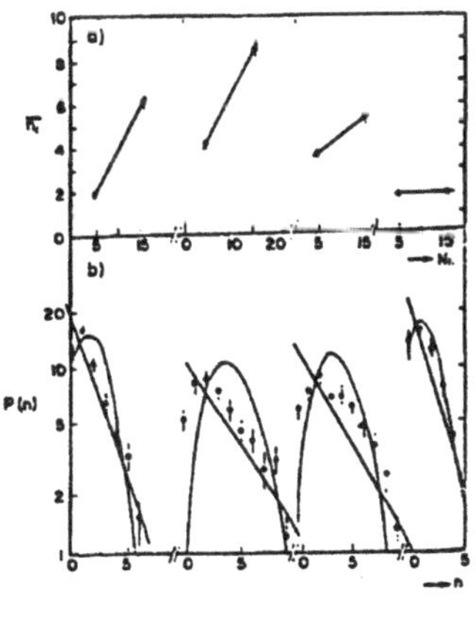


Fig. 7

Multi-Component Convolutions

A reasonable interpretation of the decrease of the apparent degree of chaoticity when y increases towards $Y_c$ can be found [25] by assuming that in each collisions two sources are at play, one of which is purely chaotic (thermal fireball due e.g. to gluon-gluon interactions), while the other behaves coherently ( as would be expected, e.g., from valence quark bremsstrahlung). The resulting MD of the total number n of produced secondaries would then be given by a convolution,

$$P_{2\ comp.}(n|m_1,m_2,k) = \sum_{n_1+n_2=n} P_{NB}(n_1|m_1,k)\cdot P_{Poiss}(n_2|m_2) \qquad (15)$$

where the subscripts 1 and 2 refer to the actual and the mean multiplicities of the chaotic and the coherent sources, respectively. Since the coherent source does not contribute to the factorial cumulants ( Eq.(14 )) it is easy to estimate $m_1$ and $m_2$ from the observed moments of the overall MD. Estimated values for $m_1$ and $m_2$ are shown in Fig..8.1 for $\sqrt{s}$ ranging from about 20 to 900 GeV. The chaotic component is well described by a $s^{1/4}$ – law, as expected from a thermal source [4] ( QGP ?) . The coherent component varies rather like log(s) consistent with a bremsstrahlung origin. Fig.8.2.a shows the estimated rapidity distribution of the two sources at $\sqrt{s}$=546 GeV. The chaotic one is concentrated at low cms rapidities while the coherent one is more or less flatly spread out towards high y. Fig.8. b then shows the estimated apparent chaoticity

$$p^* = \frac{m_1}{m_1+m_2} \qquad (16)$$

as a function of rapidity; $p^*$ decreases from ≈ 80% to ≈ 60% over the observed rapidity range.

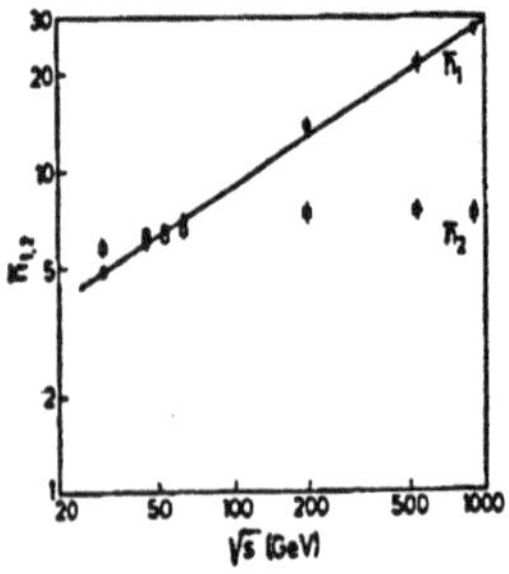

Fig. 8.1

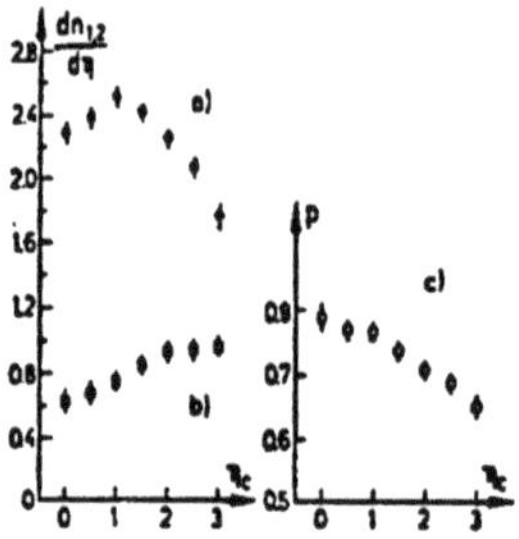

Fig. 8.2

Partially Coherent Sources with Finite Correlation Lengths

When the coherence length $\xi$ of the chaotic component is finite, two things happen: the bad news is that it is not possible to get a closed expression for P(n); the good news, however, is that it becomes possible to predict the behavior of finite rapidity windows ( at least the "cuts") from information deduced from the global distribution P(n). We shall now address these two problems in more detail.

Though no closed formula exists for P(n), numerical approximations have been worked out recently [26] , [27]. Even for low <n> the computational effort is enormous and for the large r <n> occurring at large existing or projected colliders the logistics become prohibitive. What is available is a set of closed expressions for the normalized factorial cumulants $\mu_q$ of P(n):

$$\mu_q \equiv \frac{f_q}{f_1^q}, \qquad f_q = \left.\frac{d^q \log(G_F(z))}{dz^q}\right|_{z=1} \qquad (17)$$

where $G_f(z)$ is the factorial moment generating function:

$$G_F(z) = \sum_{n=0}^{\infty} z^n P(n) . \tag{18}$$

They are given [28][19] by :

$$\mu_q = (q-1)! \left(\frac{p}{k}\right)^{q-1} \left[ p B_q(\beta) + q(1-p) \overline{B}_q(\beta) \right] \tag{19}$$

The B's are a set of elementary (but, for larger values of q, increasingly nasty !) functions of the ratio $\beta = \Delta Y/\xi$ only (where $\Delta Y$ is the width of the rapidity interval counted); the first three pairs are :

$$B_1(\beta) = \overline{B}_1(\beta) = 1 , \tag{20}$$

$$B_2(\beta) = \left(\frac{1}{2}\right)\left(e^{-2\beta} + 2\beta - 1\right) \beta^{-2} , \ \overline{B}_2(\beta) = 2\left(e^{-\beta} + \beta - 1\right) \beta^{-2} , \tag{21}$$

and

$$B_3(\beta) = \left(\frac{3}{2}\right)\left((\beta+1)e^{-2\beta} + \beta - 1\right) \beta^{-3} , \ \overline{B}_3(\beta) = \left(-e^{-2\beta} + 2(\beta+4)e^{-\beta} + 4\beta - 7\right) \beta^{-3} . \tag{22}$$

The normalized factorial cumulants $\mu_q$ can be easily connected with experimentally measured quantities, such as the normalized moments $C_q$ or the equivalent k of a fitted negative binomial. For q=2 the simple relationships are:

$$\mu_2 = \Phi_2 - 1 = C_2 - 1 - \frac{1}{\langle n \rangle} \rightarrow \frac{1}{\tilde{k}} \tag{23}$$

The fact that for a given p the $\mu_q$ are functions of the ratio $\beta$ only, allows one to scale cumulants (in a more rudimentary analysis, equivalent k - values) from one window width $\Delta Y$ (e.g. the full Y range) to a more restricted one, as long as $\xi$ remains unchanged. This property, which we call beta-scaling explains qualitatively why the MD's of symmetric windows ("cuts") become systematically wider ( in "normalized" parlance, i.e. in terms of $\mu_q$ or k) as $\Delta Y$ decreases [30] . The same formalism can be applied to explain the increase with $\sqrt{s}$ of the Forward-Backward multiparticle correlations [29] , because the width of the MD in, say, the forward cms hemisphere is determined by that of the global MD except for the change $\beta \rightarrow \beta/2$ in Eq.(19) .

By combining the results of measurements of the moments (ergo also the cumulants ) of the MD's with the information from Forward-Backward Correlation, it becomes possible to estimate simultaneously both p and $\xi$.

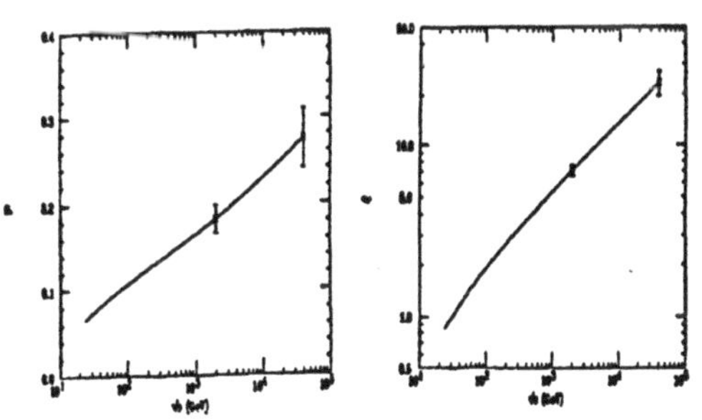

Fig. 9

It could be shown [29], [31], that both parameters increase with $\sqrt{s}$. These increases are shown in Fig.9 a) and b) for the ISR – SPPS energy range along with projection to TEVATRON and SSC energies (the latter with error bars).

Consequences for BE – Correlation (HBT) Experiments

As mentioned earlier, the fact that the two-particle correlation function $C_{(2)}$ recorded in BE-correlation experiments never reaches its maximum allowed value of 2, can be most easily interpreted in terms of partial coherence ( $p < 1$ ) of the source. As has been shown by Weiner [32], mixed (i.e. chaotic + coherent) fields lead to a correlation function

$$C_{(2)}=1+2p(1-p)\, e^{-|x_1-x_2|/\xi}+p^2\, e^{-2|x_1-x_2|/\xi} \tag{24}$$

in which even a slight admixture of coherence substantially decreases the intercept (at $|x_1 - x_2|=0$.

Very small rapidity windows: Intermittency?

It has been noticed for some time ( [33] – [36] ) that with decreasing size $\Delta y$ of the rapidity bite considered there occur strong local fluctuations in rapidity density ("spikes). A typical example observed at $\sqrt{s}$=22 GeV by the NA22 Collaboration is shown in Fig.10. The uniform distribution in azimuth illustrated in the upper half of the figure excludes the interpretation of the "spike" in y seen in the lower half as a ("hard" parton-parton) jet.

Fig. 10

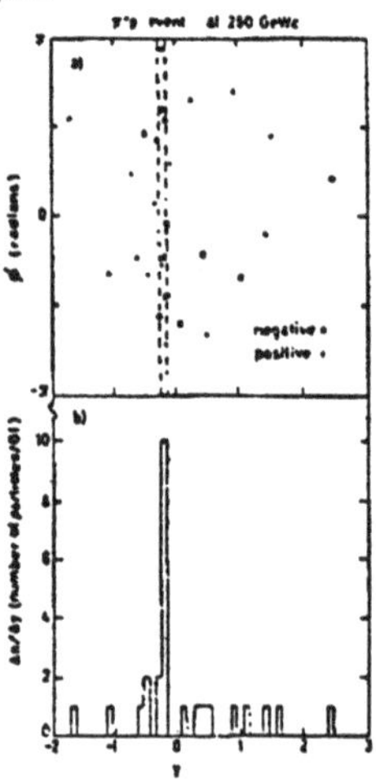

In order to deal quantitatively with this phenomenon, dubbed in analogy to hydrodynamics as intermittency, it has been proposed [37] to study the behavior of normalized factorial moments $\Phi_q$ with $\Delta y$. Certain models predict a power law behavior, whereas experimental data seem to flatten off at very small $\Delta y$. It has been shown [38] that the overall behavior of the $\Phi_q$ can be obtained directly from Eq. (19) as a consequence of the decrease of $\beta$ with $\Delta y$. Two examples of this kind of analysis are shown in Figs11 and 12 for $\sqrt{s}$ of 22 and 546 GeV, respectively. It should be noted that most of the "noisiness" noticeable for q=4 and q=5 in Fig.11 is due to the influence of the "spiky" event of Fig10 and possibly that of a few similar but less prominent extreme events ; their origin (see e.g., Appendix B) may very well lie outside the scope of QS, which otherwise

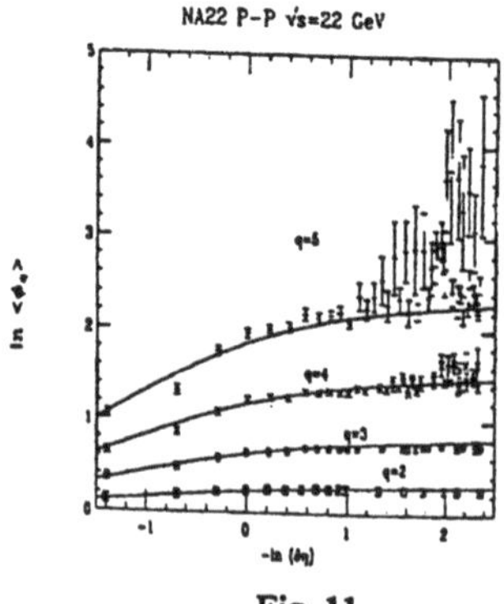


Fig. 11

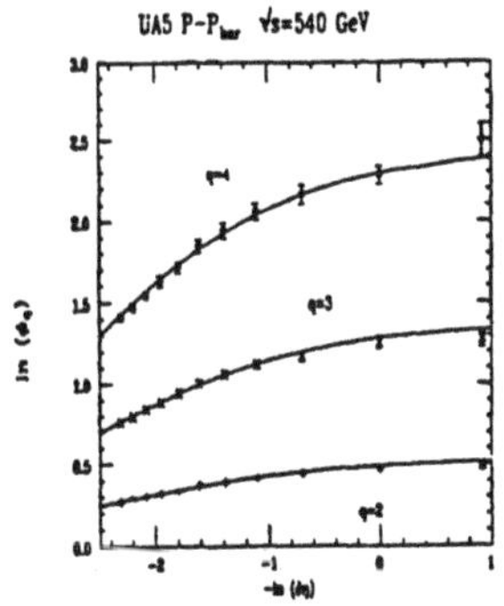


predicts the overall behavior of the $\Phi_q$ rather well (It must be stressed that in Figs.11 and 12 the parameters p and $\xi$ were fitted just to the second factorial moments and the good description of the higher moments followed automatically!).

Fig. 12

How well is P(n) *really* described by the QS formalism?

As we have seen until now, both the very fashionable negative binomial distribution and the different refinements introduced by a more detailed consideration of QS phenomena manage to give a good overall description of the shape of P(n). A legitimate question is whether "fine structures" exist which would be drowned out in the ensemble averaging reflected in the moments or be hidden in the (also very fashionable) logarithmic presentations of P(n). A suspicion that such fine structures could be present arose [39] from the consideration of possible "squeezed" quantum states, which would lead (in NB jargon) to MD's with an apparent $k < 1$. Given adequate values of the parameters of such a model one might see an oscillatory behavior superimposed on the "global" shape of the P(n).

Fig. 13 →

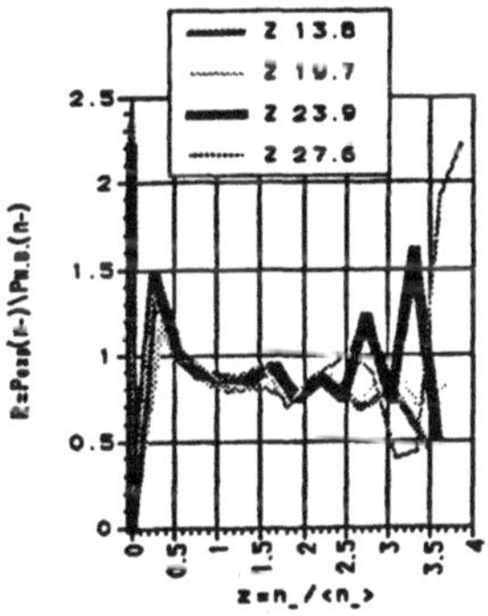


In order to "put the P(n) under a magnifying glass" we plot in Figs. 13-15 the ratios R between the experimental P(n) and the "smooth" prediction of the best fitted NB. Similar analyses have been done using as reference (instead of the NB) the PCLD or the two-component convolution (Eq.15), with essentially similar results. For better clarity the data have been grouped into "low" ($\sqrt{s}$ < 28 GeV), "medium" ($\sqrt{s}$=30 – 60 GeV, CERN ISR), and "high" ($\sqrt{s}$ >200 GeV, CERN SPPS) energy groups. The R - values are plotted against the normalized multiplicity z=n/<n> ( of KNO fame) so as to try to detect systematic deviations from a smooth distribution. Although the mid-range, i.e. z=1 - 2 data points agree roughly with the NB prediction (best in the ISR case) it is apparent that correlated structures appear around z=0.5 and z=2.5. Although statistics deteriorate towards the extremities of the z- scale (to illustrate this, I present in Fig. 16 the 200 GeV UA5 data alone, complete with error bars which would have irremediably obscured the preceding graphs), the correlated trends of the different distributions appear too similar to be due to chance fluctuations only and deserve further, quantitative, investigation. Besides squeezed quantum states one might think of other possible explanations of these deviations, such as mixtures of events obeying quite different shapes of P(n). If such a mixture could be detected by tagging the different kinds of events by means of characteristics extraneous to the MD's it might make the "good fit" by the NB distribution to appear as an accident of numbers (just as the analysis by the UA5 Collaboration had shown the apparent KNO-behavior in the ISR range be accidental).

**Fig. 14**

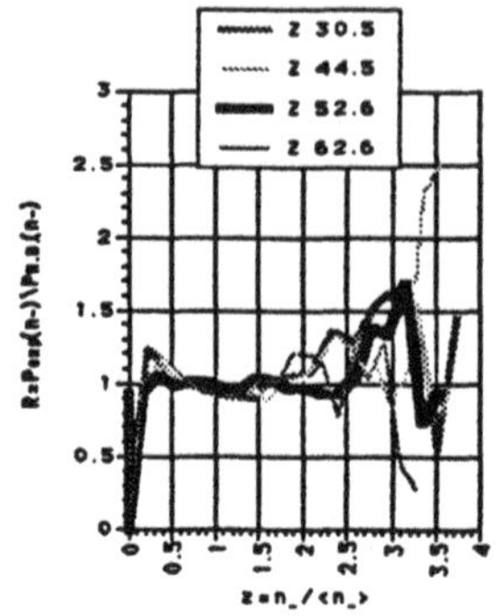


**Fig. 15**

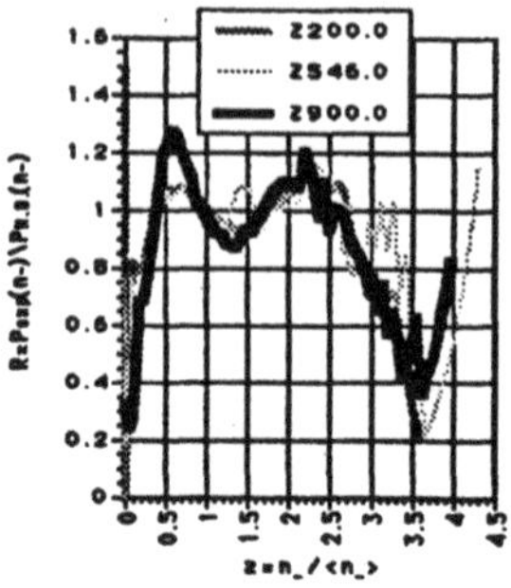


**Fig. 16**

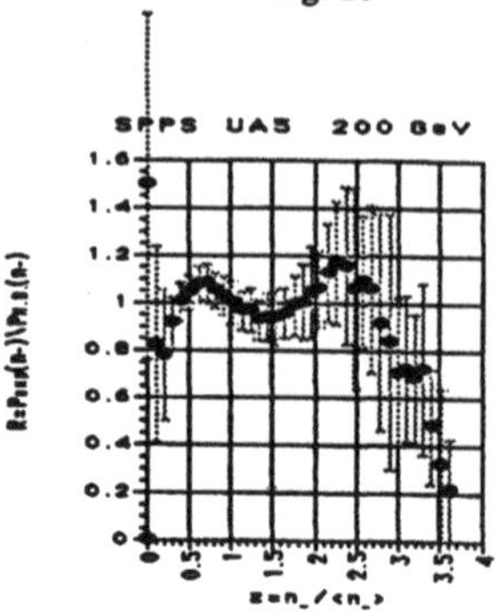


## Concluding Remarks

It goes without saying that many if not all the physical characteristics of multiplicity distributions, *taken individually*, could be explained by a host of other models (clusters, parton cascading, minijets, to name just a few). On the other hand – from quite general principles – the processes leading to multi-particle production just *have* to obey the laws of Quantum Statistics; with this in mind it is amazing how easily QS leads to a quite consistent picture of most if not all the observed phenomena. Maybe this is why it deserves further careful consideration in future investigations.

On the theoretical side these must include the effects of non-stationarity in Y, simplifications in the assumptions about the boost variable, and a better understanding of the effects of inelasticity. (See Appendix A.)

On the experimental side considerably more accurate determinations of both P(n) and the BE-correlations, with careful filtering out of systematic effects appear very desirable. Whether or not this will remain just a pious wish will depend to a large extent on the concrete programs of future accelerators as the good reliable ones ( like the ISR and the SPPS ) succumb to the looming budget crunch.

The ideas and results presented in this talk came about through the efforts and collaboration of the following colleagues, [ grouped and re-grouped in almost all n!/k! (n-k)! combinations..! ] all of which have contributed very much to whatever understanding I have of the subject and none of which bears any responsibility for my presentation thereof : M.Biyajima, P.Carruthers, G.N.Fowler, C.X.He, F.S.Navarra, U.Ornik, M.Pluemer, F.W.Pottag, C.C.Shih, I.Stern, A.Vourdas, R.M.Weiner, J.M.Wheeler and G.Wilk.

# APPENDIX A.

All Collisions are Equal, but some are More Equal than others. . .

An essential ingredient in any single-source model of particle production is the recognition of the fact that the incident particles deposit only a fraction K of their total energy into the "source" so as to make available only an "effective" energy $W=K\sqrt{s}$ for particle production. In the treatment of multi-particle production W is the only relevant energy variable. The quantity K called the coefficient of inelasticity (or, for short, the inelasticity) fluctuates from event to event ($0 < K < 1$) so that events produced by beam-target combinations producing different $\sqrt{s}$ may end up with the same physical initial conditions.

As a consequence, if observations are made at a fixed cms energy $\sqrt{s}$ and some model predicts a multiplicity distribution $P(n \mid W(K))$ then the observed multiplicity distribution $P'(n)$ will be given by

$$P'(n \mid \sqrt{s}) = \int_{K=0}^{1} P(n \mid W(K))\, \chi(K)\, dK \quad \text{(A1)}$$

Fig. A1

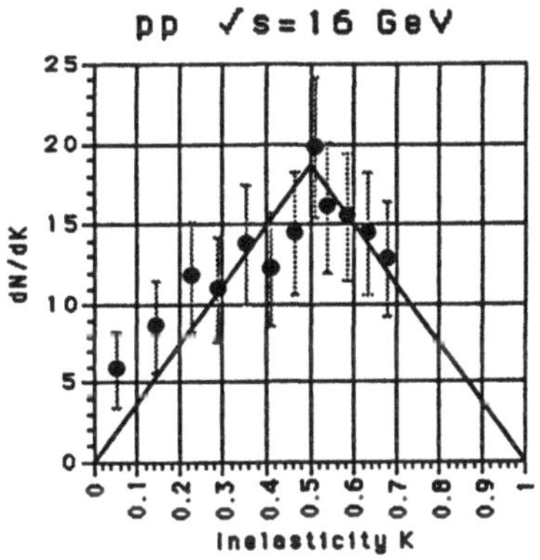


where $\chi(K)$ is the Inelasticity Distribution.

This probability distribution (which can be intuitively, if not necessarily adequately, connected with an impact parameter representation) is hardly known from experiment. It was recognized as early as the fifties (again from cosmic-ray measurements) that $\langle K \rangle$ is of the order of 1/2, i.e. that leading particles (usually the surviving baryons) carry away about half the initial energy. The only direct measurement of $\chi(K)$ [40] from collisions of an accelerator beam at $\sqrt{s}=16$ GeV, yielded a rather broad distribution centered about K=0.5.

What is better known is the Feynmann x-spectrum of individual leading protons, which is close to a flat distribution, almost from x=0 to 1. However, the concrete spectrum of K depends on the total energy loss of both incident particles and hence on the degree of correlation of the fractional energy contents x1 and x2 of the leading secondaries. It can be easily shown that if x1 and x2 are totally correlated, then K is uniformly distributed between 0 and 1, whereas if cov(x1,x2)=0 the distribution of K should be triangular and centered about $\langle K \rangle = 1/2$ (this is more or less what is seen in Fig. A.1).

It has been noted [41] that the relative widths of the rapidity distributions at $\sqrt{s}=53$ and 546 GeV, together with the relatively weak change in transverse momentum spectra at the two energies, imply a decrease of $\langle K \rangle$ with $\sqrt{s}$. Within the framework of QCD it has been shown [42] that assuming gluon-gluon interaction as the vehicle for energy deposition in the "source" leads to K-distributions similar to those observed hitherto and predicts a shrinking of $\chi(K)$ with $\sqrt{s}$ as depicted in Fig. A.2 ( Here A, B, C and D refer to $\sqrt{s}\approx 50$, 500, 2000 and 40000 GeV, respectively). With this kind of information at hand, it becomes possible to correct for the effects of $\chi(K)$ in P(n) by means of Eq.(A1) or similar procedures [22],[30].

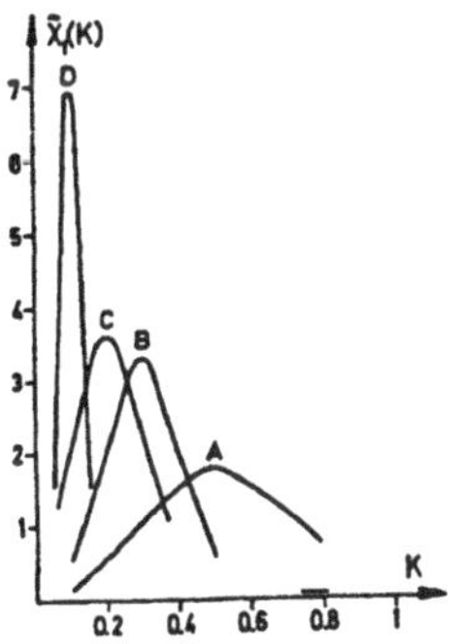


Fig. A2

# APPENDIX B

Fractal substratum of rapidity spikes? A toy model

It s well known that for certain critical values of their parameters many non-linear dynamical systems with a very deterministic behavior turn chaotic. The most trivial example is the logistic map in which a variable x defined recursively via

$$x' = \lambda\, x\ (1-x) \qquad \text{(B1)}$$

turns chaotic (in other words with a quasi - stochastic behavior) near $\lambda=4$. It seemed amusing to see what the apparent probability distribution of x - values produced by such a pseudo-random number generator is like. A histogram of a set of 200 x-values produced in this way is shown in Fig. B.1. This distribution shows an uncanny similarity to a $\cos^n \theta$ angular distribution once x is re-mapped to the "$\cos\Theta$" interval -1 to +1 .

A simple re-mapping to

$$\eta \equiv \tanh^{-1}(2x-1) \qquad \text{(B2)}$$

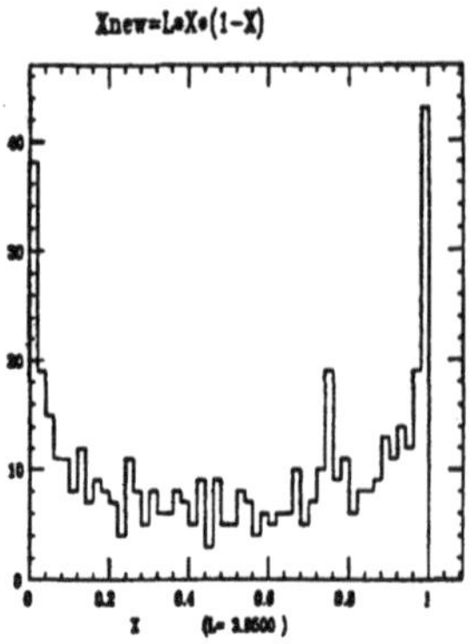


Fig. B1

turns this into a "pseudo-pseudo-rapidity distribution" shown in Fig. B.2. It displays a typical spike quite similar to those observed in( e.g. heavy ion-)experiments. One such spike turns up in a few hundreds of simulated "events" of comparable "multiplicity" . No comment !

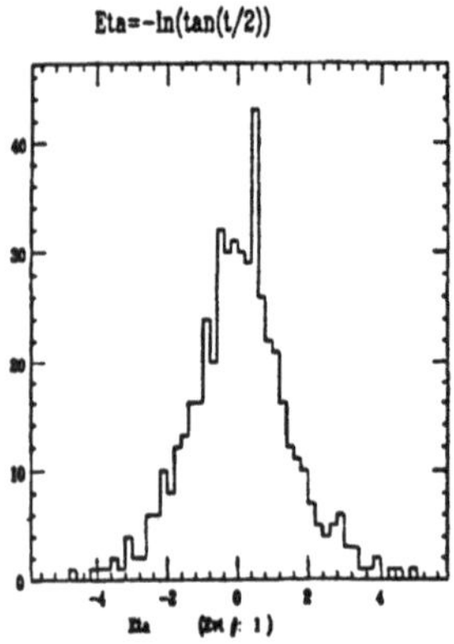


Fig. B2

## REFERENCES

[1] L.Janossy and G.D.Rochester, Proc.Roy.Soc. **A183**, 183 (1944)
[2] W.Heisenberg Kosmische Strahlung, Springer,Berlin (1953), p.148 ff.
[3] W.Heitler and L.Janossy, Proc.Phys.Soc. **62**,374 (1949)
[4] E.Fermi, Progr.Theor.Phys. **5**, 270 (1950)
[5] L.Landau, Izv.A.N.SSSR, s.fiz., **17**, 1 (1953)
[6] V.Ammossov *et al.*, Phys.Rev.Lett.,**42B**, 519 (1972)
[7] A.Firestone *et al.*, Phys.Rev. **D10**, 2080 (1974)
[8] G.J.Alner *et al.*, Phys.Lett. **160B**, 199 (1985)
[9] R.E.Ansorge *et al.*, Zs.f.Phys.C– Particles and Fields, **43**,357 (1989)
[10] G.N.Fowler,E.M.Friedlander,R.M.Weiner, Phys.Lett. **104B**, 239 (1981)
[11] J.Alner *et al.*, Nucl.Phys. **B291**, 445 (1987)
[12] M.Adamus *et al.*, Zs.f.Phys. **C37**, 215 (1988)
[13] G.J.Alner *et al.*, Phys.Rep. **154**, 247 (1987)
[14] G.Goldhaber *et al.*, Phys.Rev. **120**, 840 (1960)
[15] G.J.Kopylov and M.I Podgoretsky, Yad.Fiz. **18**, 656 (1973)
[16] G.Cocconi, Phys.Lett. **49B**, 459 (1974)
[17] M.Deutschmann *et al.*, Nucl.Phys. **103B**, 198 (1976)
[18] G.N. Fowler and R.M. Weiner, Phys.Lett. **70B**, 201 (1977)
[19] G.N.Fowler and R.M.Weiner, Phys.Rev. **D17**, 3118 (1978)
[20] T.F.Hoang , Phys.Rev. **D10**,1043 (1974)
[21] E.M.Friedlander and I.Stern, to be published
[22] G.N.Fowler *et al.*, Phys.Rev.Lett. **56**, 14 (1985)
[23] J.Cohen *et al.*, Lett.Nuovo Cim., **9**, 337 (1974)
[24] S.L.McCall and E.L.Hahn, Phys.Rev.Lett. **18**, 908 (1967)
[25] G.N.Fowler *et al.*, ibid.,**57**, 2119 (1986)
[26] C.C. Shih, Phys.Rev. **A40**, 1456 (1989)
[27] E.M.Friedlander and M.Pluemer, LBL Report 27082 (1989)
[28] A.K.Jaiswal and C.L.Mehta, Phys.Rev.**A 2**,168 (1970)
[29] G.N.Fowler *et al.*, Phys.Rev.**D37**, 3127 (1988)
[30] G.J.Alner *et al.*, Phys.Lett. **B160**, 199 (1985)
[31] G.N.Fowler *et al.*, LBL-28956,April 1990 and Journ. of Phys., under press
[32] R.M.Weiner, Proc. Hadron Matter In Collision, Tucson (1988) p.187
[33] T.H.Burnett *et al.*, Phys.Rev.Lett., **50**, 2062 (1983)
[34] E.M.Friedlander, Proc. LESIP I Conference Bad Honnef, 1984, p.105
[35] W.Kittel Proc.XVIII Int.Symp. on Multiparticle Dynamics, Tashkent (1987) p.3
[36] E.M.Friedlander *et al.*, Proc. Hadron Matter In Collision, Tucson (1988) p.210
[37] A.Bialas and R.Peschanski, Nucl.Phys. **B273**, 703 (1986)
[38] P.Carruthers *et al.*, Phys.Lett., **B222**, 487 (1989)
[39] A.Vourdas and R.M.Weiner , Phys.Rev. **D38**, 37 (1988)
[40] D.Brick *et al.*, Phys.Lett. **103B**, 242 (1981)
[41] E.M. Friedlander and R.M. Weiner, Phys.Rev. **D28**, 2903 (1983)
[42] G.N. Fowler et al., Phys.Rev. **D35**, 870 (1987)

# SCALING PROPERTIES of the $e^+e^-$ MULTIPLICITY DISTRIBUTIONS and the LOGNORMAL DISTRIBUTION

**Roman Szwed**
*Institute of Experimental Physics*
*Warsaw University*
*PL-00-681 Warsaw, ul. Hoża 69*

**Abstract**

The scaling properties of $e^+e^-$ hadronic multiplicity distributions are discussed. It is shown that the $e^+e^-$ multiplicity data obey KNO–G scaling with a scaling function of lognormal shape.

Scale invariant branching is postulated as a mechanism within which the scaling properties and the lognormal shape of the scaling function can be derived. The dependence of average multiplicity on energy derived within the postulated framework is found to be in agreement with the data.

Finally, the universal character of the lognormal distribution is illustrated by many examples of its occurrences in various branches of science.

# 1 Introduction

The aim of this paper is to review the results of a recent analysis of the scaling properties of multiplicity distributions observed in $e^+e^-$ collisions. The analysis has been made by the author of this paper in collaboration with G.Wrochna and A.K.Wróblewski [1, 2].

The most important observations of this study are the following:

- The $e^+e^-$ multiplicity distributions obey KNO–G scaling [3, 4].
- The scaling function is of lognormal shape [1, 2].
- The origin of the scaling and the lognormal shape of the scaling function were found to be well explained as a consequence of scale invariant branching [1, 2].
- The scaling properties and the lognormal shape of the scaling function have universal character and are observed in many different phenomena in nature.

## 2 Short reminder on KNO–G scaling

Before discussing the main topics of this paper let us recall the original formulation of KNO scaling [5]:

$$\langle n \rangle \, P_n = \psi(z)[1 + O\left(\frac{1}{\langle n \rangle}\right)] \ , \quad z = \frac{n}{\langle n \rangle} \ . \tag{2.1}$$

$$P_n = \frac{1}{\langle n \rangle} \psi \left(\frac{n}{\langle n \rangle}\right) \ , \tag{2.2}$$

where $\psi(z)$ is an energy–independent function, $\langle n \rangle$ is the average multiplicity, and $n$ is the actual multiplicity, $n = 0, 1, 2, 3, \ldots$ .

It should always be remembered that this formulation is valid only at asymptotic energies and that it is not self–consistent at finite energies (small $\langle n \rangle$) [6, 3, 4, 7].

Golokhvastov proposed to reformulate KNO scaling essentially by changing the old formulation of the definition of $P_n$ (Eq. 2.1) to a new one which makes KNO–G scaling self–consistent at any energy [3, 4]:

$$P_n = \int\limits_{n/\langle \tilde{n} \rangle}^{(n+1)/\langle \tilde{n} \rangle} \psi(z) dz \ , \quad \text{where} \quad z = \tilde{n}/\langle \tilde{n} \rangle \quad \text{and} \quad \langle \tilde{n} \rangle = \int\limits_0^\infty \psi(z) dz \ , \tag{2.3}$$

where $\psi(z)$ is an energy–independent function.

It is convenient to use as a scaling function the $\phi(z)$ function defined by the following formula

$$\phi(z) = \int\limits_0^z \psi(z) dz - 1 \ . \tag{2.4}$$

In this way the definition of $P_n$ (Eq. 2.3) has the simple non–integral form

$$P_n(\langle n \rangle) = \phi\left(\frac{n+1}{\langle \tilde{n} \rangle}\right) - \phi\left(\frac{n}{\langle \tilde{n} \rangle}\right) \ . \tag{2.5}$$

The usual way to test graphically for KNO scaling is to plot the multiplicity distributions presented in the KNO–variables $\langle n \rangle \, P_n$ against $n/\langle n \rangle$. However, this graphical test could be used only at asymptotic energies, at which the KNO formulation is mathematically self consistent. It does not have much sense at finite energies where it is usually performed [7].

The graphical check of the KNO–G scaling in the whole energy range follows from the relation

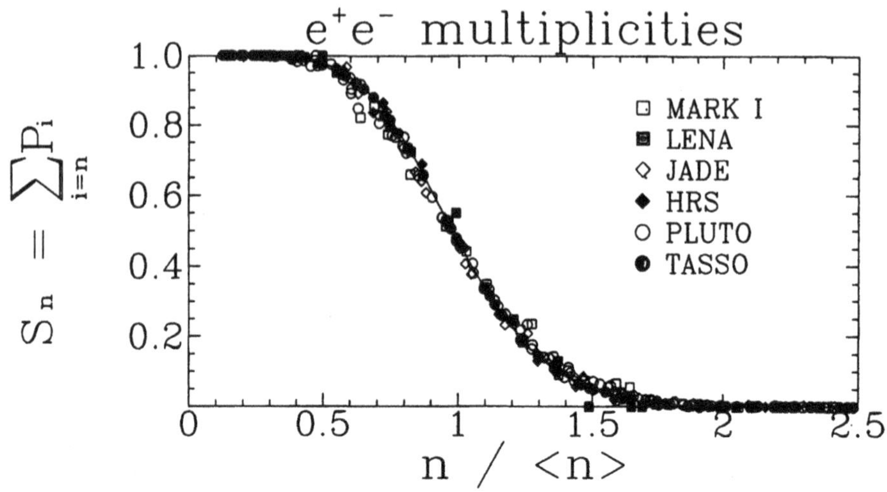


Figure 1: *Graphical test of KNO–G scaling. $S_n$ plotted as a function of reduced multiplicity $z = \tilde{n}/\langle\tilde{n}\rangle$, for the whole collection of $e^+e^-$ data.*

$$S_n = \sum_{i=n}^{\infty} P_i = \int_{n/\langle\tilde{n}\rangle}^{\infty} \psi(z)dz = -\phi(z) \quad \text{where} \quad z = \tilde{n}/\langle\tilde{n}\rangle \ . \tag{2.6}$$

Therefore all experimental points should lie on a single curve $\phi(z)$ when plotted as $S_n$ versus $z$. Such a plot gives also a straightforward suggestion for the analytical shape of the scaling function $\phi(z)$.

An extensive discussion of KNO–G scaling can be found in [3, 4, 7].

# 3 KNO–G scaling and the $e^+e^-$ data

In the recent publications we have shown that the $e^+e^-$ data obey KNO–G scaling and the scaling function has been found to be of the lognormal shape [1, 2].

Figure 1 presents a graphical test of KNO-G scaling for 25 experimental $e^+e^-$ multiplicity distributions (see Ref. [1] for references). We recall that the experimental data, plotted as $S_n$ versus $z$, should follow a single curve if the data obey KNO-G scaling [3, 4]. As seen from Fig. 1, the data obey KNO-G scaling very well.

An equivalent way to check the scaling properties of multiplicity distributions is by plotting the dependence of the $k$–order dispersions on the mean multiplicity [4, 1]. Such a plot is presented in Fig. 2. The dispersions are indeed linear functions of the mean multiplicity, which again confirms the fact that the data obey KNO–G scaling.

The scaling function has been found to be of lognormal shape [1]:

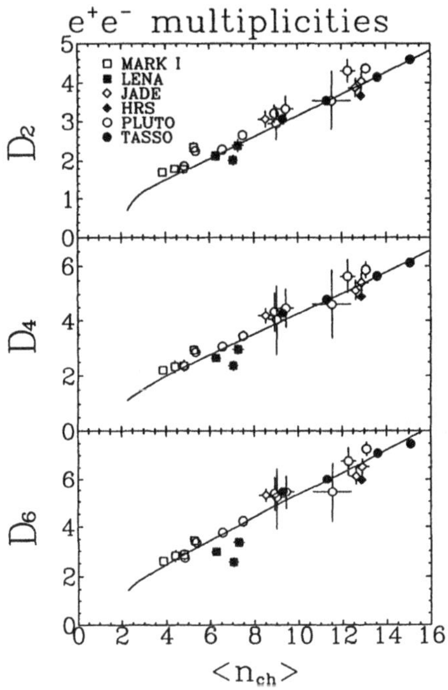


Figure 2: $D_2$, $D_4$ and $D_6$ *plotted as functions of mean multiplicity for a collection of* $e^+e^-$ *data.*

$$\psi(z) = \frac{N}{\sqrt{2\pi}\sigma} \cdot \frac{1}{z+c} \exp\left(-\frac{[\ln(z+c)-\mu]^2}{2\,\sigma^2}\right), \tag{3.1}$$

where $N$ is given by the first normalization condition [1]:

The solid curve in Fig. 1 corresponds to the lognormal distribution or more exactly to the primitive function of the lognormal distribution (function $\phi(z)$ Eq. 2.4). The following parameters have been obtained [1] as the result of a fit to the TASSO data [8]:

$$\mu = 0.581 \pm 0.001\ , c = 0.811 \pm 0.002\ , \sigma = 0.155 \pm 0.001\ , \chi^2/NDF = 41/68.$$

We believe that it is safer to fit the scaling function only to the recently published high quality TASSO data, than to all data, because the TASSO data are corrected in a uniform way and a consistent error analysis has been performed (for details the reader in directed to Ref. [1]).

The shape of the lognormal scaling function with the above parameters is presented in Fig. 3.

[1]We assume that $c > 0$. If $c \leq 0$ then $N = 1$.

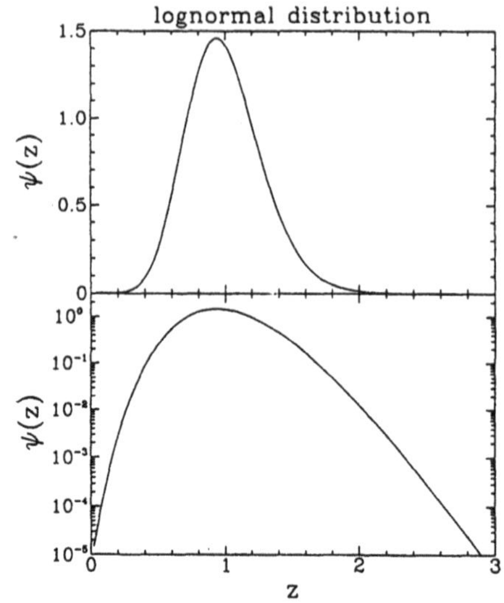


Figure 3: *Shape of the lognormal scaling function $\psi(z)$ (Eq. 3.1) for the $e^+e^-$ data.*

# 4 Genesis of the lognormal distribution

The lognormal distribution is one of the most fundamental in statistical mathematics. It appears as a result of the Central Limit Theorem in its multiplicative version (see e.g. [9]).

To get an intuitive understanding of its origin, one can consider the following reasoning. If a process is controlled by a random variable $x$, which can be changed by random impulses independent of each other, and the effects of random impulses are added, then the variable $x$ should be distributed normally due to the Central Limit Theorem

$$f(x) = \frac{1}{\sqrt{2\pi}\sigma} \exp\left\{-\frac{[x-\mu]^2}{2\,\sigma^2}\right\} . \tag{4.1}$$

Similarly, if the effects of random impulses are multiplied we obtain the lognormal distribution due to the Central Limit Theorem in its multiplicative version [9]:

$$f(x) = \frac{1}{\sqrt{2\pi}\sigma\, x} \exp\left\{-\frac{(\ln x-\mu)^2}{2\,\sigma^2}\right\} . \tag{4.2}$$

In our recent paper [2] we have assumed scale invariant branching as a mechanism through which all observed properties of the $e^+e^-$ multiplicity data, including the lognormal shape of the scaling function, could be derived.

The idea which we are intensely using in our approach comes from the fundamental papers of Polyakov [10].

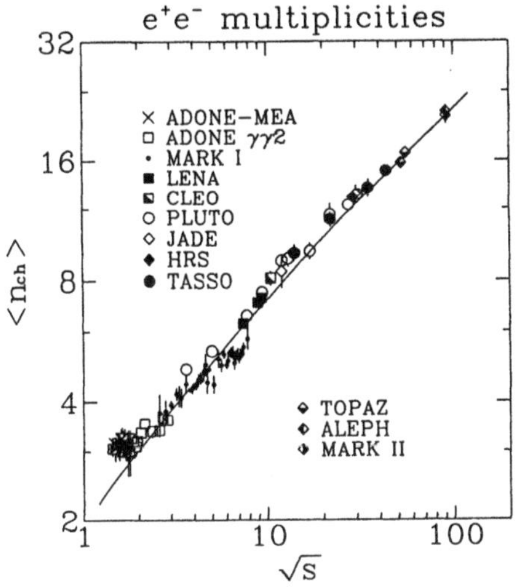


Figure 4: *The average multiplicity $\langle n_{ch} \rangle$ plotted as a function of centre of mass energy.*

The physical picture can be sketched in the following way. The process starts with a quark–antiquark pair coming from the reaction $e^+e^- \to q\bar{q}$. The kinetic energy of colliding electrons, $\sqrt{s}$, is converted into the potential energy of a string between this pair. The string can break forming a second pair of a quark and an antiquark. The sum of the potential energy of both pairs is smaller than the energy of the primary string because a fraction of the latter has been converted into the kinetic energy of the two pairs. Each of the strings can break again and the process is continued until the potential energy of the strings is of the order of the pion mass. The number of particles created in the $i$-th generation is proportional to the number of particles in the $(i-1)$-th generation. Thus the multiparticle production is a multiplicative process. We have shown in our paper [2] that in such a case the lognormal shape of the scaling function can be obtained analytically with the help of the Central Limit Theorem.

Within the postulated framework it was possible to evaluate the dependence of the average multiplicity on the energy [2]. The obtained formula is

$$\langle n_{ch} \rangle = \beta\, s^{\alpha} - 1 \ . \tag{4.3}$$

The validity of this formula is seen in Fig. 4, where the data obtained by the indicated collaboration teams are presented. The solid line presents the fit to the data with formula (Eq. 4.3). The fit gives $\alpha = 0.221$ and $\beta = 2.96$. As seen in Fig. 4 there are some systematic differences between data obtained in various experiments. Therefore, one can conclude that the derived dependence of average multiplicity on energy is confirmed by the data, but the values of the fitted parameters should be taken with caution.

## 5 Universal character of the lognormal distribution

It has been known for years that the lognormal distribution can be used to describe statistical properties of various natural phenomena [11]. There is a vast literature on the subject and references could be found, e.g., in [9, 11].

In the following we quote some examples – from various branches of science – of distributions which were found to be of lognormal shape [9, 11]:
- distribution of size of particles coming from grinding rocks (geology),
- distribution of incomes and family budgets (economy),
- distribution of size of organisms (biology),
- distribution of abundance of species, number of lesions in plants and animals (biology),
- and distribution of age of marriage (sociology).

We have presented the evidence for the similarity between the $e^+e^-$ multiplicity distributions and the distributions of the age of marriage in our recent paper [2]. It is fascinating to observe very similar scaling properties in the distributions depending on completely different "dynamics".

## 6 Conclusions

It has been found that the $e^+e^-$ multiplicity distributions obey KNO–G scaling with a scaling function of the lognormal shape.

It seems that a scale invariant branching process provides an explanation of the observed properties of the multiplicity distributions.

The fact that the statistical descriptions of various phenomena in nature exhibit similar properties as the $e^+e^-$ multiplicity distributions is attributed to the universality of the Central Limit Theorem.

## Acknowledgements

I would like to thank the organizers of the Marburg meeting and especially Prof. Richard Weiner for their generous hospitality and perfect organization of the workshop.

# References

[1] R.Szwed and G.Wrochna: Z. Phys. C – Particles and Fields 47 (1990) 449.

[2] R.Szwed, G.Wrochna and A.K.Wróblewski: "Genesis of the lognormal multiplicity distribution in the $e^+e^-$ collisions and other stochastic processes", Warsaw University Preprint IFD/1/1990, to be published in Modern Physics Letters A (1990).

[3] A.I.Golokhvastov: Sov. J. Nucl. Phys. 27 (1978) 430; 30 (1979) 128.

[4] R.Szwed and G.Wrochna: Z. Phys. C – Particles and Fields 29 (1985) 255.

[5] Z.Koba, H.B.Nielsen and P.Olesen: Nucl. Phys. B40 (1972) 317.

[6] A.Chodos, M.H.Rubin and R.L.Sugar: Phys. Rev. D8 (1973) 1620.

[7] R.Szwed: "Multiplicity distributions in the pp and nucleus–nucleus collisions", Warsaw University Preprint IFD/6/1989.

[8] W.Braunschweig et al.: Z. Phys. C – Particles and Fields 45 (1989) 1939.

[9] J.Aitchison and J.A.C.Brown: "The Lognormal Distribution", Cambridge University Press 1957.

[10] A.M.Polyakov: Sov. Phys. JETP 32 (1971) 296; 33 (1971) 850 (in Russian: Zh. Eksp. Teor. Fiz. 59 (1970) 542; 60 (1971) 296).

[11] J.H.Gaddum: Nature 156 (1945) 463.

# SECOND MOMENTS OF MULTIPLICITY DISTRIBUTIONS IN HIGH ENERGY NUCLEUS-NUCLEUS COLLISIONS

J. Bartke, E. Gładysz-Dziaduś and P. Stefański

Institute of Nuclear Physics
High Energy Physics Laboratory
ul.Kawiory 26A
30-055 Cracow, Poland

NA35 Collaboration CERN
Propane Bubble Chamber Collaboration JINR

ABSTRACT

Second moments of multiplicity distributions of negatively charged secondaries from C+C and S+S collisions at Dubna and CERN energies are analyzed and compared with Monte Carlo models. Entropy of the multiplicity distribution is obtained for central S+S collisions at 200AGeV.

## 1. INTRODUCTION

In the present paper we analyze multiplicity distributions of negatively charged secondaries coming from $^{12}C + ^{12}C$ collisions at 4 A GeV and $^{32}S + ^{32}S$ collisions at 200 A GeV. The first experiment was done at the Dubna synchrophasotron by the Alma-Ata - Baku - Belgrade - Bucharest - Budapest - Cracow - Dubna - Moscow - Prague - - Sofia - Tashkent - Tbilisi - Ulan Bator - Varna - Warsaw - Yerevan Collaboration using the propane bubble chamber of 1 m effective length, the second one was done at the CERN SPS by the NA35 Collaboration (Athens - Bari - Berkeley - CERN - Cracow - Darmstadt - Frankfurt - Freiburg - - Marburg - Munich - Warsaw - Zagreb) using the 2 m streamer chamber. Multiplicity of secondary particles has been determined by visual scanning of the film, momenta and angles of charged secondaries have been partly measured. The choice of colliding nuclei of the same mass enables us to study particles in a restricted rapidity region in which the detection efficiency is the best and then extrapolate to full phase space. Various corrections have been applied to the multiplicity data, their full description will be given in a separate paper [1]. The experiment at 4 A GeV was described in earlier publications [2].

At 4 A GeV the "inclusive" sample contains about 90% of all inelastic C+C interactions with the most peripheral collisions excluded. The "central" sample has been selected allowing for at most one spectator proton from the projectile ($Q_s \leqslant 1$) , it corresponds to about 20% of $\sigma_{inel}$.

At 200 A GeV the "minimum bias" sample contains some 90% of all inelastic S+S interactions, with the most peripheral collisions excluded by the trigger. The "hard veto" trigger selects "central" events by requiring at most five non-interacting nucleons from the projectile ($E_{veto} \leqslant 1.1$ TeV what corresponds to about 2% of $\sigma_{inel}$.

## 2. SECOND MOMENTS OF MULTIPLICITY DISTRIBUTIONS IN CENTRAL RAPIDITY INTERVALS

In Figs 1 a,b the second-order correlation integral for negatively charged secondaries $f_2^- = D_-^2 - \langle n_- \rangle$ is plotted as a function of the half-width of the rapidity interval for intervals centered about the centre-of-mass rapidity $y_c$ ($y_c = 1.1$ at 4 GeV and 3.0 at 200 GeV), for the four event samples studied. One can see that in the narrow central rapidity interval $f_2^- = 0$ for all samples, i.e. the multiplicity distribution has Poissonian shape. This could be related to the formation of a thermalized fireball at midrapidity as found from pion interferometry [3]. The half-width of the least interval plotted is 0.4 at 4 A GeV and 0.1 at 200 A GeV, but at 200 A GeV we find $f_2^- = 0$ also for much more narrow intervals, down to $\Delta y = 0.02$ . The data are compared to Monte-Carlo events generated according to the Dubna Cascade Model DCM [4] at 4 A GeV and the FRITIOF model [5] version 1.6 with the JETSET version 6.2 at 200 A GeV.

For non-impact-parameter-selected events ("inclusive"

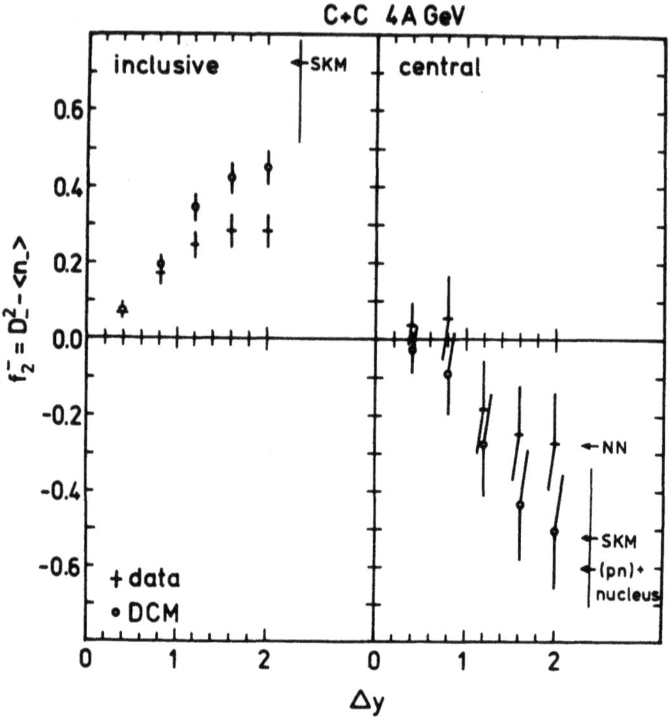


Fig.1a Correlation integral $f_2^-$ versus $\Delta y$ at 4 A GeV

or "minimum bias" samples) $f_2^-$ increases with increasing width of the rapidity interval. This is in qualitative (but not quantitative!) agreement with model predictions: DCM overestimates, while FRITIOF underestimates the $f_2^-$. For events representing central collisions the situation is different: at 4 A GeV with increasing width of the rapidity interval $f_2^-$ becomes negative and increases in absolute value, this behaviour being only approximately described by DCM. At 200 A GeV $f_2^-$ increases with increasing $\Delta y$, while the values from FRITIOF fluctuate about zero, in strong disagreement with the data. This drastic change in behaviour of $f_2^-$ between 4 and 200 A GeV can be understood by invoking the characteristics of nucleon-nucleon collisions at the same energy per nucleon. At 4 GeV the multiplicity dis-

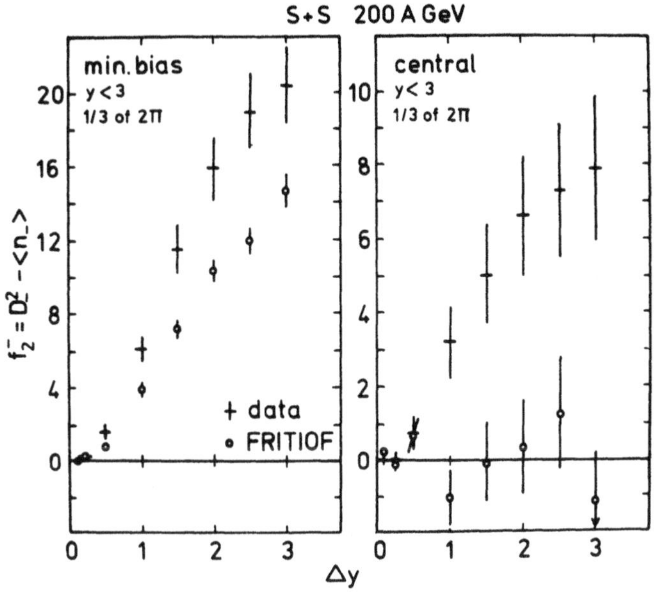


Fig.1b Correlation integral $f_2^-$ versus $\Delta y$ at 200 A GeV

tribution in N-N collisions is narrower than Poissonian, i.e. $f_2^- < 0$ [6], while at 200 GeV it is wider than Poissonian, $f_2^- > 0$ . Thus, multiplicity distributions in central collisions reflect some features of elementary N-N collisions, but it seems difficult to describe them quantitatively by simple superposition models.

In Figs 2 a,b we have plotted the value of the expression $C_2^- \langle n_- \rangle - 1$ where $C_2$ is the second normalized moment $C_2 = \langle n^2 \rangle / \langle n \rangle^2$ , as a function of the mean multiplicity in the rapidity intervals of increasing width, again centered about $y = y_c$ . According to Hwa [7] a linear dependence $C_2 \langle n \rangle - 1 = A \langle n \rangle$ should be observed. We find for central collisions the slope A compatible with 1.0 , for inclusive or minimum bias collisions slope greater than one and increasing with projectile energy. If we attribute this difference to geometrical correlations which should be important for non-impact-parameter-selected col-

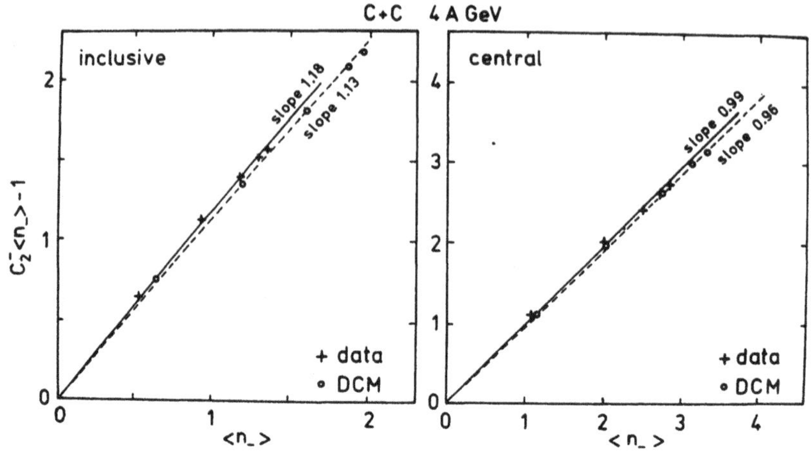


Fig.2a Normalized moment $C_2^-\langle n_-\rangle-1$ vs $\langle n_-\rangle$ at 4 A GeV

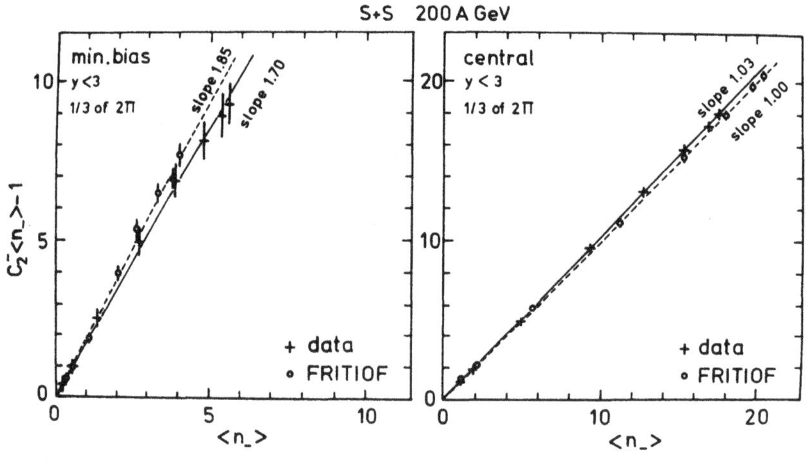


Fig.2b Normalized moment $C_2^-\langle n_-\rangle-1$ vs $\langle n_-\rangle$ at 200 A GeV

lisions, then in view of [7] it seems to be no room for dynamical correlations.

A similar study has been made in ref. [8] for nearly--central $^{16}O$ and $^{32}S$ interactions in nuclear emulsion.

## 3. ENTROPY

Šimak, Šumbera and Zborovský [9] called attention to the the entropy S of the multiplicity distribution, defined as

$$S = - \sum P(n) \ln P(n) \qquad \text{with } \sum P(n) = 1 ,$$

and demonstrated its remarkable scaling properties. For various high energy hadron-hadron collisions the value of entropy $S_{max}$ calculated for the full available phase space increases linearly with maximum possible rapidity $y_{max} = \ln(\sqrt{s}/m_{\pi})$ . Also, entropy calculated in rapidity windows centered at the c.m. rapidity $y_c$ seems to be a universal function of the window width:

$$S/y_{max} = A[1 - \exp(-By/y_{max})]$$

with A= 0.41 ± 0.01 and B= 8.6 ± 1.0 .

Ghosh et al.[10] calculated entropy for $^{12}C$ and $^{24}Mg$ nuclei interacting with heavy nuclei Ag/Br of photographic emulsion, at 4.5 A GeV/c incident momentum. They used pseudorapidity windows centered about the central value $\eta_c$ . Data for both types of projectile nuclei show the same dependence on $\eta/\eta_c$ and can be described by a similar function with A= 0.50 ± 0.05 and B= 6.8 ± 0.2 (we scaled the coefficient B by a factor of two as $\eta_c \cong y_c \cong y_{max}/2$).

We have calculated entropy for our sample of central S+S collisions at 200 A GeV in pseudorapidity windows centered about $\eta_c$=3.0 , for both all charged and negatively charged secondaries. Results for $n_{ch}$ are shown in Fig.3, together with predictions of the FRITIOF model. One can see that the shape of $S/\eta_{max}$ as function of $\eta/\eta_{max}$ is similar to that found in earlier studies [9, 10]. For the central pseudorapidity window ±1.5 units wide, corresponding to $\eta/\eta_{max} \cong 0.5$ , S reaches already about 90% of its limiting value, which means that the main contribution to S comes from the central region, fragmentation regions contributing very little. Similarity of the observed features of S to those found for elementary collisions and for the lightest nuclei, a compatible limiting value of

$S/\eta_{max}$ (0.45 ± 0.02 for all charged, 0.37 ± 0.02 for negatively charged secondaries), and a reasonable description of the data by the FRITIOF model are worth to be mentioned.

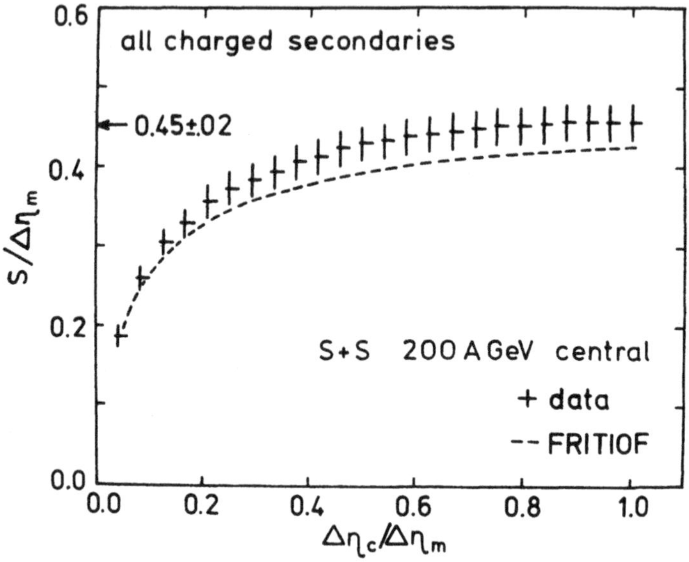


Fig.3 Entropy versus $\Delta\eta$ for S+S central at 200 A GeV

REFERENCES

1. J.Bartke et al., NA35 Collab., to be published.
2. E.O.Abdrahmanov et al., Z.Phys.C5, 1 (1980) ; H.N.Agakishiev et al., Z.Phys.C16, 307 (1983) ; C27, 177 (1985).
3. A.Bamberger et al., Phys.Lett. B203, 320 (1988).
4. V.D.Toneev and K.K.Gudima, Nucl,Phys.A400, 73c (1983).
5. B.Andersson et al., Nucl.Phys. B281, 289 (1987).
6. A.I.Golokhvastov, Ph.D.Thesis, Dubna 1989.
7. R.C.Hwa, Phys.Rev. D37, 2451 (1988).
8. K.Sengupta, G.Singh and P.L.Jain, Phys.Lett. B213, 548 (1988).
9. V.Šimak, M.Šumbera and I.Zborovský, Phys.Lett. B206, 159 (1988).
10. D.Ghosh et al., Proc.21st Int.Cosmic Ray Conference, Adelaide, 6-19 January 1990, paper HE 1.3-6.

# CHARGED PARTICLE MULTIPLICITY DISTRIBUTIONS AND FLUCTUATIONS IN IN 60 AND 200 *A* GEV $^{16}$O + NUCLEUS REACTIONS

F. Berger[4], R. Albrecht[1], T. C. Awes[5], M. Bahr[4], P. Beckmann[4,a], R. Bock[1], G. Claesson[3], G. Clewing[4], L. Dragon[4], A. Eklund[3], R. L. Ferguson[5], A. Franz[5,b], S. Garpman[3], R. Glasow[4], H. Å. Gustafsson[3], H. H. Gutbrod[1], G. Hölker[4], J. Idh[3], P. Jacobs[2], K. H. Kampert[4], B. W. Kolb[1], P. Kristiansson[3], H. Löhner[4,a], I. Lund[1], F. E. Obenshain[5,b], A. Oskarsson[3], I. Otterlund[3],T. Peitzmann[4], S. Persson[3], F. Plasil[5], A. M. Poskanzer[2], M. Purschke[4], H. G. Ritter[2], S. Saini[5], R. Santo[4], H. R. Schmidt[1], R. Schmidt[4], S. P. Sørensen[5,b], K. Steffens[4], E. Stenlund[3], D. Stüken[4], M. L. Tincknell[5] and G. R. Young[5]

**WA80 Collaboration**

1. *Gesellschaft für Schwerionenforschung, D-6100 Darmstadt, Fed. Rep. of Germany*
2. *Lawrence Berkeley Laboratory, Berkeley, California 94720, USA*
3. *University of Lund, S-22362 Lund, Sweden*
4. *University of Münster, D-4400 Münster, Fed. Rep. of Germany*
5. *Oak Ridge National Laboratory, Oak Ridge, Tennessee 37831, USA*

a. *now at: KVI, University of Groningen, NL-9747 AA Groningen, Netherlands*
b. *University of Tennessee, Knoxville, Tennessee 37996, USA*

**Abstract**

Multiplicities of charged particles in ultrarelativistic heavy ion reactions have been measured in the WA80 experiment with streamer tube detectors in the pseudorapidity range $1.2 \leq \eta \leq 4.2$. The dependence of the mean value, and the dispersion as a measure of the fluctuations of the distributions on the target mass, beam energy, pseudorapidity window, and the centrality will be discussed. A geometrical explanation of the multiplicity distributions and a statistical extrapolation from larger to smaller parts of phase space which are both able to reproduce the predictions of the Monte-Carlo version of the string model FRITIOF will be compared to the experimental results. The origin for the slight deviations between the experimental data and the model of independent nucleon-nucleon collisions might be caused by target fragments which are not taken into account by the applied event generators.

# 1 Introduction

First results from ultrarelativistic heavy ion experiments at the CERN SPS [1,2] have shown that in 200 *A* GeV $^{16}$O + Au collisions high energy densities are created which approach the critical values where hadronic matter could be melted to form a quark-gluon-plasma (QGP). This phase transition could lead to *unusually* large multiplicity fluctuations [3] where unusual means significant deviations of the experimental data from the models of independent nucleon-nucleon collisions like, for example, FRITIOF [4].

# 2 Experimental Setup

The complete WA80 experimental setup and minimum bias trigger conditions are described in Ref. [5]. The streamer tube detectors [6] have been used for the measurement of charged particles in the pseudorapidity range $1.2 \leq \eta \leq 4.2$ with a resolution in $\eta$ of better than 0.1. These electronic detectors feature a high data rate for accumulating very good statistics and low energy thresholds for pions (14 MeV) and protons (25 MeV) which means that not only the fast secondary produced particles are registered but also slow target fragments. First results with applied data corrections for reactions of $^{16}$O-projectiles with nuclear targets at 60 and 200 $A$ GeV are presented in [7]. The forward energy, measured by the zero-degree calorimeter (ZDC), allows to determine the centrality of the collisions.

# 3 Multiplicity Distributions

Before any quantitative discussion of the multiplicity fluctuations of heavy ion reactions can begin, a qualitative understanding of the shape of the multiplicity distributions $P(n)$ is important. The broad plateau and therefore large dispersion of the distribution for collisions averaged over different impact parameters is mainly explained by the clean cut geometry of the reaction zone between target and projectile. The peripheral (central) events with few (many) sources of participants contribute to the lower (higher) multiplicities. This interpretation is supported by many experimental data [1,2] and by the comparison to models which assume the independent superposition of nucleon-nucleon collisions. The same conclusion has been drawn for the transverse energy $E_T$ which is closely related to the multiplicity if the transverse energy per particle is nearly constant. The dominant role of the geometry and the assumption that only the particle production is dependent on the beam energy should in turn lead to distributions which, after proper scaling with the mean multiplicity, have the same shape [8]. This is shown in Fig. 1 which confirms with much better statistics the results of streamer chamber and emulsion experiments [9,10]. Such a scaling in energy is known from hadron-hadron reactions as KNO-scaling [11] although the underlying mechanism is probably different.

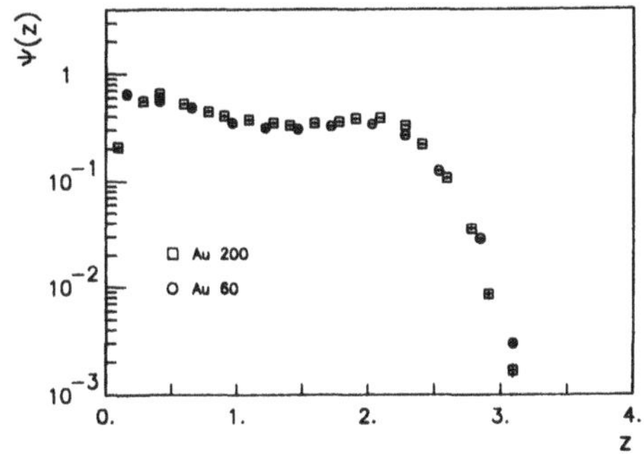


Figure 1: Minimum bias multiplicity distribution of charged particles in KNO variables $\Psi(z) = <n> P(n)$ $(z = n/<n>)$ of WA80 data for $^{16}O$ reactions at 60 (○) and 200 (□) $A$ GeV in the pseudorapidity range $1.2 \leq \eta \leq 4.2$.

# 4 Multiplicity Fluctuations

As a measure of the fluctuations, the dispersion $D = \sqrt{<n^2> - <n>^2}$ will be taken which will be always related to the mean $<n> = \sum_{n=1}^{N} nP(n)$ for $N$ events. In Fig. 1 the dispersion $D$ is plotted versus $<n>$ for $^{16}O$ collisions with various targets and energies recorded under minumum-bias conditions in the pseudorapidity range $1.2 \leq \eta \leq 4.2$. All values lie on a line, as observed in [9], showing no obvious deviation for the 200 $A$ GeV $^{16}O + Au$ system where the probability for a phase transition should be largest. This behaviour can be also connected to the geometrical picture [9,12].

To overcome the problem of separating the dynamical fluctuations from the geometrical part it was suggested [3] to study the dependence of normalized moment, for example $\gamma_2 = D^2/<n>^2$, on the size of the pseudorapidity window. A steep rise of $\gamma_2$ for windows getting smaller and smaller would be a indication for large non-geometrical fluctuations. This has been done for 200 $A$ GeV $^{16}O + Au$ minimum bias reactions and, in order to reduce the impact parameter variations, also for central data samples (relative beam energy $E_{ZDC}/E_{beam} \leq 20\%$). The center for the symmetric pseudorapidity windows $\Delta\eta$ has been chosen from the position of the maximum of the pseudorapidity distribution. In Fig. 3 the results are shown for the experimental data together with FRITIOF simulations. The minimum bias data indeed show a slow rise at smaller window values which is strongly enhanced for the central reactions. But, the same behaviour is reproduced by the string model. Also the choice of the center for the pseudorapidity window does not change this correspondence [13] of model and data.

A possible explanation for this effect is the purely statistical assumption that

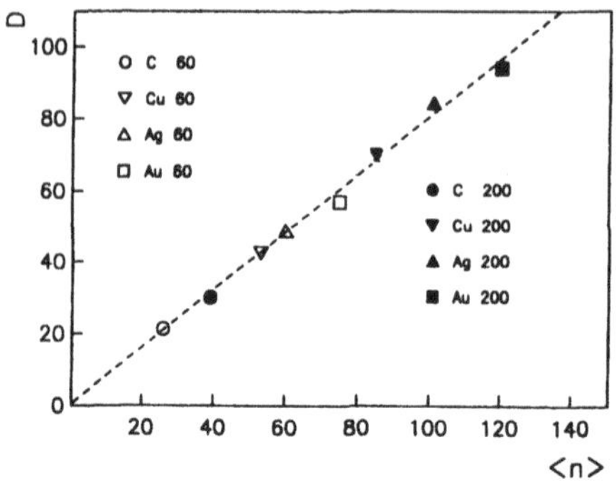


Figure 2: Dispersion D versus the mean $< n >$ extracted from minimum bias multiplicity distributions for $^{16}$O collisions with different targets (C (∘), Cu (▽), Ag (▽) and Au (△)) and energies (open symbols for 60 $A$ GeV and filled symbols for 200 $A$ GeV). The dashed line is to guide the eye.

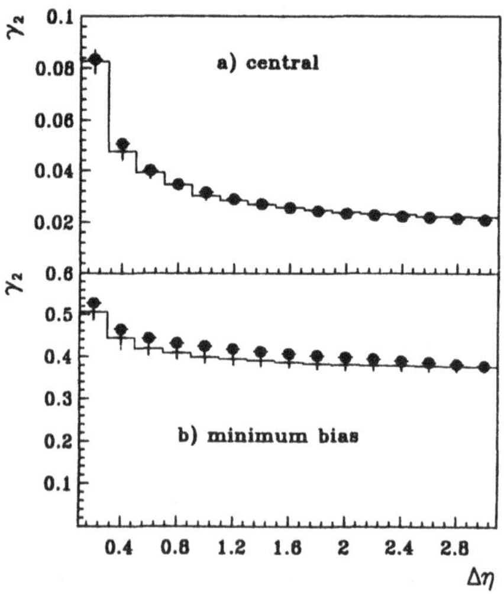


Figure 3: $\gamma_2$ as function of the pseudorapidity window $\Delta\eta$ for $^{16}$O+Au reactions determined from WA80 data (∘) and FRITIOF simulations(histogram). The upper graph (a) shows central ($E_{ZDC}/E_{beam} \leq 20\%$) and the lower graph shows (b) minimum bias data.

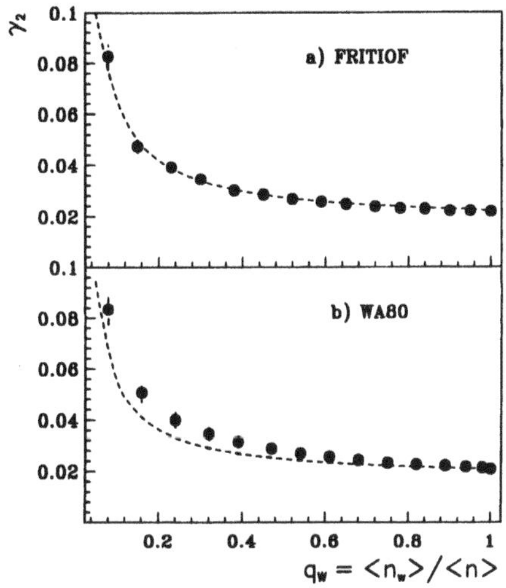


Figure 4: $\gamma_2$ as function of $q_W$ for central ($E_{ZDC}/E_{beam} \leq 20\%$) $^{16}O + Au$ reactions (•) and a binomial extrapolation (dashed line). The upper graph (a) shows results from the FRITIOF calculations and the lower graph (b) shows WA80 data. For the extrapolations in each case the full distributions $P(n)$ were approximated by the multiplicity distributions in $1.2 \leq \eta \leq 4.2$ for FRITIOF and WA80, respectively.

if the multiplicity distribution $P(n)$ with mean $< n >$ is known in the full phase space, the corresponding distributions $P_W(n_W)$ with $< n_W >$ in certain windows can be calculated by :

$$P_W(n_W) = \sum_n P(n) \binom{n}{n_W} q_W^{n_W} (1 - q_W^{n - n_W}). \tag{1}$$

and are only dependent on the probability $q_W = < n_W > / < n >$ to observe particles in the subspaces. Such extrapolation are, for example, used in the multi-source model [14,15]. Fig. 4 shows for the central reactions perfect agreement with FRITIOF and to a large extent with the WA80 data. The small deviations from the experimental sample might be due to target fragments which are present at low pseudorapidity values [16], and are therefore included to determine $P(n)$. By going to smaller windows the target fragments are more and more exclude [17] until only the secondary particles are left. This might limit the validity of the simple statistical extrapolation in the case of the experimental data whereas in FRITIOF the projectile and target spectators are always excluded.

# 5 Summary

The analysis of ultrarelativistic heavy ion collisions has shown that a geometrical explanation of the multiplicity distributions and a statistical extrapolation from larger to smaller parts of phase space are both able to reproduce the predictions of the Monte-Carlo versions of the string models and a large fraction of the experimental findings. The origin of the slight deviations between the experimental data and the models of independent nucleon-nucleon collisions might be caused by the target fragments, which are not taken into account by the event generators used or by other phenomena, for example intermittency. The new version of the string model VENUS [18] which implements the effect of rescattering in the target will be compared with the data in the near future.

# References


[1] H. Satz, H. Specht, and R. Stock, editors, *Proceedings of the sixth International Conference on Ultra-Relativistic Nucleus-Nucleus Collisions*, Z. Phys. C – Particles and Fields 38, 1988.

[2] G. Baym, P. Braun-Munzinger, and S. Nagamiya, editors, *Proceedings of the seventh International Conference on Ultra-Relativistic Nucleus-Nucleus Collisions*, Nucl. Phys. A 498, 1989.

[3] R. Hwa, Enhanced multiplicity fluctuations as a possible signature of quark matter, In Satz et al. [1], 277–281.

[4] B. Nilsson-Almqvist and E. Stenlund, Comput. Phys. Commun. **43** (1987) 387.

[5] R. Albrecht et al., Study of Relativistic Nucleus-Nucleus Collisions at the CERN SPS, Aug. 1985, preprint:CERN-EP/85-39.

[6] R. Albrecht et al., Nucl. Instr. and Meth. **A276** (1989) 131–139.

[7] R. Albrecht et al., Phys.Lett. **B202** (1988) 596–602.

[8] C. Wei-qin and L. Bo, Z. Phys. C – Particles and Fields **42** (1989) 337–340.

[9] A. Bamberger et al., Phys.Lett. **B205** (1988) 583–589.

[10] M. Adamovich et al., Phys.Lett. **B223** (1989) 262–266.

[11] Z. Koba, H. Nielsen, and P. Olesen, Nucl. Phys. **B40** (1972) 317–334.

[12] F. Berger, *Multiplizitäts- und Pseudorapiditätsverteilungen geladener Teilchen in ultrarelativistischen Schwerionenreaktionen*, Dissertation, Institut für Kernphysik, Münster, 1990.

[13] R. Albrecht et al., Z. Phys. C – Particles and Fields **45** (1989) 31–37.

[14] C. Xu, C. Wei-qin, and M. Ta-chung, Phys.Rev. **D36** (1987) 2009–2018.

[15] M. Ta-chung, Fireball models and statistical methods in multiparticle production processes, 1987, Preprint FUB-HEP/87-13.

[16] R. Albrecht et al., Z. Phys. C – Particles and Fields **45** (1990) 529–537.

[17] T. Csörgő et al., Fragmentation of target spectators in ultrarelativistic heavy ion collisions, 1989, KFKI-1989-34/A.

[18] K. Werner and P. Koch, Phys.Lett. **B242** (1990) 251–256.

# SUM RULES FOR CLUSTER PARAMETERS*

Pan Ji–cai
*Institut für Theoretische Physik*
*Freie Universität Berlin*
*Berlin, Germany*

ABSTRACT

A set of sum rules for hadronic cluster parameters is derived with statistical methods. With the help of these sum rules the cluster properties can be determined successively. It is found that the clusters produced in $e^+e^-$ annihilation decay into less charged hadrons than the clusters produced in hadron–hadron collisions do.

Since the early 1970s there has been persistent interest in studying hadronic clusters in high energy collisions[1]. It is assumed that the final state hadrons are the decay products of such clusters. This assumption is supported by the observation of short range correlations[1] and the rapidity gap distributions[2].

In recent years many collaborations[3] have studied the multiplicity distributions in various rapidity windows. Chao, Meng and myself[4] have studied these data with statistical methods and found that also these data indicate strongly the existence of clusters. Some authors[5] asserted recently that the power law behavior of factorial moments in small rapidity regions can be accounted for in terms of short–range correlations. This may imply further that the study of intermittency first proposed by Bialas and Peschanski[6] reveals the existence of hadronic clusters.

Although the hadronic clusters have been discussed extensively in the past, not very much about their nature are known. This is because that most cluster parameters, (e.g. size, charge, distribution in rapidity space, etc.), are not directly associated with measurable quantities. It would, therefore, be helpful to establish more direct links between these cluster parameters and the experimentally measurable quantities.

Based on the early works[4] Chao, Gao, Meng and myself[7; 8] derived a set of sum rules for such cluster parameters with which the characteristic properties of hadronic clusters can be studied in a systematical manner.

To begin with, let us consider a ideal case. Suppose $N$ clusters are produced independently in a high energy collision in the full rapidity space and the probability for any one of the $N$ clusters to be inside a rapidity window $W$ is $q_w$. Let $P'(N)$

* Supported in part by Deutsche Forschungsgemeinschaft (DFG: ME 470/5-1 and 5-2)

be the distribution of clusters in the full rapidity space, the distribution of clusters $P'_w(N_w)$ in the rapidity window $W$ is then given by a sum of weighted binomial distributions. The weight is nothing else, but the probability distribution of clusters $P'(N)$ in the full rapidity space. That is

$$P'_w(N_w) = \sum_N P'(N) \binom{N}{N_w} q_w^{N-w} (1 - q_w)^{N-N_w} \quad . \tag{1}$$

If every cluster decays into two hadrons locally, we have

$$n_w = 2 \cdot N_w \quad . \tag{2}$$

The multiplicity distribution in the rapidity window $W$ is then given by

$$P_w(n_w/2 = N_w) = \begin{cases} P'_w(N_w), & \text{if } n_w = 2 \cdot N_w; \\ 0, & \text{otherwise.} \end{cases} \tag{3}$$

We note that the $P_w(n_w/2)$ is measurable, $P'(N)$ and $q_w$ are the cluster parameters. Through this equation the measurable quantities and cluster parameters are connected.

We consider now a more general situation. We assume that every cluster may decay into 0, 1, 2, $\cdots c$ charged hadrons. Let $N_w(l|N)$ be the number of clusters, each of which contributes $l$ charged hadrons to the rapidity window $W$ in an N–cluster event. Then, the number of charged hadrons $n_w$ in $W$ is given by[7]

$$n_w = \sum_{l=0}^{c} l N_w(l|N) \quad , \tag{4}$$

while

$$N = \sum_{l=0}^{c} N_w(l|N) \tag{5}$$

is the number of cluster in that event. We note that $n_w$ is a function of the rapidity window, while $N$ is independent of $W$.

Let $P(N)$ be the probability of the emitting system to produce $N$ clusters, and $\alpha_w(l)$ be the average probability for any one of the $N_w(l|N)$ clusters to contribute $l$ charged hadrons in the window $W$, then the probability for observing $n_w$ charged hadrons in the rapidity window $W$ is given by the sum of weighted multinomial distributions. The weight is determined by the $P(N)$:

$$P_w(n_w) = \sum_N{}' \sum P(N) \left[ N! / \prod_{l=0}^{c} N_w(l|N)! \right] \times \prod_{l=0}^{c} \alpha_w(l)^{N_w(l|N)} \tag{6}$$

where the sum is first taken over $N_w(l|N)$. The prime on the summation sign indicates that the conditions given in Eqs. (4) and (5) should be satisfied.

Eq. (6) connects the characteristic properties of the hadronic clusters and the experimentally direct measurable quantities. It is the basis of sum rules for cluster parameters, with the help of which $\alpha_w(l)$ and $P(N)$ can be estimated from the experimental data for $P_w(n_w)$.

It should be noticed that the formula for $P_w(n_w)$ can be readily generalized to include cases in which more than one independent emitting systems contribute to the given window.

It should be also noticed that the average probability $\alpha_w(l)$ for having a cluster which contributes exactly $l$ charged hadrons in a given rapidity window $W$ is a useful quantity. On the one hand, it is closely related with the intrinsic properties of the clusters and, on the other hand, it reflects the relative importance of the different kinds of clusters contributing to the given rapidity window. For example, extraordinary large clusters or clusters with extremely wide rapidity distributions may or may not exist in the high energy collisions, but the role such clusters play statistically is not important one. Hence, these clusters can be neglected in the low order approximation.

From Eq. (6) following relations can be readily shown:

$$\langle n_w \rangle = \langle N \rangle \sum_l^c l\alpha_w(l) \quad , \tag{7}$$

$$\langle n_w(n_w-1)\rangle = \langle N(N-1)\rangle \left[\sum_l^c l\alpha_w(l)\right]^2 + \langle N\rangle \sum_l^c l(l-1)\alpha_w(l) \quad , \tag{8}$$

$$\langle n_w(n_w-1)(n_w-2)\rangle = \langle N(N-1)(N-2)\rangle \left[\sum_l^c l\alpha_w(l)\right]^3 + 3\langle N(N-1)\rangle$$
$$\times \left[\sum_l^c l\alpha_w(l)\right]\left[\sum_l^c l(l-1)\alpha_w(l)\right] + \langle N\rangle \sum_l^c l(l-1)(l-2)\alpha_w(l). \tag{9}$$

These as well as similar equations which include higher moments of the $n_w$ and those of the $N$ distributions are the ***sum rules for cluster parameters***. These sum rules can be used successively. In practice, the $P_w(n_w)$ data are always taken for a series of rapidity windows $W$, including that for the entire rapidity space. One simple and convenient way to carry out the successive approximation is to start with an ansatz for $\alpha_{wmax}(l)$ in the largest rapidity window, and see whether the corresponding solution for $\alpha_w(l)$ and $P(N)$ is consistent with the experimental distribution $P_w(n_w)$ for all other rapidity windows.

We[7; 8] used the sum rules to study the cluster parameters successively. As an example let us consider the $e^+e^-$ annihilation data of Derrick et al.[9]. We assume that $\alpha^{(0)}_{wmax}(2) = 1.$ for the full rapidity space, as the zero-order approximation. In this case we have:

$$\langle n_{wmax}\rangle = 2\langle N\rangle^{(0)} \qquad , \langle n^2_{wmax}\rangle = 4\langle N^2\rangle^{(0)} \tag{10}$$

and

$$P_{wmax}(n_{wmax}) = P(N = n_{wmax}/2). \tag{11}$$

We can now calculated $q^{(0)}_w(l)$ for every rapidity window $W$ by inserting the Eq. (10) into the first two sume rules (Eqs. 7 and 8). The results are shown in Fig. 1. It can be shown that[7] that a low boundary for the ratio $\alpha_w(1)/\alpha_w(2)$ is given by

$$\frac{P_w(1)}{P_w(2)} < \frac{\alpha_w(1)}{\alpha_w(2)} \quad . \tag{12}$$

Here $P_w(1)$ and $P_w(2)$ are the probability of observing one and two particles in the window $W$ respectively. The low limit of $\alpha^{(0)}_w(1)$ is shown in hatched area in Fig. 1. The boundaries are determined by the errors of $P_w(1)$ and $P_w(2)$. As we can see some of the calculated $\alpha^{(0)}_w(1)$ are lower than the low boundary determined by Eq. (12).

In Fig. 2 we show the results of first order approximation which we obtained from the ansatz $\alpha^{(1)}_{wmax}(2) = 0.75$ and $\alpha^{(1)}_{wmax}(1) = 0.25$. This means that 25% of the produced "hadronic clusters" are nothing else, but the charged hadrons themselves. Compare the Figs. 1 and 2 we see that the later is in better agreement with the low boundary. The quantity of this approximation can be further tested by inserting the experimental data of moments $\langle n_w\rangle \cdots \langle n^3_w\rangle$ into the left-hand side, $\alpha^{(1)}_w(1)$, $\alpha^{(1)}_w(2)$ and the $\langle N\rangle \cdots \langle N^3\rangle$ into the right-hand side of the third sum rule (Eq. 9). The difference between the left-hand and the right-hand side is shown in Fig. 3. If the error bars shown in Fig. 3 were so small such that we could conclude that this approximation is definitely insufficient, we would proceed one step further. That is, we would try to determine $\alpha_{wmax}(3)$ and so on.

We[8] also used the successive approximation to determine $\beta(1)$ and $\beta(2)$ for negative particles from the measured multiplicity distributions for negatively charged hadrons inside different rapidity windows. (Here $\beta(l)$ is the average probability for cluster, which contributes $l$ negatively charged hadrons, to exist). The only available data of this kind are those given by the NA22 collaboration[10] for $\pi^+p$ and $pp$ collisions at $p_{lab}$=200 GeV/c. The corresponding results we obtained are shown in Figs. 4 and 5. From the figures we see that there is a considerable large percentage of neutral clusters which decay into *more than two* charged hadrons in the $\pi^+p$ and $pp$ collisions. This result differs very much from that in $e^+e^-$ annihilation where the clusters decay dominantly into *only two* oppositely charged hadrons.

In summary, we have derived a set of sum rules for cluster properties. With the help of such sum rules cluster properties can be determined successively. We found

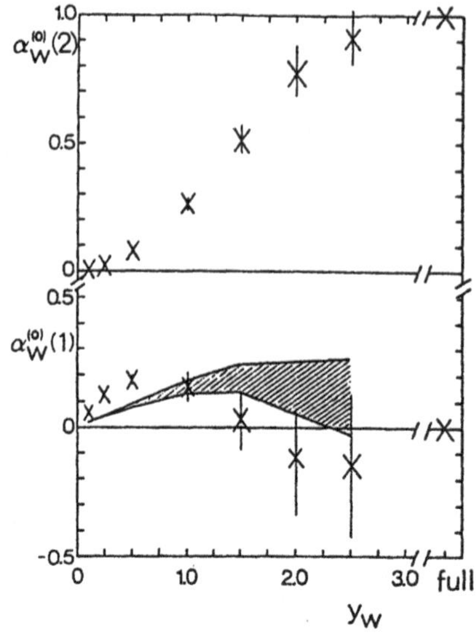


Fig. 1. The calculated probabilities $\alpha_w^{(0)}(1)$ and $\alpha_w^{(0)}(2)$ for charged hadrons in $e^+e^-$ annihilation as functions of the size of rapidity window $y_w$ ($|y| < y_w$). The values and the corresponding error bars are calculated from the data of [9]. The value for the full rapidity space is used as input. The low limit for $\alpha_w^{(0)}(1)$ is obtained from the inequality (12) taken together with the above-mentioned data. The hatched area shows the possible values for the low limit. The boudaries of the hatched area are determined by the errors of data for $P_w(1)$ and $P_w(2)$. The figure is reproduced from [7].

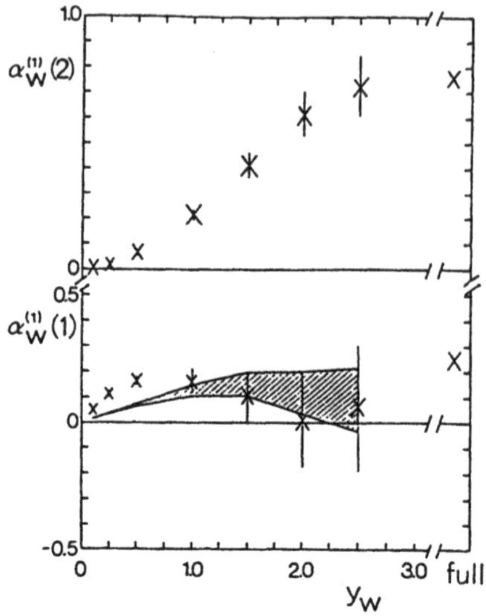


Fig. 2. The same as in Fig. 1., but for $\alpha_w^{(1)}(1)$ and $\alpha_w^{(1)}(2)$. The figure is reproduced from [7].

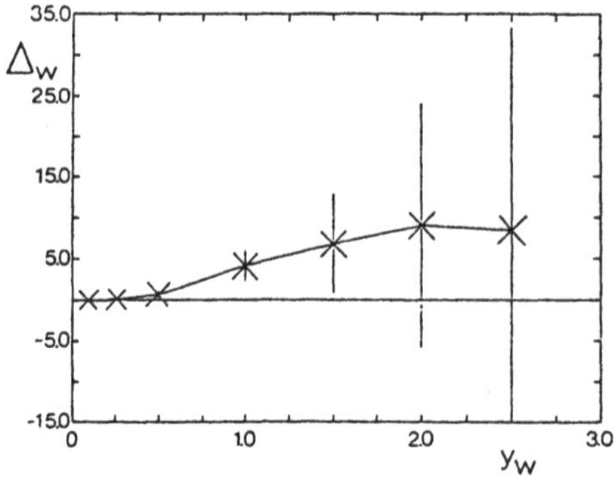


Fig. 3. $\Delta_w$, the difference between the left–hand side and the right–hand side of the third sum rule (Eq. 9), as a function of the size of the rapidity window $y_w$. The values and error bars of the left–hand side are obtained from the data of [9] for $\langle n_w \rangle$, $\langle n_w^2 \rangle$ and $\langle n_w^3 \rangle$. The right–hand side is obtained from the calculated values for $\langle N \rangle^{(1)}$, $\langle N^2 \rangle^{(1)}$, $\langle N^3 \rangle^{(1)}$, $\alpha_w^{(1)}(1)$ and $\alpha_w^{(1)}(2)$. The figure is reproduced from [7].

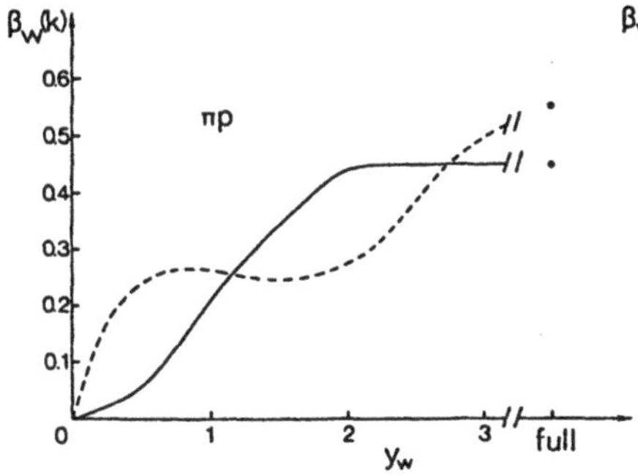


Fig. 4. The probabilities $\beta_w(1)$ (the dashed curve) and $\beta_w(2)$ (the solid curve) calculated by using the measured multiplicity distributions for negatively charged hadrons in different rapidity windows for $\pi^+p$ collisions at $p_{lab}$=200 GeV/c. The input: $\beta_{wmax}(1)$=0.55, and $\beta_{wmax}(2)$-=0.45. The figure is reproduced from [8].

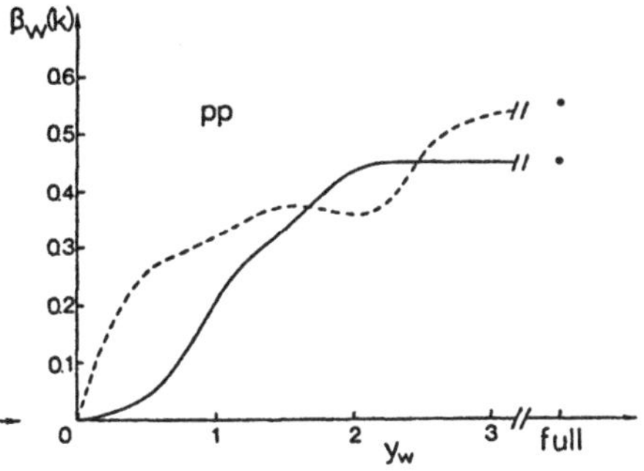


Fig. 5. The same as Fig. 4, but for *pp* collisions at the same energy. The figure is reproduced from [8].

that in hadron–hadron collision a considerable large percentage of neutral clusters decay into *more than two* charged hadrons, while those in $e^+e^-$ annihilation decay mainly into *two* oppositely charged hadrons.

The results we get here are different from those of Derrick et al.[11]. They found that the multiplicity distribution in the full rapidity space is remarkably close to a poisson distribution which could be expected for the random emission of single particles. Our results seem to be in contradiction with the experiments[12]. I think the reason for this may be the following. The cluster has typically a size of one rapidity unit which can be estimated from short-range correlations. If the rapidity window is larger than the size of clusters the cluster effects should be less significant. This can be very clearly seen from Fig. 4 in the second reference of [4]. In that figure we compared the calculated multiplicity distributions in different rapidity windows *with and without* cluster assumption with the data of Derrick et al.[9]. In that figure we see that in the rapidity windows between $|y| = 0.5$ and $|y| = 1.0$ the results with cluster assumption are in good agreement with data and the results without cluster assumption fall too fast. In large rapidity windows, however, the results with and without cluster assumption are very close to each other. Both of them are in good agreement with experimental data. This indicates that the cluster effects in large rapidity window are suppressed. It is therefore better to study clusters in the rapidity window of the size comparable with the size of cluster, e.g. around one rapidity unit. Further, experiments at LEP should be very helpful in clarifying this problem.

**References**

[1] A. Bialas, in Proceedings of IV International Symposium on Multiparticle Hydrodynamics, Pavia, Italy, 1973, eds. F. Dulmin et al., p93; L. Foa, Phys. Rep. **22**, 1 (1975); J. Whitmore, ibid. **27**, 187 (1976); G. Giacomelli and M. Jacob, ibid. **55**, 1 (1979).
[2] ABCDHLLW collaboration, W. Bell at al., Z. Phys. **C32**, 335 (1986).
[3] UA5 Colllaboration, G.J. Alner et al., Phys. Lett. **160B**, 193 (1985); M. Derrick et al., ibid. **168B**, 299 (1986). NA22 Collaboration M. Admus et al., ibid. **177B**, 239 (1986); The European Muon Collaboration, M. Arneodo et al., Z. Phys. **C35**, 335 (1987). ABCDHW collaboration, G. Giacomelli, Invited paper at Les Rencontres de Physique de la Vallee d'Aoste, La Thuile, Aosta Valley, Italy (1987).
[4] Chao Wei–qin, Meng Ta–chung, and Pan Ji–cai, Phys. Lett. **176B**, 211 (1986); Phys. Rev. **D35**, 152 (1987).
[5] P. Carruthers and I. Sarcevic, Phys. Rev. Lett. **63**, 1562 (1989); A. Capella, K. Fialkowski and A. Krzywicki, Phys. Lett. **230B**, 149 (1989).
[6] A. Bialas and R. Peschanski, Nucl. Phys. **B273**, 703 (1986); **B308**, 803 (1988).
[7] Chao Wei–qin, Gao Chong–shou, Meng Ta–chung, and Pan Ji–cai, Phys. Rev. **D36**, 2702 (1987).
[8] Chao Wei–qin, Gao Chong–shou, Meng Ta–chung, and Pan Ji–cai, Phys. Rev. **D41**, 838 (1990).
[9] M. Derrick et al., in reference [3].
[10] NA22 Collaboration, M. Admus et al., in reference [3].
[11] M. Derrick et al., Phys. Rev. **D34**, 3304 (1986); and Derrick at el. in [3].
[12] The author thanks C.C. Shih for valuable comments and discussions.

# SINGLE $b(\bar{b})$ SHOWER PROPERTIES IN THE QCD PARTON SHOWER MODEL

A. Giovannini *and* R. Ugoccioni

*Dipartimento di Fisica Teorica, Università di Torino and INFN, Sezione di Torino, Italy*

L. Van Hove

*CERN - Geneva*

**ABSTRACT**
The hadronic properties of $e^+e^-$ annihilation in the QCD Parton Shower Model can be understood in terms of the partonic properties by means of Local Parton Hadron Duality (LPHD). The partonic properties have been shown to result from the superposition of the two single – parton showers produced by the "ancestor" quark and antiquark. Continuing our previous work we have investigated the rapidity and multiplicity distributions of single $b(\bar{b})$ showers in the $b\bar{b}$ system at c.m. energies $\sqrt{s} = 91$ GeV, $\sqrt{s} = 200$ GeV and $\sqrt{s} = 2000$ GeV. The evolution of ancestors $b(\bar{b})$ has been studied from the initial to the final state at parton level and then to $B$-hadrons at hadron level. The emerging structures are remarkably simple at high c.m. energy ($\sqrt{s} = 2000$ GeV). At c.m. energy $\sqrt{s} = 91$ GeV ($Z^0$ peak, present LEP energy), due to the relatively low virtuality of the process the heavy flavor ancestor gives origin to unexpected features, which are of particular interest.

*Work supported in part by Ministero della Pubblica Istruzione (Italy) under Grant 1989.*

## 1. Introduction

The simple properties which we found in the QCD Parton Shower Model when describing multiplicity distributions (MD's) in $gg$ and $q\bar{q}$ systems (with all flavors included) in symmetric rapidity windows at different energies at parton and at hadron level [1],[2], stimulated our research on single $g$–jets and $q(\bar{q})$–jets resulting by separating the above two systems in their elementary substructures [3]. Continuing this program we focused our attention on the channel $e^+e^- \to b\bar{b}$ and on its single $b(\bar{b})$–jet substructure.

We decided to investigate beauty flavors rapidity and multiplicity distributions and to follow the $b(\bar{b})$ ancestor evolution from initial state to final state at parton level and then, after hadronization, to $B$-hadrons and eventually to their decay products. Accordingly we generated a sample of 2000 events in the $b\bar{b}$ system at c.m. energies $\sqrt{s}$ = 91 GeV ($Z^0$ peak, present LEP energy), 200 GeV and 2000 GeV by using the QCD Parton Shower Model with Lund string fragmentation prescription for electron–positron annihilation as implemented in the program JETSET version 6.3.[4] As in our previous work[1],[2],[3] the values of the shower evolution parameters are $\Lambda$ = 0.4 GeV (QCD scale in LLA) and $Q_0$ = 1 GeV (virtuality cutoff for parton evolution). The $b\bar{b}$ system has been separated in single $b(\bar{b})$ forward and $b(\bar{b})$ backward jets (or showers) after grouping partons originated by the ancestor parton with positive (forward) and negative (backward) rapidity. Rapidity $y$ is defined as usual with respect to the linear sphericity axis in the $b\bar{b}$ c.m. frame.

The motivation of the present work is to be traced back to our view on electron – positron annihilation which is summarized in the following. The evolution of the parton shower at perturbative level is given by the Altarelli – Parisi equations (AP), which are based on standard QCD and are the common feature of the most successful parton shower models.[4],[5] We assume then that hadronization is controlled by Local Parton Hadron Duality (LPHD)[1],[2], which is indeed contained in the QCD Parton Shower Model with string fragmentation when applied to $gg$ and $q\bar{q}$ systems at c.m. energies $\geq$ 200 GeV. In order to see whether heavy flavors behave differently, it is of interest to study the $b\bar{b}$ system above and below such energy.

The paper proceeds, in Section 2, by discussing in some detail rapidity distributions of single forward and backward $b(\bar{b})$ showers both in the $b\bar{b}$ c.m.

system and in the forward (backward) ancestor frame. The evolution of the $b(\bar{b})$ ancestors from initial to final state and to $B$-hadrons is also studied. In Section 3 MD's are presented once again in the $b\bar{b}$ c.m. system and in the forward (backward) ancestor frame. An attempt is made to fit them with negative binomial (NB) distributions. In Section 4 we analyse the event structure in the single $b(\bar{b})$ ancestor rest frame. Conclusions are added at the end.

## 2. Rapidity distributions in single ancestor $b(\bar{b})$–jet

The final parton rapidity distribution in the c.m. $b\bar{b}$ system at $\sqrt{s} = 91$ GeV is the superposition of the single forward and backward $b(\bar{b})$–jet (see Fig. 1, heavy dashed and continuous lines, respectively); this property has been already verified in $gg$ and $q\bar{q}$ systems (with all flavors)[3]. In addition a new feature appears in the parton distribution of the $b\bar{b}$ system: on top of the expected shape two peaks show up for $y$ in the rapidity intervals $2.75 \div 3.00$ (forward $b(\bar{b})$–jet) and $-3.00 \div -2.75$ (backward $b(\bar{b})$–jet). They are due to the forward and backward ancestors $b(\bar{b})$ in the final state, as can be seen in Fig. 2 (dashed line), where the same distributions plotted in Fig. 1 are presented after excluding the $b(\bar{b})$ ancestor. In fact, as will be discussed later in more detail, $b(\bar{b})$ ancestors are in general located at large rapidities and 10.4 percent of them are on the mass–shell. One may wonder whether the presence of extra $b\bar{b}$ and $c\bar{c}$ pairs may influence our findings: we counted in our sample one extra $b\bar{b}$ pair and 26 $c\bar{c}$ pairs and therefore their contribution is negligible. The two peak structure in the final parton distribution is slightly reflected in the final hadron rapidity distribution (Fig. 1, thin line). This distribution does not include the $B$-hadrons but includes their charged metastable decay products (hadrons and leptons). The structure is washed out by subtracting $B$-hadron decay products (Fig. 2, thin line). The situation is summarized in Fig. 3, where the two distributions (continuous lines) are plotted together with the rapidity distribution of the subtracted fraction of final particles (dashed line).

In view of the above considerations on the role of the ancestors $b(\bar{b})$ partons we decided to study their evolution from the initial state to the final state at partonic level and further down to $B$-hadrons. The rapidity distribution of the ancestors in the initial state (see Fig. 4a) indicates a two bump structure, the forward contained in the rapidity interval $0 < y < 3$, the backward being in the interval $-3 < y < 0$.

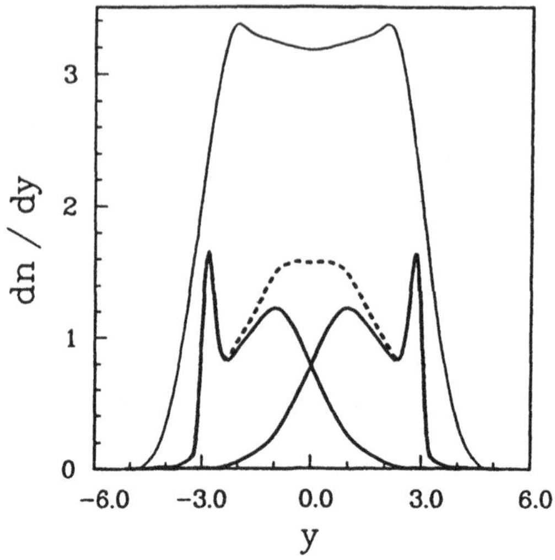


**Figure 1.** Final parton rapidity distribution of forward and backward $b(\bar{b})$–jet with respect to the $b\bar{b}$ c.m. system (heavy lines) at $\sqrt{s} = 91$ GeV vs rapidity $y$; final total parton (dashed line) and final hadron (thin line) rapidity distribution vs rapidity $y$. The mean numbers of partons and hadrons are $\langle n \rangle = 8.0$ and $\langle n \rangle = 21.18$, respectively. The final hadron distribution is as defined in the text.

It shows – as already anticipated – that $b(\bar{b})$ ancestors are predominantly located at large rapidities.

The rapidity distribution of the ancestors in the final state (see Fig. 4b) has two very sharp peaks at $y = 2.75 \div 3.00$ (forward ancestor $b(\bar{b})$–jet) and at $y = -3.00 \div -2.75$ (backward ancestor $b(\bar{b})$–jet). The cut at $y \approx \pm 3$ in Fig. 4b is an effect of the phase space limit. At $\sqrt{s} = 91$ GeV the maximum allowed value of rapidity for a $b(\bar{b})$ parton of mass equal to 5 GeV is $\pm 2.9$. At higher energies, where phase space is larger (e.g., at 2000 GeV in the $b\bar{b}$ c.m. system) a spread in rapidity is clearly visible (see Table I).

It is striking to remark that the above mentioned peaks are found again in the rapidity distribution of $B$-hadrons (see Fig. 4c): hadronization of the ancestors goes by picking up light quarks with very little displacement in rapidity.

After considering rapidity distributions in the $b\bar{b}$ c.m. system we move to study the corresponding distributions in the $b(\bar{b})$–ancestor frames.

In Fig. 5 we plot the rapidity distribution of final partons in the rest frame of the

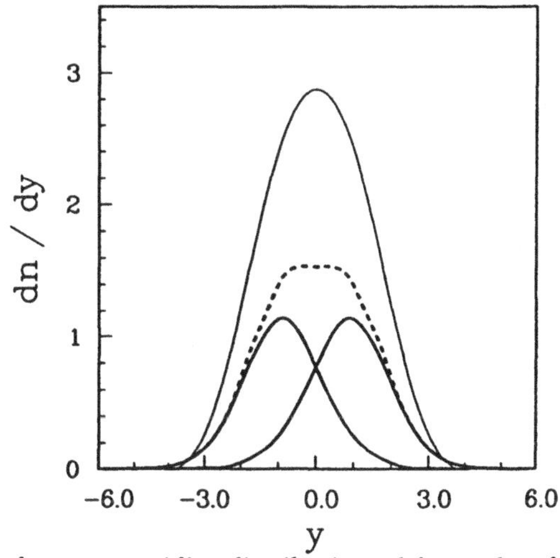


**Figure 2.** Final parton rapidity distribution of forward and backward $b(\bar{b})$–jet after excluding $b(\bar{b})$ ancestors in the final state (heavy lines) at $\sqrt{s} = 91$ GeV vs rapidity $y$ and their sum (dashed line). Final hadron rapidity distribution after suppressing $B$-hadron decay products as defined in the text (thin line) vs rapidity $y$. The mean numbers of partons are $\langle n \rangle = 3.0$ and $\langle n \rangle = 4.0$ and the mean number of hadrons is $\langle n \rangle = 10.36$.

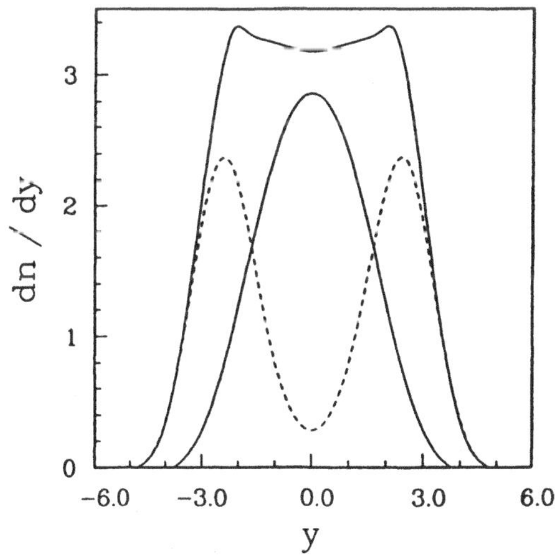


**Figure 3.** Rapidity distribution of $B$-hadron decay products in the final state (dashed line), final hadron rapidity distribution with (thin line – upper curve) and without (thin line – lower curve) $B$-hadron decay products at $\sqrt{s} = 91$ GeV vs rapidity $y$. The mean number of hadrons is $\langle n \rangle = 10.82$. $B$-hadron decay products are defined in the text.

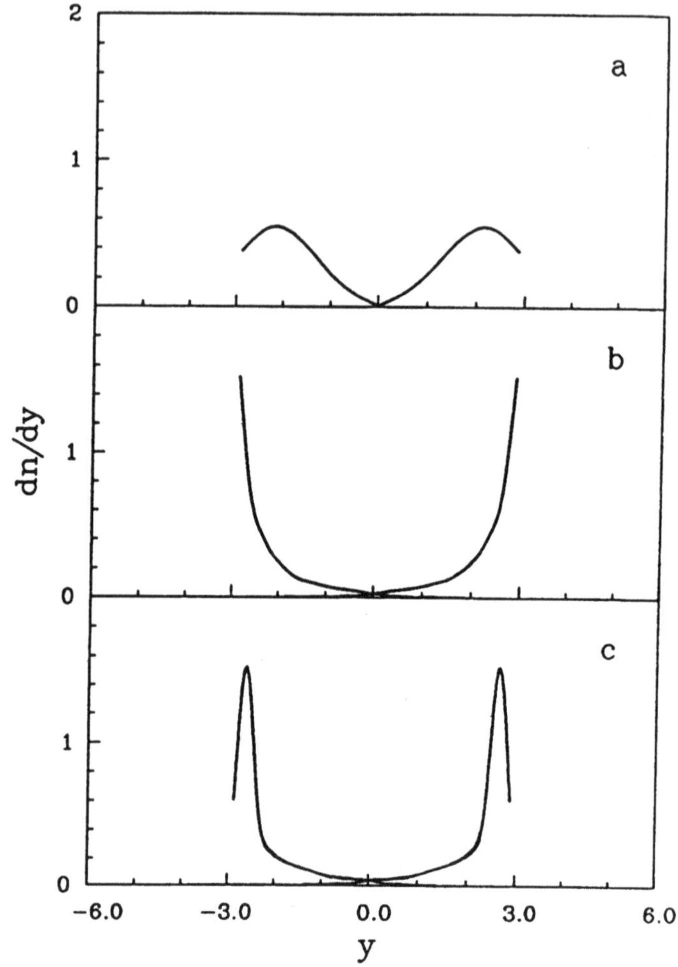


**Figure 4.** **a)** Rapidity distribution of forward (on the right) and backward (on the left) $b(\bar{b})$ ancestor in its initial state (i.e., with its original virtuality) in the $b\bar{b}$ c.m. frame at c.m. energy $\sqrt{s} = 91$ GeV vs rapidity $y$.
**b)** Rapidity distribution of forward (on the right) and backward (on the left) $b(\bar{b})$ ancestor in its final state (the virtuality is then the mass of 5 GeV) in the $b\bar{b}$ c.m. frame at c.m. energy $\sqrt{s} = 91$ GeV vs rapidity $y$.
**c)** Rapidity distribution of forward (on the right) and backward (on the left) $B$-hadrons (before decay) in the $b\bar{b}$ c.m. frame at c.m. energy $\sqrt{s} = 91$ GeV vs rapidity $y$.

**Table Ia.** Forward final partons mean rapidity, $\langle y\rangle$, in the $b\bar{b}$ c.m. frame and dispersion, $D_y = \sqrt{\langle (y-\langle y\rangle)^2\rangle}$, are compared at c.m. energies $\sqrt{s} = 91$ GeV, 200 GeV and 2000 GeV. The mean numbers of partons $\langle n\rangle$ are also given.

| $b(\bar{b})$–jet | Final Partons | | |
|---|---|---|---|
| $\sqrt{s}$ | $\langle y\rangle$ | $D_y$ | $\langle n\rangle$ |
| 91 GeV | 1.23 | 1.21 | 4.02 |
| 200 GeV | 1.42 | 1.31 | 5.90 |
| 2000 GeV | 1.95 | 1.54 | 14.42 |

**Table Ib.** Forward $b(\bar{b})$ ancestor mean rapidity in the initial state and dispersion and forward $b(\bar{b})$ ancestor mean rapidity in the final state and dispersion are compared at different energies.

| $b(\bar{b})$–jet | $b(\bar{b})$ ancestor, initial state | | $b(\bar{b})$ ancestor, final state | |
|---|---|---|---|---|
| $\sqrt{s}$ | $\langle y\rangle$ | $D_y$ | $\langle y\rangle$ | $D_y$ |
| 91 GeV | 1.85 | 0.66 | 2.33 | 0.72 |
| 200 GeV | 2.09 | 0.74 | 2.82 | 0.87 |
| 2000 GeV | 2.54 | 0.94 | 3.74 | 1.36 |

**Table Ic.** Forward $B$-hadrons (before decay) mean rapidity and dispersion are compared at different energies.

| $b(\bar{b})$–jet | Beauty Hadrons | |
|---|---|---|
| $\sqrt{s}$ | $\langle y\rangle$ | $D_y$ |
| 91 GeV | 2.26 | 0.68 |
| 200 GeV | 2.77 | 0.85 |
| 2000 GeV | 3.68 | 1.43 |

forward ancestor $b(\bar{b})$ in its initial state, $dn/dY$ with $Y = y - y_{anc}$; $b(\bar{b})$–ancestors are included in the upper curve, and excluded in the other. As already discussed in the $b\bar{b}$ c.m. system, the peak disappears when ancestor $b(\bar{b})$ is excluded.

We follow then the evolution of ancestors $b(\bar{b})$'s to their final state at partonic level and to $B$-hadrons after hadronization. The rapidity distribution of forward ancestor $b(\bar{b})$ in its final state is shown in Fig. 6a; we notice at $Y = 0$ a delta function contribution on top of the smooth curve. Since $Y = 0$ corresponds to saying that the rapidities of the ancestor in the initial and final state are equal, this delta function is due to initial on mass–shell $b(\bar{b})$ ancestors, which do not

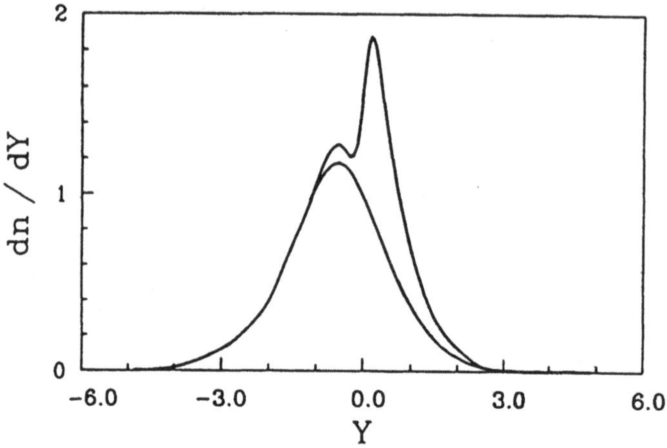


**Figure 5.** Forward final parton rapidity distribution including (upper curve) and excluding (lower curve) $b(\bar{b})$ ancestor in the final state at c.m. energy $\sqrt{s} = 91$ GeV vs rapidity $Y = y - y_{anc}$ with respect to forward ancestor in its initial state.

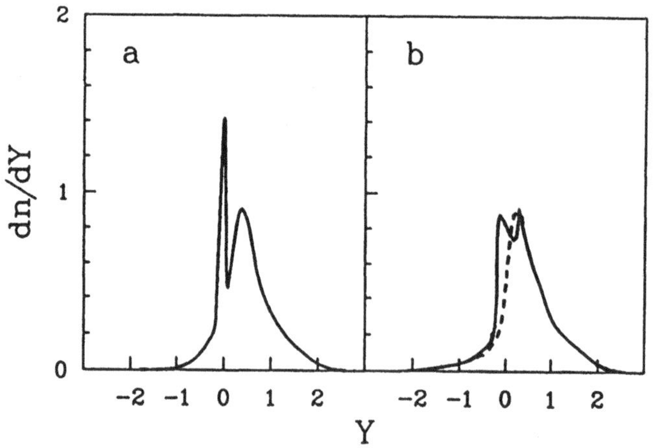


**Figure 6.** a) Rapidity distribution of forward $b(\bar{b})$ ancestor in the final state at c.m. energy $\sqrt{s} = 91$ GeV vs rapidity $Y = y - y_{anc}$ with respect to the forward ancestor in its initial state. By suppressing ancestors on mass-shell in their initial state the peak at $Y = 0$ disappears and the smooth curve remains.
b) Rapidity distribution of forward $B$-hadrons (before decay) at c.m. energy $\sqrt{s} = 91$ GeV vs rapidity $Y = y - y_{anc}$ with respect to the forward ancestor in its initial state (full curve). The dashed curve is obtained by suppressing $B$-hadrons generated by ancestors on mass-shell in their initial state.

radiate. This structure is found back in the rapidity distribution of $B$-hadrons after hadronization (Fig. 6b). In fact, the sharp structures disappear at partonic and hadronic levels when initial on mass–shell $b(\bar{b})$'s are removed in the distributions (see again Fig.s 6a and 6b).

We also studied how these results are modified by increasing the c.m. energy of the $b\bar{b}$ system (see Table I). The most relevant finding in this context is the weakening of the two peaks at 200 GeV and their disappearance at 2000 GeV. It should be noticed also that the average rapidity of the ancestor in the final state is growing from 2.33 at 91 GeV to 3.74 at 2000 GeV and that the corresponding spread in rapidity distribution, goes from 0.72 to 1.36. An interesting point is that at each energy the average rapidity and dispersion of $B$-hadrons are very close to the corresponding quantities of the ancestors in the final state (compare Table Ic with the two last columns of Ib).

## 3. Multiplicity distributions in single–ancestor $b(\bar{b})$–jet

We first studied MD's of final partons and hadrons at $\sqrt{s}$ = 91, 200 and 2000 GeV in the $b\bar{b}$ c.m. system in symmetric rapidity windows, $|y| < y_0$ $(y_0 = 0.5, 1, 2, 3, 4)$ and in full phase space. For MD's in full phase space at partonic level we consider as usual [3] the distribution of newly created partons, i.e., of final partons after subtracting the ancestor(s) parton(s). When the two $b(\bar{b})$–jets are not yet separated in the $b\bar{b}$ c.m. system we find that MD's are described well by NB's and that the corresponding $\chi^2/_{\rm NDF}$ are in general quite good. Results for $\sqrt{s}$ = 91 and 2000 GeV are listed in Table II. We also confirmed Local Parton–Hadron Duality for $\sqrt{s} \geq 200$ GeV ($\bar{n}_h \simeq 2\bar{n}_p$ and $k_h \simeq k_p$, see the $\sqrt{s}$ = 2000 GeV results of Table II). Regarding the energy independence of the mean number $\bar{N}$ of clans in fixed symmetric rapidity windows, it here holds only for $|y| < 1$ and it is much better at partonic than at hadronic level. The windows $|y| < 1$ at partonic level are the only ones not to be occupied by $b(\bar{b})$'s with appreciable probabilities.

The presence of the $b(\bar{b})$ partons has more drastic effects on the partonic MD's of single forward and backward $b(\bar{b})$–jets. For symmetric windows in the c.m. frame of the $b\bar{b}$ system NB distributions are found only for $|y| < 0.5$ at $\sqrt{s}$ = 91 GeV, whereas at $\sqrt{s}$ = 2000 GeV they occur up to $|y| < 3$. NB properties are altogether absent in symmetric windows $|Y| < Y_0$ in the ancestor frame of the jet.

**Table II.** Standard NB parameters $(\bar{n}, k)$, and corresponding average number of clans, $\bar{N}$, and average clan multiplicity, $\bar{n}_c$, for multiplicity distributions of partons and hadrons in symmetric windows $|y| < y_0$ and in full phase space (FPS) in the $b\bar{b}$ c.m. system, at c.m. energies $\sqrt{s} = 91$ GeV and $\sqrt{s} = 2000$ GeV. $\chi^2$/number of degrees of freedom (NDF) for NB fits are given in the last column.

| PARTONS c.m. system | $\bar{n}$ | $k$ | $\bar{N}$ | $\bar{n}_c$ | $\chi^2$/NDF |
|---|---|---|---|---|---|
| $\|y\| < 0.5$ | | | | | |
| 91 GeV | 1.56 ± 0.03 | 3.61 ± 0.42 | 1.29 ± 0.03 | 1.20 ± 0.04 | 10.81/8 |
| 2000 GeV | 3.83 ± 0.12 | 0.71 ± 0.03 | 1.32 ± 0.04 | 2.91 ± 0.12 | 15.31/25 |
| $\|y\| < 1.0$ | | | | | |
| 91 GeV | 3.06 ± 0.05 | 4.55 ± 0.38 | 2.34 ± 0.05 | 1.31 ± 0.04 | 21.01/12 |
| 2000 GeV | 7.58 ± 0.18 | 1.14 ± 0.05 | 2.32 ± 0.06 | 3.27 ± 0.11 | 38.50/40 |
| $\|y\| < 2.0$ | | | | | |
| 91 GeV | 5.53 ± 0.08 | 6.19 ± 0.42 | 3.95 ± 0.08 | 1.40 ± 0.03 | 32.94/19 |
| 2000 GeV | 15.04 ± 0.25 | 2.36 ± 0.09 | 4.72 ± 0.11 | 3.19 ± 0.09 | 63.26/57 |
| ... | ... | ... | ... | ... | ... |
| FPS | | | | | |
| 91 GeV | 6.01 ± 0.08 | 7.29 ± 0.56 | 4.38 ± 0.09 | 1.37 ± 0.03 | 9.913/18 |
| 2000 GeV | 26.55 ± 0.23 | 9.73 ± 0.42 | 12.81 ± 0.25 | 2.07 ± 0.04 | 93.66/63 |

| HADRONS c.m. system | $\bar{n}$ | $k$ | $\bar{N}$ | $\bar{n}_c$ | $\chi^2$/NDF |
|---|---|---|---|---|---|
| $\|y\| < 0.5$ | | | | | |
| 91 GeV* | 3.07 ± 0.08 | 2.01 ± 0.17 | 1.86 ± 0.06 | 1.65 ± 0.07 | 34.62/18 |
| 2000 GeV+ | 7.36 ± 0.29 | 0.72 ± 0.04 | 1.74 ± 0.07 | 4.22 ± 0.23 | 73.16/48 |
| $\|y\| < 1.0$ | | | | | |
| 91 GeV | 6.44 ± 0.11 | 2.88 ± 0.14 | 3.38 ± 0.08 | 1.90 ± 0.05 | 49.24/27 |
| 2000 GeV+ | 14.74 ± 0.40 | 1.11 ± 0.06 | 2.95 ± 0.11 | 5.00 ± 0.23 | 82.13/73 |
| $\|y\| < 2.0$ | | | | | |
| 91 GeV | 13.07 ± 0.18 | 4.34 ± 0.17 | 6.03 ± 0.12 | 2.17 ± 0.05 | 62.20/40 |
| 2000 GeV | 31.40 ± 0.53 | 2.62 ± 0.08 | 6.71 ± 0.14 | 4.68 ± 0.12 | 115.2/119 |
| ... | ... | ... | ... | ... | ... |
| FPS | | | | | |
| 91 GeV | 21.18 ± 0.15 | 24.48 ± 1.65 | 15.26 ± 0.27 | 1.39 ± 0.03 | 28.98/19 |
| 2000 GeV | 63.36 ± 0.50 | 11.36 ± 0.44 | 21.39 ± 0.46 | 2.96 ± 0.07 | 111.9/127 |

* (first bin excluded)  + (first two bins excluded)

We found that the NB properties are generally improved when one considers MD's excluding the $b(\bar{b})$ at partonic level and excluding the $B$-hadron decay products at hadronic level. That the properties of single $b(\bar{b})$–jets are complicated is related to the occurrence of sizable correlations between the forward and backward jets. These correlations can be studied theoretically and experimentally. They should provide interesting tests of the QCD Parton Shower Model, especially the dynamics of the ancestor partons, but their discussion is beyond the scope of the present paper.

## 4. Structure of the single $b(\bar{b})$–jet

In the ancestor $b(\bar{b})$ frame with the $z$ axis made to coincide with the rapidity axis we studied the final parton distribution in the polar angle $\theta$ ($\cos\theta = p_z/p$), $dn/d\cos\theta$ (see Fig. 7a). The resulting distribution at $\sqrt{s} = 91$ GeV shows an elongated shape as in Ref. [3]. In addition the backward peak ($\theta \approx \pi$) is now higher than the forward peak ($\theta \approx 0$), an asymmetry induced by the heavy ancestor parton. As the energy increases ($\sqrt{s} = 2000$ GeV) the distribution becomes more symmetric around $\theta \approx \pi/2$ but remains asymmetric in the $\theta \approx 0$ and $\theta \approx \pi$ regions. This situation should be compared with our previous work on light partons where the distribution was symmetric.

We looked also at the final parton distribution in transverse momentum $p_T$ ($p_T = \sqrt{p_x^2 + p_y^2}$), $dn/dp_T^2$ (see Fig. 8). Low $p_T$ final partons are favored, confirming the tendency of their distribution to be elongated in shape along the rapidity axis. It is to be pointed out that for $p_T \leq 0.6$ GeV the final parton distribution in transverse momentum depends on the c.m. energy (it is growing) contrary to what has been noticed in Ref. [3], where it is energy independent.

It is interesting to analyse the effect of $b(\bar{b})$ exclusion on the above structures. We see that $b(\bar{b})$ partons at both energies are generated at small $\theta$ (compare Figs. 7a and 7b). The distribution in $p_T$ shows no appreciable difference after $b(\bar{b})$ subtraction. At $\sqrt{s} = 91$ GeV there is a concentration of 10 percent of the total number of $b(\bar{b})$'s in the region of very small $p_T$ ($p_T < 0.25$ GeV).

## 5. Conclusions

Studying the $e^+e^- \to b\bar{b}$ annihilation channel in the Parton Shower Model, we have analysed rapidity distributions in terms of single $b(\bar{b})$ forward and back-

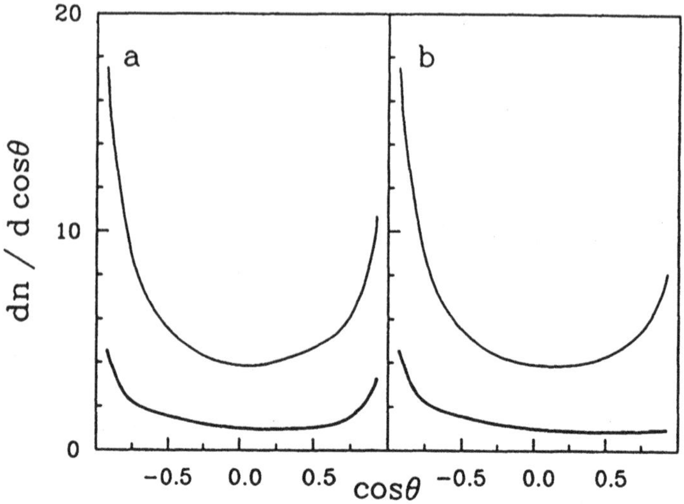


**Figure 7.** a)Forward final parton distribution in polar angle $\theta$ at $\sqrt{s} = 91$ GeV (heavy line) and $\sqrt{s} = 2000$ GeV (thin line) with respect to the forward ancestor frame vs $\cos\theta$. The $z$ axis coincides with the rapidity axis, $\cos\theta = p_z/p$.
b) The same as a), after excluding $b(\bar{b})$ ancestors in the final parton state.

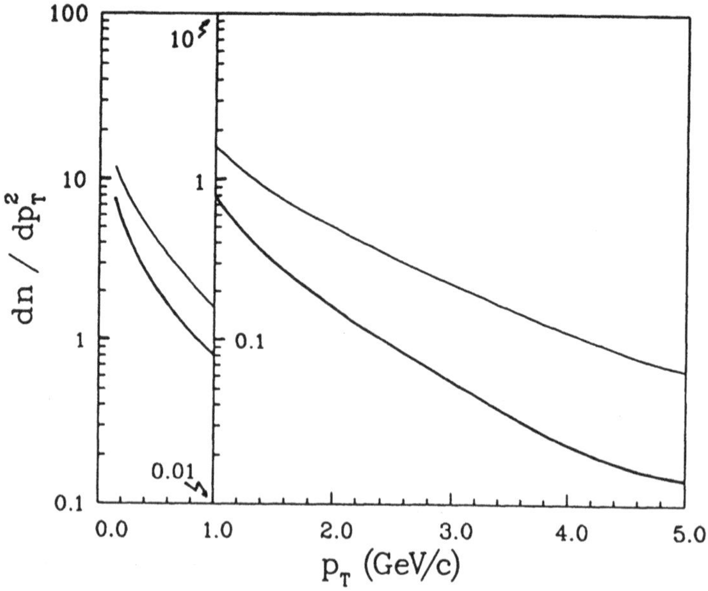


**Figure 8.** Final parton distribution in transverse momentum squared at $\sqrt{s} = 91$ GeV (heavy line) and 2000 GeV (thin line).

ward ancestor jets in the $b\bar{b}$ c.m. frame. At c.m. energy $\sqrt{s} = 91$ GeV, the rapidity distribution of the total system has two peaks, one forward and one backward, which are already present in the single forward and backward $b(\bar{b})$–jet rapidity distribution and are found back at hadron level. They are due to $b(\bar{b})$ ancestors. They disappear when excluding the $b(\bar{b})$ ancestors and the $B$-hadron decay products from the final state. These peaks strongly affect the overall hadronic distribution (Fig. 3). At $\sqrt{s} = 2000$ GeV there are no such peaks at our level of statistics.

The $b(\bar{b})$ ancestors have on the average larger rapidity than the other final partons and hadronization of the heavy flavor occurs with little displacement in rapidity. This fact outlines a peculiar configuration in rapidity for the heavy flavor with respect to light flavors, e.g., the forward $b(\bar{b})$–jet should be thought as a leading $b(\bar{b})$ followed by a well separated tail of light partons. The same situation is expected to occur at hadron level, where decay products of $B$-hadrons populate in general the large rapidity regions.

In our study on MD's at parton and hadron level in symmetric rapidity windows and in full phase space we found that the MD's of the total $b\bar{b}$ system are well described by NB fits. In single forward and backward $b(\bar{b})$–jet with respect to the ancestor frame in $|Y| < Y_0$ they are not fitted by NB's. In the $b\bar{b}$ c.m. system, for single forward and backward $b(\bar{b})$–jet, the domain of validity of the NB regularity in $|y| < y_0$ is growing with energy (it is limited to $y_0 = 0.5$ at $\sqrt{s} = 91$ GeV and it extends up to $y_0 = 3.0$ at $\sqrt{s} = 2000$ GeV).

The event structure in the single ancestor $b(\bar{b})$–jet frame shows an elongated asymmetric shape with respect to the rapidity axis, the $b(\bar{b})$ ancestor being forward ($\theta \approx 0$). This picture is in agreement with the general finding that heavy flavors are leading in rapidity with respect to the light partons.

One final comment. We found that heavy flavor properties are quite different from those of light flavors in the LEP energy range. This should open the possibility of new tests of the QCD Parton Shower Model.

## References

[1] L. Van Hove and A. Giovannini, Acta Phys. Pol. **B19**, (1988) 917; L. Van Hove and A. Giovannini, Acta Phys. Pol. **B19**, (1988) 931.

[2] M. Garetto, A. Giovannini, T. Sjöstrand and L. Van Hove, in Proceedings of the Perugia Workshop on Multiparticle Production, eds. G. Pancheri et al., (World Scientific, Singapore, 1989), p. 181.

[3] A. Giovannini, R. Ugoccioni and L. Van Hove, to appear in the Proceedings of the XXVth Rencontres de Moriond at Les Arcs, March 12–16, 1990.

[4] M. Bengstsson and T. Sjöstrand, Nucl. Phys. **B289**, (1987) 810; T. Sjöstrand and M. Bengstsson, Computer Physics Commun. **43**, (1987) 367.

[5] G. Marchesini and B. R. Webber, Nucl. Phys. **B238**, (1984) 1; B. R. Webber, Nucl. Phys. **B238**, (1984) 492; HERWIG version 3.2, G. Marchesini and B. R. Webber, Nucl. Phys. **B310**, (1988) 461.

# MULTIFRACTAL PROPERTIES OF MULTIPARTICLE PRODUCTION

RUDOLPH C. HWA
Institute of Theoretical Science and Department of Physics
University of Oregon, Eugene, OR 97403

## ABSTRACT

The language of multifractals is reviewed for the study of multiplicity fluctuations in high-energy collisions. Connection with intermittency is discussed. Analysis using the G-moments is introduced. Various issues related to the multifractal analysis are briefly mentioned. They include the size of resolution width, statistical background, and the application to the diagnostics of quark-gluon plasma.

In the last two years the subject of intermittency[1] has gained considerable interest both experimentally and theoretically.[2-6] Although there is still controversy on whether some of the data ( mainly on hadronic collisions) require a physical mechanism other than the usual short-range correlation to explain,[4] I shall take the point of view here that intermittency is a new effect that reveals self-similarity over a range of resolution scale.[6] The subject of this talk is then a description of how this new effect can be presented in the language of multifractals.[7,8]

As proposed by Bialas and Peschanski,[1] intermittency is the power-law behavior of normalized factorial moments

$$F_q = \frac{\langle k(k-1)\ldots(k-q+1)\rangle}{\langle k\rangle^q} \propto \delta^{-a_q} \qquad (1)$$

where k is the multiplicity in a bin and $\delta$ is the bin width; $a_q$ is the intermittency index. The advantage of $F_q$ over the standard normalized moments

$$C_q = \frac{\langle k^q \rangle}{\langle k \rangle^q} \tag{2}$$

is that $F_q$ filters out the statistical fluctuation,[1] so that in a plot of log $F_q$ vs log $1/\delta$ one can by inspection infer from a persistent linear behavior the existence of a dynamical self-similar property that is contained in $C_q$. The drawback of using $F_q$ is that it cannot readily be generalized to include negative values of q. While positive q emphasizes the peaks in rapidity fluctuation, negative q emphasizes the valleys, which contain equally interesting information about the fluctuation. The multifractal analysis overcomes that drawback. Moreover, it provides an interpretation of the intermittency indices in terms of the fractal properties of the multiparticle system.

The multifractal analysis can best be done by using the G moments,[8,9] which are defined by

$$G_q = \sum_{j=1}^{M}{}' p_j^q \,, \qquad p_j = k_j/n \tag{3}$$

where $k_j$ is the multiplicity in the $j$th bin and $n = \sum_j k_j$ is the total multiplicity in a chosen rapidity interval, which is divided into M bins, each of width $\delta$. The prime on the summation sign in (3) signifies that only the non-empty bins are summed. Thus q may be negative. The standard moments $C_q$ can also be extended to negative q if the empty bins are left out in the horizontal analysis, i.e.

$$C_q = \frac{\sum_{k=1}^{\infty} k^q Q_k}{\left(\sum_{k=1}^{\infty} k Q_k\right)^q} \left(\sum_{k=0}^{\infty} Q_k\right)^{q-1} \tag{4}$$

where $Q_k$ is the number of bins with multiplicity k. Note that k=0 is included in the normalization of $P_k = Q_k / \sum_{k=0}^{\infty} Q_k$. From (3) and (4) we get

$$G_q = C_q M^{1-q} \tag{5}$$

Clearly, if $F_q$ possesses a power-law behavior as shown in (1), then so does $C_q$ and $G_q$. Assuming that $C_q$ has indeed the same dynamical dependence on $\delta$ as $F_q$, then a parameterization of the $\delta$-dependence of $G_q$ by

$$G_q \sim \delta^{\tau(q)} \tag{6}$$

implies that

$$\tau(q) = -a_q + q - 1 \tag{7}$$

for $q > 1$.

The conventional theory of multifractals considers the limit of infinitely high resolution. In high energy collisions that limit creates a unique problem. Experimental resolution is not a problem in principle. The problem is that at finite energy the number of particles produced is finite. When $n<<M$, k in (1) is small and may not be greater than q. But even that is not a fatal problem, because a rare event may have $k > q$ and gives rise to observable $F_q$. Indeed, intermittency is precisely a measure of those rare events.[1] It is only a matter of statistics to have enough events that contain those peaks. That is for positive q. For negative q the situation is different. In the limit of infinitely good resolution, all bins have either k=0 or 1. Then according to (3) we have $G_q = n^{1-q}$, which is independent of M. This saturation takes place earlier for $q<0$ than for $q>0$, since the negative moments are insensitive to the peaks but receive large contribution from the k=1 bins. Good statistics does not help. Thus self-similarity in both positive and negative values of q cannot be expected in multiparticle production over a wide range of M at present energies. Nevertheless, we shall study the multifractal structure at finite $\delta$.

From the plot of $\ell n G_q$ vs $-\ell n \delta$ one can determine the slope $\tau(q)$ from (6) over any region of $\delta$, either from the data or from the results of Monte Carlo calculation. We say that a multifractal structure exists if $\tau(q)$ satisfies the following properties.

1. Defining $\alpha_q = d\tau(q)/dq$, there exists a range of q in which $\alpha_q < \alpha_{q'}$, if $q>q'$.

2. Defining

$$f(\alpha_q) = q\alpha_q - \tau(q) \tag{8}$$

there exists a range of $\alpha_q$ in which $d^2f(\alpha)/d\alpha^2<0$.

These are the properties of multifractals in problems where the limit $\delta \to 0$ can be taken.[7] Here we take these properties to select the region of $\delta$ in which multifractals are defined.

A number of relations between intermittency and multifractality now follow. From (7) we obtain

$$\alpha_q = 1 - b_q \;, \qquad\qquad b_q = da_q/dq \tag{9}$$

Since (8) implies $df(\alpha_q)/d\alpha_q = q$, one also has[10]

$$d^2f(\alpha_q)/d\alpha_q{}^2 = -[d^2a_q/dq^2]^{-1} \tag{10}$$

Furthermore, from (7) - (9) follows[10]

$$f(\alpha_q) = -qb_q + a_q + 1 = -q^2 \frac{d}{dq}\left[\frac{a_q+1}{q}\right] \tag{11}$$

Since $a_q$ is defined for q>1, Eqs. (9) - (11) are therefore also for q>1, although they can be generalized to all real q, when $a_q$ is continued to negative q through the use of $C_q$. Eq. (10) relates the negative curvature of $f(\alpha)$ to the positive curvature of $a_q$. Thus strong intermittency (large $d^2a_q/dq^2$) corresponds to a broad $f(\alpha)$ spectrum. As indicated by (9), $\alpha_q$ deviates from 1 to the extent of the slope of $a_q$. Note that if

$$a_q = (q-1)\,\bar{d} \tag{12}$$

where $\bar{d}$ is independent of q, then $\alpha = 1 - \bar{d}$ and $f(\alpha) = \alpha$. This is the condition for a monofractal, as in the case of the Cantor set, for which $\alpha = \ell n2/\ell n3$. In general, when the limit $\delta \to 0$ can be taken, the Renyi dimension[11] is $D_q = \tau(q)/(q-1)$, so the q-dependent $\bar{D}_q$ [defined as in (12) in place of $\bar{d}$] is $\bar{D}_q = 1 -$

$D_q$, an expression that gives justification to referring to $\bar{D}_q$ as the Renyi codimension. In usual hadronic and nuclear collisions $\bar{D}_q$ depends on q, and we say that the system has multifractal structure. Something special must happen when $\bar{D}_q$ becomes $\bar{d}$, independent of q. Bialas and I have proposed using it as a place to look for signals of quark-gluon plasma.[12]

Efforts have begun on the multifractal analysis of experimental data; in particular, Sugano has reported his work on $e^+e^-$ annihilation (HRS)[13] and Rimondi on $\bar{p}p$ collision (CDF).[14] In the plots of $\ell nG_q$ vs $-\ell n\delta$ the saturation effects become prevalent before $\delta$ gets to be very small. Thus self-similarity can only be considered for low values of M. One may ask what the resultant $f(\alpha)$ has to do with the usual notion of fractals, when only a few bins are used to determine the exponent $\tau(q)$ in (6). If one is optimistic, one may regard the initial slopes in the log-log plot as indicative of the universal asympotic slopes when s is increased, just as Sarcevic and Satz[15] have found phenomenological evidence for such universality in the factorial moments. More interestingly, it would be good if one can establish a form of ergodic theorem that can relate vertical averaging of the fractal properties of a few not-so-small bins at low energy to the horizontal averaging of the same properties of many, very small bins at much higher energies. Florkowski and I have made some progress in our recent study of this subject.[16] With that relationship established, we can then expect a kind of scaling that describes the shapes of the curves in the log-log plot (and not just the straight portion) for all energies.

In a different direction one can also ask whether the statistical contribution to multiplicity fluctuation can be filtered out of $f(\alpha)$. The answer is yes. Chiu, Fialkowski and I have looked into this problem and found a way to define a new set of $\tilde{G}$ moments, which reduces to the trivial moments when the fluctuation is purely statistical, but retains nontrivial multifractal structure when the G moments have both dynamical and statistical components. The spectral function derived from $\tilde{G}_q$, denoted by $\tilde{f}(\tilde{\alpha})$, then represents the filtered component. The details of this work is reported elsewhere.[17]

In conclusion I remark that the multifractal description of multiplicity fluctuation is an excellent arena for confrontation between theory and experiment. It not only contains information about the spikes, as summarized by intermittency behavior, but also gives a measure of the valleys in rapidity (and/or $\phi$) distribution. It makes connection with the usual language of self-similarity in other fields, and offers a possibility of diagnosing self-similar systems, such as phase transitions between quark and hadron matters.

I am grateful to many colleagues for beneficial discussions that advance my understanding of this subject. They include A. Bialas, P. Carruthers, C.B. Chiu, K. Fialkowski, W. Florkowski, L.S. Lu, V. Simak, and K. Sugano. This work was supported in part by the A. S. Department of Energy and Grant No. DE-FG06-85ER-4022.

## REFERENCES

1. A. Bialas and R. Peschanski, Nucl. Phys. **B273**, 703 (1986); **B308**, 803 (1988).
2. *Multiparticle Dynamics*, edited by A. Giovannini and W. Kittel (World Scientific, Singapore, 1989).
3. W. Kittel and R. Peschanski, in Proc. Europhys. Conf.. on High Energy Physics 1989 (to be published); B. Buschbeck and Lipa, Mod. Phys. Lett. **A4**, 1871 (1989).
4. Proceedings of the Santa Fe Workshop on Intermittency in High Energy Collisions (1990).
5. R.C. Hwa, OITS-440, an overview to be published in the Proc. of Santa Fe Workshop (Ref. 4).
6. A. Bialas, in Proc. of Quark Matter '90, Nucl. Phys. A (to be published).
7. J. Feder, *Fractals* (Plenum Press, New York, 1988).
8. R.C. Hwa, Phys. Rev. **D41**, 1456 (1990).
9. C.B. Chiu and R.C. Hwa, OITS-431 (1990).
10. Ph. Brax and R. Peschanski, Saclay SPhT/89-203.
11. A. Renji, *Probability Theory* (North-Holland, Amsterdam, 1970) The relevant material is supposedly contained in the Appendix that was left out in the English edition.
12. A. Bialas and R.C Hwa, CERN-TH-5754/90 or OITS-444 (1990).

13. K. Sugano, ANL-HEP-CP-90-37, to be published in the Proceedings of the Santa Fe Workshop (loc. cit); also talk given at CAMP Workshop, Marburg, 1990.
14. F. Rimondi, talk given at CAMP Workshop, Marburg, 1990.
15. I. Sarcevic and H. Satz, Phys. Lett. **B233**, 251 (1989).
16. W. Florkowski and R.C. Hwa, OITS-xxx (1990).
17. C.B. Chiu, K. Fialkowski and R. C.Hwa, OITS-441 (1990).

# FRACTAL STRUCTURE IN MULTIPARTICLE PRODUCTION IN $e^+e^-$ ANNIHILATIONS AT 29 GeV

K. Sugano
Argonne National Laboratory, High Energy Physics Division
Argonne, IL 60439, USA

## ABSTRACT

The fractal structure in multiparticle production in $e^+e^-$ annihilations at 29 GeV has been studied using the HRS detector at PEP. Very high fractal moments have been measured and their implications are discussed in terms of an $\alpha$-model and a thermodynamical model. Two-dimensional fractal moments in the rapidity and azimuthal angle space have also been measured and the results are compared to the Lund string model predictions.

The theory of fractal geometry[1] has revealed the structure of self-similarity in nature and has been applied successfully in various fields such as physics, astronomy, chemistry, computer graphics, and so on. The application to the multiparticle production in high energy physics has just begun. If quarks and gluons become jets of particles through a cascading process, the particles inside a jet should show a self-similar behavior which leads to a fractal/multifractal structure. Therefore, the analysis techniques developed in fractals are powerful tools to study the underlying physics of the hadronization process.

In this talk we report a study of fractal behavior of the charged particle multiplicity in $e^+e^-$ annihilations, using a high statistics data sample corresponding to a total integrated luminosity of 300 $pb^{-1}$ obtained with the High Resolution Spectrometer (HRS) at PEP. The HRS detector was a solenoidal spectrometer that measured charged particles and electromagnetic energy over 90% of the solid angle. The details of the detector and the event selection criteria can be found elsewhere.[2] A rapidity of each particle is calculated with respect to a thrust axis of all charged particles. The resolution of the rapidity is estimated to be about 0.1 including an ambiguity in determining the true jet axis which is the initial quark direction. The rapidity distribution, after detector corrections for the inclusive data sample is relatively flat in the range between $y = -2$ and $y = +2$, but the uncorrected distribution has a valley near $y =$

0 mainly because of the low tracking efficiency of slow particles and tracks near the beam line. In the following analysis, we use uncorrected variables.

A rapidity interval of total length $Y_o$ between y = -2 and y = +2 is divided into $M_o$ bins of width $\delta = Y_o/M_o$ and let $k_i$ be the number of particles in the $i^{th}$ bin. Since there may be bins that have no particles, we define M to be the number of non-empty bins, which constitute a fractal set. A multifractal moment is defined as follows:[3]

$$G_q = \sum_{j=1}^{M} p_j^q$$

where $p_j = k_j/n$ with $n = k_1 + k_2 + \ldots + k_M$, q is a real number, and the summation is carried over non-empty bins only. If the particle production process exhibits self-similar behavior, the moments show a power law relation of $G_q \propto \delta^{\tau(q)}$. This relation does not necessarily occur in the limit of $\delta \to 0$. Therefore, in calculating the power $\tau(q)$, we use the first two points in $G_q(M_o = 2$ and $3)$ and an average of $\ell n\, G_q$ over all events. Once $\tau(q)$ is determined from $G_q$, we can apply the theory of multifractals to calculate the singularity spectrum function $f(\alpha)$ by Legendre transform:

$$\alpha_q = \frac{d\tau(q)}{dq}$$

$$f(\alpha) = q\alpha_q - \tau(q)$$

The spectrum function $f(\alpha)$ for the inclusive data sample is shown in Fig. 1. The left-hand side of the curve corresponds to multifractal moments with $q > 0$ and the edge points are from high moments with q up to 20. The right-hand side corresponds to the negative moments with q down to -20. The statistical error bars are less than the mark size except near the endpoints. The $f(\alpha)$ stays above zero up to very large moments of $q = \pm 20$, which has thermodynamical implications.

In fluid-dynamics of fully developed turbulence, a geometrical model called "α-model" has been used successfully. It is based on a random cascading process with a two-level probability distribution at each stage. The fractal structure of this model has been studied theoretically and the spectrum functions $f(\alpha)$ are sketched in Fig. 2 according to four different regions in the parameter space.[4] The regions I, II, III, and IV are respectively named a peak transition region, a hole transition region, a full transition region, and a no phase transition region. This is because $f(\alpha)$ can be regarded as a local entropy in a thermodynamical interpretation. Therefore, if

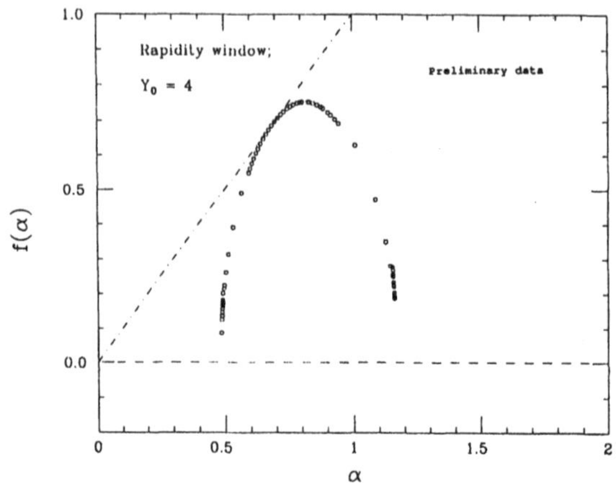


**Fig. 1**

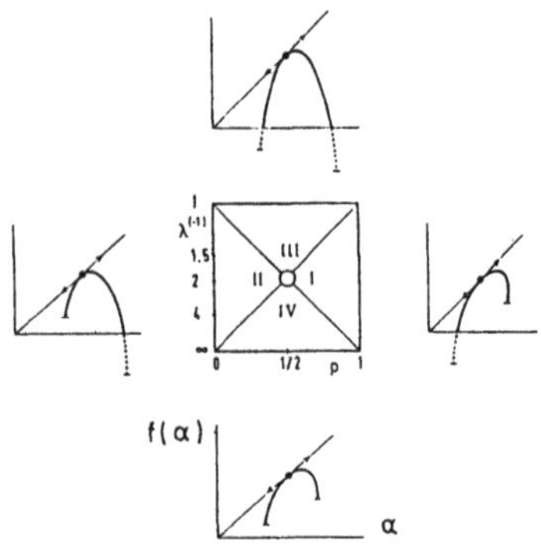


**Fig. 2**

it reaches zero, the free energy is frozen at the maximum and the system undergoes a phase transition of the spin-glass type. The regions I, II, and III have each corresponding models in nature such as growing crystals and a percolating cluster in a random resistor network. The region IV, however, has few corresponding models so far and it seems that the rapidity distribution of multiparticle production shown in Fig. 1 is in this category. The underlying physics of this process needs to be studied further. The Lund string model[4] reproduces this limiting behavior of f(α) well, although there are some discrepancies for the negative moments where coherence effects are important in the valley regions among jets. More detailed comparisons between data and the Lund model for two-jet and inclusive samples can be found elsewhere.[6]

The multifractal moments of azimuthal angles around the beam axis ($\phi_{beam}$) have been measured. The distribution of $\phi_{beam}$ averaged over many events is flat, although the azimuthal angles in each event are clustered in narrow phase space regions due to the jet structure of events. The resulting $f(\alpha)$ is shown in Fig. 3. The curve seems to reach zero for large q values near ±20, the interpretation of which is still unclear. The shape of $f(\alpha)$ is much wider than that in the rapidity space reflecting the nature of narrowly collimated jets. The fractal dimension given at the peak of the curve is larger than the one in the rapidity space.

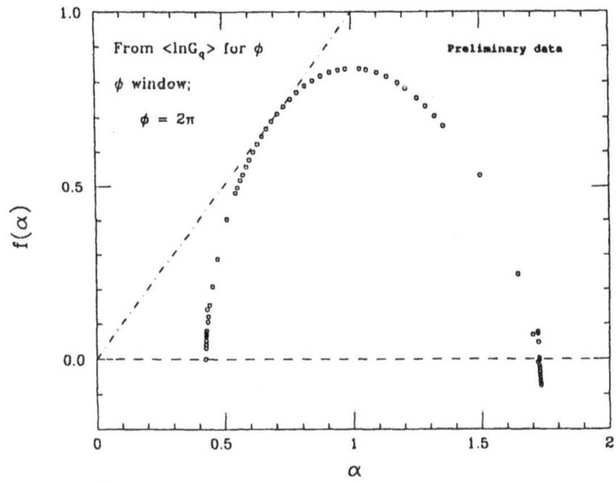


**Fig. 3**

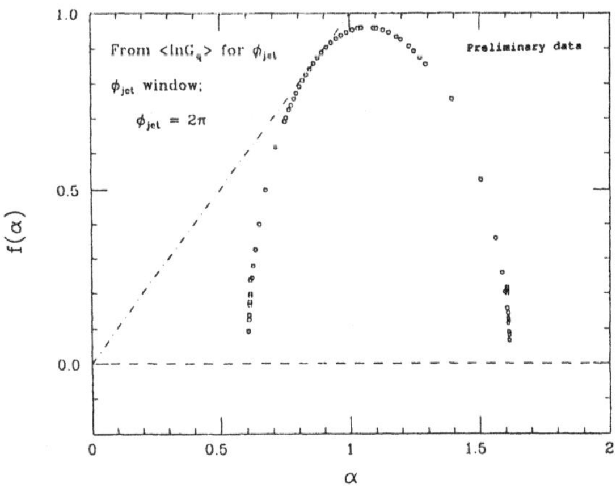


**Fig. 4**

On the other hand, the $f(\alpha)$ from the multifractal moments of azimuthal angle around the jet axis ($\phi_{jet}$) has a different behavior as shown in Fig. 4. The curve stays positive for moments up to q = ±20 as in the case of rapidity. The shape is narrower than that in the case of $\phi_{beam}$, and the fractal dimension is close to 1. This is consistent with an observation that particles inside a jet are almost uniformly distributed around the jet axis.

In order to study the fragmentation function of jets, the multifractal moments in the two-dimensional space of y and $\phi_{jet}$ have been measured. The total lengths of $-2 < y < +2$ and $0 < \phi_{jet} < 2\pi$ are divided into the equal number of internals, n, resulting in the total number of grids $M = n^2$ with n = 1,2,3,...40. The moments with various q values are shown in Fig. 5, where the abscissa is basically the number of grids with $\nu_2 = \ln M/\ln 4$. The factor $\ln 4$, instead of $\ln 2$ was used because the size of each rectangular grid is proportional to $\sqrt{M}$ in the two-dimensional case. All the moments saturate faster than in the one-dimensional case due to a faster rise in moments. The singularity spectrum $f(\alpha)$ is calculated using the two points of M = 1 and M = 4 in $G_q$ and shown in Fig. 6. As expected, the fractal dimension is about 1.7 reflecting the nature of two-dimensional analysis. The Lund string model after detector simulations, shown in Fig. 7, predicts a remarkable agreement with the data.

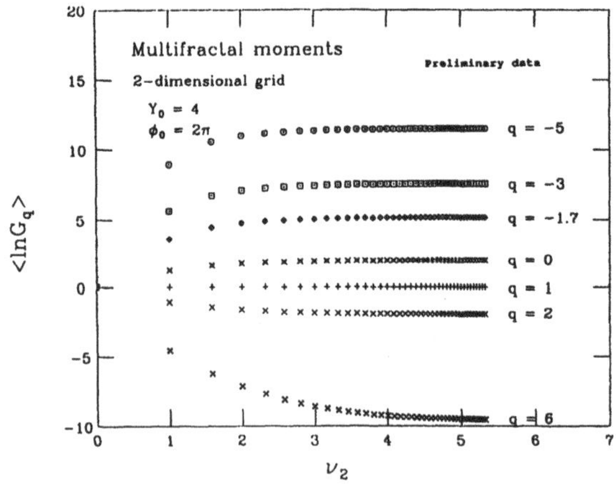


**Fig. 5**

In conclusion, very high multifractal moments up to q = ±20 have been measured. The singularity spectrum functions $f(\alpha)$ in rapidity and in azimuthal angle around the jet axis stay positive up to high moments, which is interpreted as a no phase transition in the $\alpha$-model. On the other hand, the $f(\alpha)$ in azimuthal angle around the beam axis reaches zero for very high moments, which is interpreted as

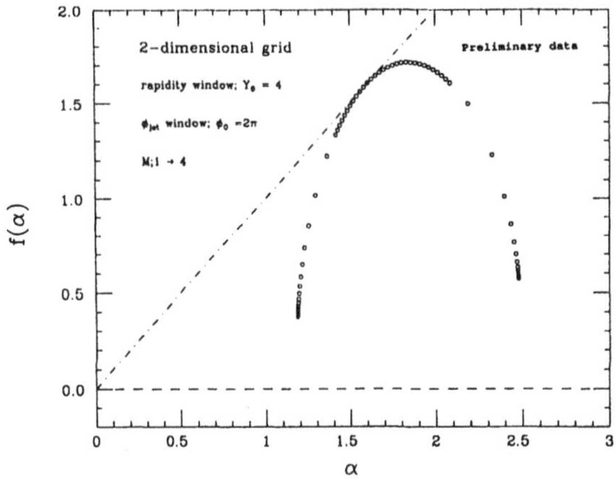


**Fig. 6**

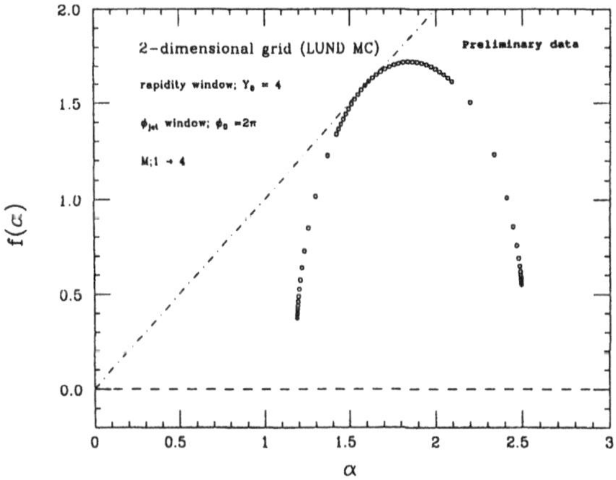


**Fig. 7**

full transition. More theoretical studies are needed to get a coherent physical picture of these data. The fractal dimensions and the widths of f(α) in two different kinds of azimuthal angles clearly show the difference in the event-by-event distributions of particles in jets. As expected, the particles are narrowly collimated in azimuthal angle around the beam axis, but they are almost uniformly distributed in azimuthal angle around the jet axis.

The two-dimensional multifractal moments in rapidity and azimuthal angle around the jet axis have been measured. All the moments saturate faster than in the one-dimensional case because of a faster rise. The fractal dimension is abut 1.7 reflecting the two-dimensional nature. The Lund string model after detector simulations agrees remarkably well with the data.

I would to like thank the members of the organizing committee under Richard Weiner for this lively workshop. I am indebted to the stimulating and enlightening conversations throughout the conference. This work was supported in part by the U.S. Department of Energy, Division of High Energy Physics, under contract W-31-109-ENG-38.

1. B. Mandelbrot, "The Fractal Geometry of Nature", (Freeman, New York, 1982); J. Fluid Mech. 62, 331 (1974).
2. D. Bender et al., Phys. Rev. D31, 1 (1985); M. Derrick et al., Phys. Lett. 168B, 299 (1986).
3. C. Chiu and R. Hwa, Oregon preprint OITS-431 (1990).
4. Ph. Brax and R. Peschanski, Saclay preprint, SPhT/89-203 (1989).
5. B. Andersson et al., Phys. Rep. 97, 32 (1983). (JETSET 5.3 was used.)
6. K. Sugano, Argonne preprint ANL-HEP-CP-90-37; Proc. Santa Fe Workshop on Intermittency in High Energy Collisions, Santa Fe, NM (1990).

# RELATIVE ENTROPY OF MULTIPLICITY DISTRIBUTIONS IN HIGH ENERGY NUCLEON-NUCLEON AND NUCLEUS-NUCLEUS COLLISIONS

**CAI XU, ZHOU DAICUI, ZHOU JINGCHEN**

*Institute of Particle Physics, Huazhong Normal University*

*Wuhan 430070, China*

and

**ZHOU ZHUOMEI**

*Department of Physics, Huazhong University of Science and Technology*

*Wuhan 430074, China*

ABSTRACT

A new characteristic quantity, called relative entropy, is introduced to study multiplicity distributions of produced particles consistently both in the cases of nucleon-nucleon and nucleus-nucleus collisions at high energies.

At the moment the best theoretical tool in strong interaction field might be the Quantum Chromodynamics (QCD). Everything what can be calculated by now is in the perturbative way within this theory. The perturbative calculuses are applicable to the processes with large momentum transfers between the colliding particles or their contituents. So, the hard processes, such as the $e$ $e^+$annihilations, the deep inelastic lepton-hadron scatterings or the hadron-hadron collisions with large transfers of transverse momentum, can be successfully described by the QCD. But, the soft processes in which the final states involve small momentum transfers and a large number of produced hadrons, are responsible for the most of cross section. The situation is more complicated in the case of soft processes where the perturbative QCD calculuses cannot be directly applied. Therefore, we need to learn how to treat the unperturbative problems in the soft processes, because of the primary important understanding on these processes.

A lot of what we know at present about the dynamics of the multihadron production comes from the study of multiplicity distributions. Detailed analysis of their properties may provide information about the soft processes in the wide range of energy.

Recently, a new type of scaling law has been observed from the experimental data of multiplicity distributions in high energy hadron-hadron collisions [1]. Some authors [2] called it an entropy scaling. They suggested an information entropy of multiplicity distribution $P(n)$ as

$$S = -\sum P(n) \ln P(n), \qquad \sum P(n) = 1 \tag{1}$$

and a maximum rapidity range of a produced pion in nucleon-nucleon collisions at a given total cms energy $\sqrt{s}$ as

$$Y_{max} = \ln ( ( \sqrt{s} - 2 m_N ) / m_\pi ) \qquad (2)$$

where $m_N$ and $m_\pi$ are mass of nucleon and pion respectively. The ratio $S/Y_{max}$ seems to be independent of the energy from the data in full rapidity space at high energy. Obviously, it is difficult to generalize the analysis in the case of nucleon-nucleon collisions to the case of nucleus-nucleus. The latter is more complicated. Unlike the case of nucleon-nucleon collisions, the multiplicity distributions in the nucleus-nucleus collisions depend not only on the colliding energy, but also on the nuclei mass numbers of projectile ($B$) and target ($A$) and on the 'centrality' of the collision between projectile nucleus and target nucleus. And more difficltly, one can not define a maximum rapidity range in the case of nucleus-nucleus collisions as the Eq. (2) in the case of the nucleon-nucleon collisions.

In this report we shall discuss some questions existed in their suggestion for the case of nucleon-nucleon collisions and introduce a new characteristic quantity to replace it and apply consistently to both cases of nucleon-nucleon and nucleus -nucleus collisions.

Let us recall the definition of the information entropy [3]

$$H_\alpha = - \sum_{i=1}^{N} P_i \log_\alpha P_i \ , \qquad (3)$$

which is considered as a measure of the uncertainty with $N$ possible outcomes whose probabilities are equal to $P_i$ ( $i$ = 1, 2, ..., $N$ ). One should notice that the base of logarithm is nothing else but the unit of entropy, since for any positive $\alpha$, there is a known mathematical formula $\log_{\alpha} P = (1 / \log_{\beta} \alpha) \log_{\beta} P$. To multiply transformation modulus $1/\log_{\beta} \alpha$ is equivalent to the change of the unit of entropy from $\beta$-digit to $\alpha$-digit. Of course, the choice of the unit of entropy is of no important and allowed to be arbitrary. But it is necessary to use the same unit to make comparison between different distributions.

Now, we consider a nucleon-nucleon collision at a given energy $\sqrt{s}$, which is an event with $N$ possible final states whose probabilities are $P(n)$ of $n$ produced particles ($n = 1, 2, \ldots, N$). It should be pointed out here that $N$ is not only the finite number of possible final states of a collision, but also the maximum multiplicity of produced particles at the given energy $\sqrt{s}$. For the sake of simplicity and definition, taking all produced particles in final states as pions, the maximum multiplicity at $\sqrt{s}$ is estimated as

$$N = (\sqrt{s} - 2 m_N) / m_{\pi} \qquad (4)$$

which is also the finite number of final states. If we calculate the entropy of multiplicity distributions based on the entropy unit of $N$-digit, we have

$$H_N = - \sum_{n=1}^{N} P(n) \log_N P(n) , \qquad \sum_{n=1}^{N} P(n) = 1 \qquad (5)$$

It quite evident that the entropy defined by Eq. (5) for different multiplicity distributions at different energies are base on different units of entropy, so that such defined quantities are incomparable with each other. What is the meaning of a quantity which obtained from dividing $H_N$ by $Y_{max}$ suggested in Ref. [2]?

In the following we would like to exploit a different strategy, which was firstly discussed by Zhou Zhuomei and Cai Xu [4]. Making choise of a fixed unit of the multiplicity entropy, for example, $\gamma$-digit, we can calculate them as

$$H = -\sum_{n=1}^{N} P(n) \log_{\gamma} P(n), \tag{6}$$

for different multiplicity distributions at different energies. In the view of information theory [3], while the entropy is equal to zero if and only if a trial has a single outcome with probability 1, the trial with $N$ equiprobable outcomes whose probabilities are all equal to $1/N$ possesses the maximum entropy $H_{max}$. That is to say, $0 \le H \le H_{max}$. According to Eq.(6), it is readily to obtain the maximum entropy

$$H_{max} = \log_{\gamma} N \tag{7}$$

Now, we can naturally introduce a new characteristic quantity, called relative entropy, defined as

$$R = H / H_{max} \tag{8}$$

which obviously is an unit independent quantity and can be used to compare for different multiplicity distributions at different energies. Taking the data [1] at the energy range $\sqrt{s}$ = 22 - 900 GeV, the value of $R$ is approximately equal to 0.42 covering this wide range of energy.

The most meaningful is that the defined quantity $R$ can be directly applied to the case of nucleus-nucleus collisions by estimating the maximum multiplicity $N_{BA}$. It is known that the multiplicity of produced particles in nucleus-nucleus collisions depends on the number of participant nucleons of the colliding nuclei $B$ and $A$ and the impact parameter $b$ between them within a simple geometrical picture, except its dependence on energy. Take a function of $B$, $A$ and $b$, $W(B, A, b)$, the maximum multiplicity in nucleus-nucleus collisions can be approximately expressed as

$$N_{BA} = (N/2)\, W(B, A, b) \qquad (9)$$

where $N$ is the maximum multiplicity in nucleon-nucleon collisions at the same energy. In the first approximation of the geometrical picture, $W(B, A, b)$ is simply the number of nucleons in the overlapping volumes of the two colliding nuclei for head on collisions [5]

$$W = B + A\,(1 - (1 - (B/A)^{2/3})^{3/2}) \qquad (10)$$

In term of Eqs. (6) and (7), $H_{BA}$ and $H_{BA,max}$ can be readily calculated. Thus the relative entropy of multiplicity distributions for nucleus-nucleus collisions can be obtained by

$$R_{BA} = H_{BA} / H_{BA,max} \quad . \qquad (11)$$

The mulptilicity distributions in Eq. (6) is replaced by $P_{BA}(n)$.

**Acknowlegements**

This work is supported in part by the National Natural Foundation of China, the Distinguished Teacher Fund of the State Education Commission of China and the Fok Yin-dong Education Foundation.

**References**

[1] C. Fuglesang, Proc. 19th Int. Symp. on Multiparticle Dynamics, Arles, 1988; preprint CERN-EP/89-135(1989);
R. Szwed, Warsaw Unversity preprint IFD/6/1989 and references therein.

[2] V. Simak, Proc. 19th Int. Symp. on Multiparticle Dynamics, Arles,1988;
P.A. Carruthers, M. Plumer, S. Raha and R.M. Weiner, Phys. Lett. **B212**(1988) 369; P.M. Weiner, Z. Phys. **C38**(1988)199.

[3] L. Brillouin, *Science and Information Theory,* Academic, New York, 1962;
A. Wehrl, Rev. Mod. Phys. **50**(1978)221.

[4] Zhou Zhuomei and Cai Xu, Chinese Phys. Lett. **6**(1989)317.

[5] I. Otterlund, Nucl. Phys. **A418**(1984)87;
B. Andersson *et al.*, LUIP 8602/LUTP-86-9 (1986).

# FACTORIAL CORRELATORS FROM $\pi^+p$ AND $K^+p$ COLLISIONS AT 250 GeV/c

EHS/NA22 Collaboration

Presented by W. Kittel

## ABSTRACT

Factorial correlators are studied in 250 GeV/c $\pi^+p$ and $K^+p$ collisions as a function of the distance in rapidity. The correlators are found to increase with decreasing correlation distance, independently of the rapidity resolution. The increase follows a power law, but the power is considerably larger than expected from a log-normal approximation in the simple $\alpha$ model of intermittency.

Intermittency [1] has now been studied in $e^+e^-$ [2,3,4], $\mu p$ [5], hadron-hadron [6,7,8], hadron-nucleus [9,10] and nucleus-nucleus [9,11,12] collisions. Recent reviews are given in [13,14]. The study has been performed in terms of a power law dependence of the scaled factorial moments

$$\langle F_i \rangle = \frac{1}{M} \sum_m \frac{\frac{1}{N_{evt}} \sum_{evt} F_{i,m}}{\langle n_m \rangle^i} \quad , \tag{1}$$

where $M$ is the number of phase-space bins of size $\delta = \Delta/M$ into which an original region $\Delta$ is subdivided, $n_m$ is the multiplicity in bin $m$ ($m = 1, ..., M$), $F_{i,m} = n_m(n_m-1).....(n_m-i+1)$ is the factorial of order $i$ in bin $m$. $\langle n_m \rangle = \frac{1}{N_{evt}} \sum_{evt} n_m$ is the mean multiplicity in bin $m$ and $\langle n \rangle = \frac{1}{N_{evt}} \sum_{evt} n$ is the mean multipicity in the original region $\Delta$.

While in the early stage of the analysis the interest mainly arouse from the failure of currently used models to reproduce the effect, a large number of alternative interpretations is available now. These range from conventional short range correlations and Bose-Einstein interference, via pencil jets and extended parton cascades, to a possible signal for quark-gluon-plasma. Different data support different interpretations, so that more decisive information is needed.

While the moments defined in (1) measure local density fluctuations in phase space, additional information is contained in the correlation between these fluctuations within a given event. This correlation can be extracted by means of the factorial correlators [1]

$$\langle F_{ij}^{mm'} \rangle = \frac{\langle n_m(n_m-1)\dots(n_m-i+1)\, n_{m'}(n_{m'}-1)\dots(n_{m'}-j+1)\rangle}{\langle n_m \dots (n_m-i+1)\rangle \langle n_{m'}\dots(n_{m'}-j+1)\rangle} , \quad (2)$$

where $n_m$ is the multiplicity in bin $m$ and $n_{m'}$ that in bin $m'$ (see Fig.1). The correlators are calculated at given $\delta$ for each combination $mm'$ and then averaged over all combinations with given $D$.

According to the simple intermittency model ($\alpha$-model) described in [1], the $\langle F_{ij} \rangle$ should depend only on $D$ and not on $\delta$, and the dependence should be according to the power law

$$\langle F_{ij} \rangle \propto (\Delta/D)^{f_{ij}} \quad . \quad (3)$$

For the power $f_{ij}$ (slope in a log-log plot) the following relation has been derived [1]:

$$f_{ij} = f_{i+j} - f_i - f_j = ij f_2 \quad , \quad (4)$$

where the first equal sign is due to the $\alpha$ model, the second to the log-normal approximation. Since according to (4) $f_{11} = f_2$, this can also be written in the form

$$f_{ij} = ij f_{11} \; . \quad (5)$$

Preliminary results for pseudo-rapidity resolution $\delta\eta \geq 0.2$ have been reported by the HELIOS Collaboration [15], where, however, the multiplicities had to be obtained from the transverse energy $E_T$.

The data presented here come from the NA22 experiment performed at CERN. In this experiment the European Hybrid Spectrometer (EHS) is equipped with the Rapid Cycling Bubble Chamber (RCBC) as a vertex detector and exposed to a 250 GeV/c tagged positive, meson enriched beam. In data taking, a minimum bias interaction trigger is used. The details of the spectrometer and the trigger can be found in previous publications [16,17].

Charged particle tracks are reconstructed from hits in the wire- and drift-chambers of the two lever-arm magnetic spectrometer and from measurements in the bubble chamber. The average momentum resolution $\langle \Delta p/p \rangle$ varies from a maximum of 2.5% at 30 GeV/c to around 1.5% above 100 GeV/c.

Events are accepted for the present analysis when measured and reconstructed charge multiplicity are consistent, charge balance is satisfied, no electron is detected among the secondary tracks and the number of badly reconstructed (and therefore rejected) tracks is 0. Elastic events are excluded. Furthermore, an event is called single-diffractive and excluded from the sample if the total charge multiplicity is smaller than 8 and at least one of the positive tracks has $|x| > 0.88$. After these cuts, our "cleaned" inelastic non-single-diffractive sample consists of 59 232 $\pi^+p$ and $K^+p$ events.

For momenta $p_{LAB} < 0.7$ GeV/c, the range in the bubble chamber and/or the change of track curvature is used for proton identification. In addition, a visual ionization scan has been used for $p_{LAB} < 1.2$ GeV/c on the full $K^+p$ and 62% of the $\pi^+p$ sample. Particles with momenta $p_{LAB} > 1.2$ GeV/c are not identified in the present analysis and are treated as pions.

The experimental resolution is better than 0.1 units in the rapidity $y$ variable used in the analysis.

As mentioned above, no losses due to bad reconstruction are allowed in our sample. It has been checked by a comparison to an event sample with track losses that slopes are hardly affected by track losses. For further analysis of possible biases see [18].

In Fig.2a-d we show $\ln\langle F_{ij} \rangle$ as a function of $-\ln\ D$ for four values of $\delta y$ (corresponding to $M = 10$, 20, 30 and 40, respectively). The original region is chosen to be $-2.0 \leq \Delta \leq 2.0$. Errors are preliminary. At given $D$, the $\langle F_{ij} \rangle$ are indeed independent of the resolution $\delta y$. This can be seen from a comparison of Figs.2a-d at a fixed value of $D$ and is indicated in Fig.3 for the value $D = 0.4$.

From $-\ln\ D \approx 0$ onwards, the $\langle F_{ij} \rangle$ of Fig.2 approximately follow a straight line. The slopes $f_{ij}$ are compared to $f_2$ as a function of $ij$ in Fig.4. The slopes lie in a line $f_{ij}/f_2 = a + bij$ with $b$ much steeper than expected from (5) (dashed line in Fig.4).

We conclude that the correlators $\langle F_{ij} \rangle$ are indeed independent of the resolution $\delta y$ and increase with decreasing correlation length $D$ according to a power law. The powers $f_{ij}$ increase linearly with the product $ij$ of the orders, but are considerably larger than expected from the simple $\alpha$ model.

**REFERENCES**


[1] A. Białas and R.Peschanski, Nucl. Phys. B273, 703 (1986); B308, 857 (1988).

[2] B. Buschbeck, P.Lipa, R. Peschanski, Phys. Lett. B215, 299 (1988).

[3] W. Braunschweig et al. (TASSO), "Study of Intermittency in Electron-Positron Annihilation into Hadrons", DESY preprint 89-092.

[4] K. Sugano, "Intermittency Study in $e^+e^-$ Annihilation at 29 GeV", Proc. Workshop on Intermittency in High Energy Collisions, ed. F. Cooper (Los Alamos 1990), to be published.

[5] I. Derado, G. Jancso, N. Schmitz and P. Stopa, "Intermittency in $\mu p$ Scattering at 280 GeV", preprint MPI-PAE/Exp. El. 221 (1990).

[6] I.V. Ajinenko et al. (NA22), Phys. Lett B222, 306 (1989) and B235 (1990) 373.

[7] B. Buschbeck (UA1),"Intermittency Measured in $p\bar{p}$ Collisions at 630 GeV", Festschrift Léon Van Hove, eds. A. Giovannini and W. Kittel (World Scientific 1990), p.211.

[8] G. Ekspong (UA5), "Search for Intermittency in UA5 Data", Proc. Workshop on Intermittency in High Energy Collisions, ed. F. Cooper (Los Alamos 1990), to be published.

[9] R. Hołynski et al. (KLM), Phys. Rev. Lett. 62, 733 (1989).

[10] NA22 Coll., "Intermittency in 250 GeV/c $\pi^+$ and $K^+$ Collisions on $Al$ and $Au$ Nuclei", paper submitted to the Int. Europhys. Conf. on High Energy Physics, Madrid 1989.

[11] I. Derado (NA35), "Investigations of Intermittency in the Reaction $^{32}S + ^{32}S$ at 200 GeV per Nucleon", Festschift Léon Van Hove, eds. A. Giovannini and W. Kittel (World Scientific 1990), p.257.

[12] R.J. Wilkes (EMU01), "Scaled Factorial Moment Analysis of 200 $A$ GeV Sulfur + Gold Interactions", Proc. Workshop on Intermittency in High Energy Collisions, ed. F. Cooper (Los Alamos 1990), to be published.

[13] B. Buschbeck and P. Lipa, Mod. Phys. Lett. A4, 1871 (1989).

[14] W. Kittel and R. Peschanski, "Review on Intermittency in Particle Multiproduction", Proc. Europhys. Conf. on High Energy Physics 1989, to be published.

[15] HELIOS Collaboration

[16] M. Adamus et al. (NA22), Z. Phys. C32, 475 (1986).

[17] M. Adamus et al. (NA22), Z. Phys. C39, 311 (1988).

[18] NA22 Collaboration (presented by W. Kittel), "Low $p_T$ Intermittency in $\pi^+p$ and $K^+p$ Collisions at 250 GeV/c", Proc. Santa Fe Workshop on Intermittency in High-Energy Collisions, ed. F. Cooper (Los Alamos).

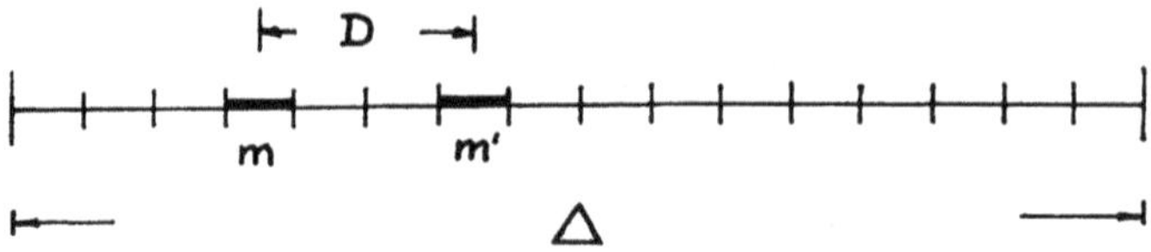


Fig.1 Definition of bin size $\delta$ and correlation distance $D$.

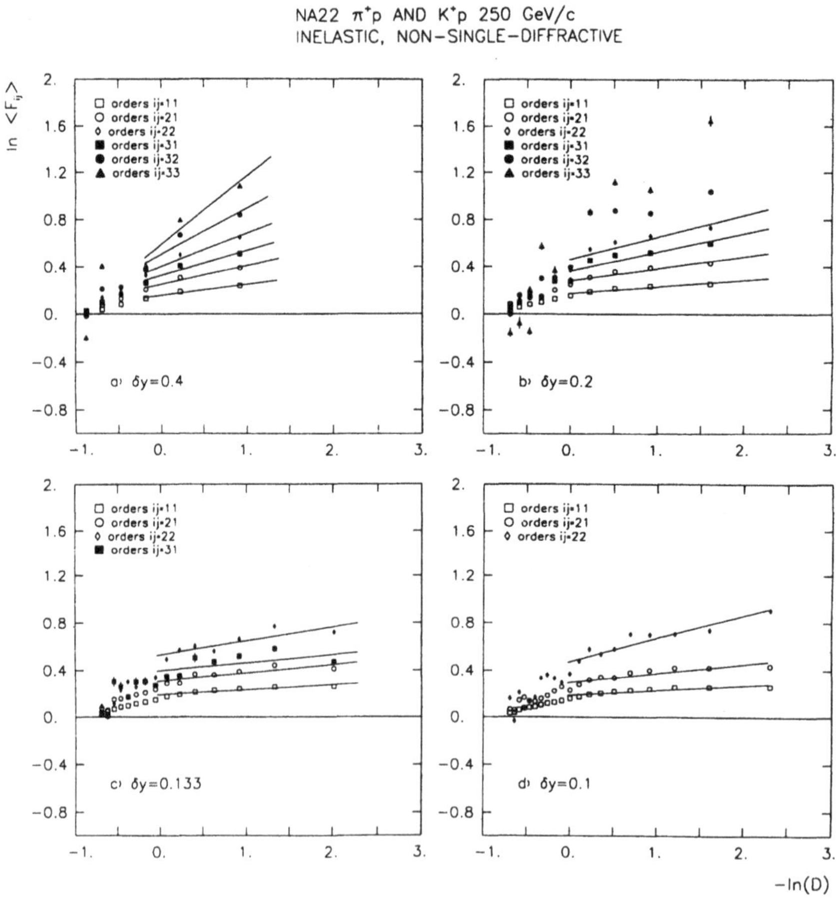


Fig.2 $\ln\langle F_{ij}\rangle$ as a function of $-\ln D$ for four values of $\delta y$, as indicated.

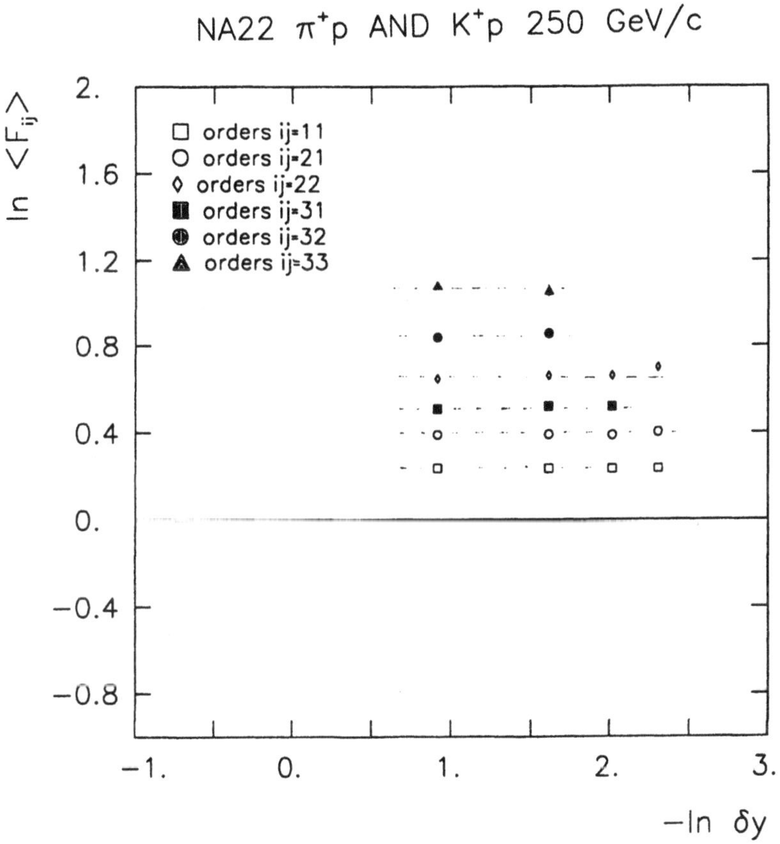


Fig.3 Dependence of $\ln\langle F_{ij}\rangle$ on the bin size $\delta y$, for a correlation distance $D = 0.4$.

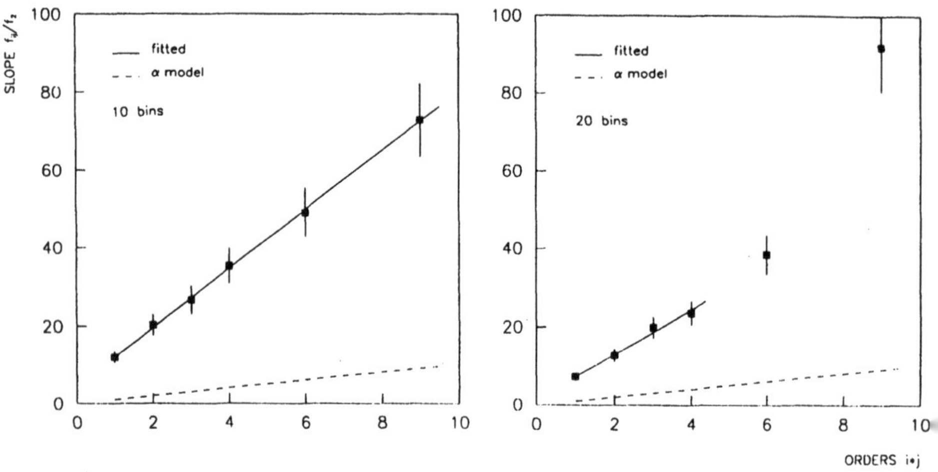


Fig.4 The increase of $f_{ij}/f_2$ with increasing order $ij$, compared to that expected from the $\alpha$ model (dashed line).

# FRACTAL STRUCTURES AND INTERMITTENCY IN QCD

Gösta Gustafson
Department of Theoretical Physics, University of Lund
Sölvegatan 14A, S-22362 Lund, Sweden

**Abstract**

New results are presented for fractal structures and intermittency in QCD parton showers. The particle momenta define a fractal curve in energy-momentum space with a dimension $1 + \sqrt{3\alpha_s/2\pi}$, i.e. one plus the anomalous dimension of QCD. For the distributions in small $y$-regions the property with jets within jets corresponds to a fractal structure with a dimension $\sqrt{3\alpha_s/2\pi}$ (for high moments). At lower energies in the PEP-PETRA range the soft hadronization is important. The experimental data can be understood if the properties of the directly produced pions are carefully taken into account.

## 1. Introduction

The earlier results discussed in this introduction are obtained in collaboration with B. Andersson and P. Dahlqvist, while the new results in sections 2 and 3 are obtained together with the students A. Nilsson and C. Sjögren. In section 2 I will discuss properties of perturbative QCD. New since the Santa Fe workshop are analytic results for moments in small rapidity windows. In section 3 I discuss intermittent effects in soft hadronization. I here concentrate on data for $e^+e^-$-annihilation from PEP and PETRA, but the results are relevant also to hadron collisions and DIS.

If we study e.g. $e^+e^-$-annihilation into hadrons, this process is often described in terms of two phases, a hard perturbative phase, described in terms of quarks and gluons, and a soft phase in which the energy of these partons is transformed into the observable hadrons. The latter phase can be described in terms of clusters or in terms of strings as in the Lund model. At high energies the properties of the hard phase are most important, while the soft phase is more essential at lower energies, as will be further discussed in the following.

To study the hard perturbative phase we use two important tools:

- The dipole formulation of QCD cascades [1]
- An infrared stable measure on parton states related to the hadronic multiplicity [2].

*Dipole Formulation*

A high energy $q\bar{q}$-system radiates gluons according to the dipole formula

$$dn = \frac{3\alpha_s}{4\pi^2}\frac{dk_\perp^2}{k_\perp^2}dy\,d\varphi \tag{1}$$

Here the phase space available is given by the relation

$$|y| \lesssim \frac{1}{2}\ln(s/k_\perp^2) \tag{2}$$

which corresponds to the triangular region in a $y - \ln k_\perp^2$ diagram as shown in fig. 1a.

If two gluons are emitted, then the distribution of the hardest gluon is described by eq. (1), while the distribution of the second, softer, gluon corresponds to two dipoles, one stretched between the quark and the first gluon, and the second between this gluon and the antiquark [3].

If the masses of the $qg_1$ and $g_1\bar{q}$ systems are $\sqrt{s_{12}}$ and $\sqrt{s_{23}}$ respectively, then the transverse momentum and rapidity of the first gluon are given by the relations

$$\begin{cases} s \cdot k_{\perp 1}^2 &= s_{12} \cdot s_{23} \\ y_1 &= \frac{1}{2}\ln\frac{s_{23}}{s_{12}} \end{cases} \tag{3}$$

Therefore the rapidity range $\Delta y$ available for the second gluon is given by

$$\begin{aligned} \Delta y &= \ln(s_{12}/k_{\perp 2}^2) + \ln(s_{23}/k_{\perp 2}^2) = \\ &= \ln s + \ln k_{\perp 1}^2 - 2\ln k_{\perp 2}^2 \end{aligned} \tag{4}$$

The phase space available for the second gluon thus corresponds to the folded surface in fig. 1b, with the constraint $k_{\perp 2}^2 < k_{\perp 1}^2$, as the first gluon is assumed to be the hardest one.

This procedure can be generalized so that the distribution of a third, still softer gluon corresponds to three dipoles. With $n$ gluons the emission of gluon number $n+1$ is given by a chain of dipoles. We note that the dipoles connect the gluons in the same way as the string in the Lund fragmentation model.

Thus, with many gluons the gluonic phase space can be represented by the multifaceted surface in fig. 1c. Each gluon adds a fold to the surface, which increases the phase space for softer gluons.

We note that in this process the recoils are neglected, as is normal in the leading log approximation. With these approximations it is possible to make analytic calculations, as we will see below. Recoil effects and kinematical constraints for hard gluons can be taken into account in a Monte Carlo simulation program (for hard gluons the constraint in eq. (2) is not exactly correct but should be replaced by $cosh(y) < \sqrt{s}/2k_\perp$). Such a program called ARIADNE is developed by U. Pettersson and L. Lönnblad [4].

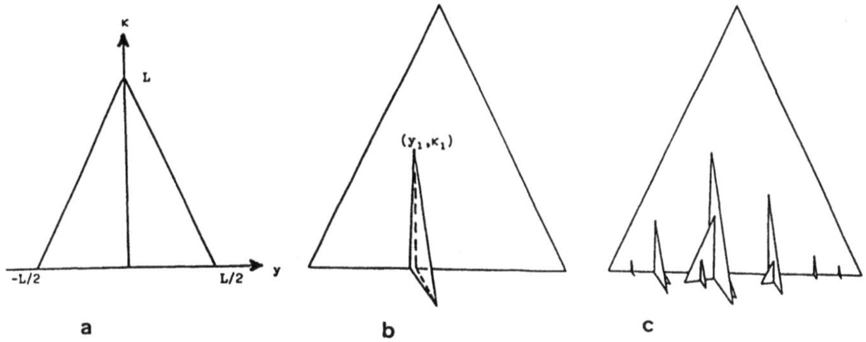


**Fig. 1a.** The phase space available for a gluon emitted by a high energy $q\bar{q}$ system is a triangular region in the $y$-$\kappa$ plane ($\kappa = \ln k_T^2/\Lambda^2$).
**Fig. 1b.** If one gluon is emitted at $(y_1, \kappa_1)$ the phase space for a second (softer) gluon is represented by the area of this folded surface.
**Fig. 1c.** Each emitted gluon increases the phase space for the softer gluons. The total gluonic phase space can be described by this multifaceted surface.

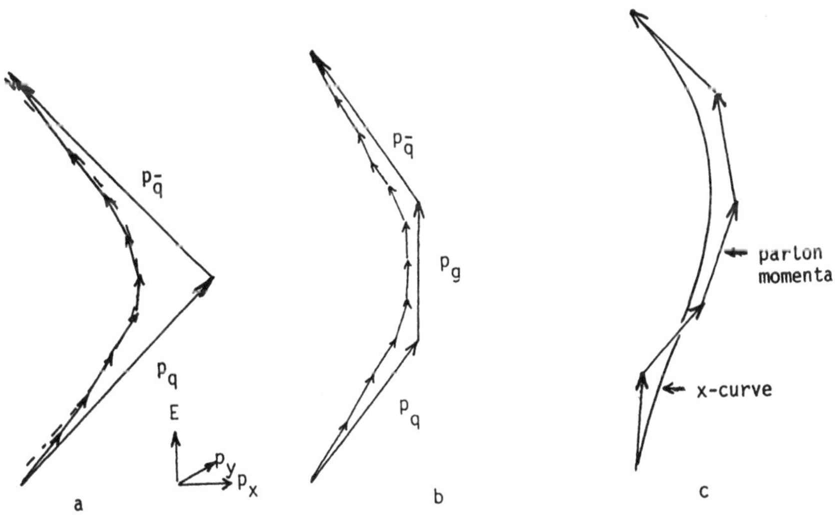


**Fig. 2a.** In a $q\bar{q}$ system the hadron momenta are distributed around a hyperbola in energy-momentum space.
**Fig. 2b.** For a $q\bar{q}g$-system the hadron momenta are distributed around two hyperbolae.
**Fig. 2c.** For a multigluon state the hadron momenta are distributed around a curve in energy-momentum space (called the x-curve) which smoothly follows the (colourordered) parton momenta.

*Multiplicity Measure*

In string fragmentation (or a longitudinal phase space model) the hadronic multiplicity for a simple $q\bar{q}$-system is proportional to $\ln s$. The hadrons are evenly distributed in rapidity, which means that their energy-momentum four-vectors are distributed around a hyperbola (cf fig. 2a).

For a $q\bar{q}g$-system, we obtain in the Lund string fragmentation model a bent string with two straight segments. The energy in the segments is $s_{12}$ and $s_{23}$, where $s_{ij} = (k_i + k_j)^2$ and $k_1, k_2$ and $k_3$ are the momenta of the $q, g$ and $\bar{q}$ respectively. Thus the average multiplicity, $n$, is given by the relation

$$\langle n \rangle \sim \ln s_{12} + \ln s_{23} \approx \ln s + \ln k_\perp^2 \tag{5}$$

Here $k_\perp$ is the transverse momentum of the gluon. We also note that this expression is equal to the length of the baseline of the surface in fig. 1b. The momentum distribution of the hadrons is in this case given by two hyperbolae (fig. 2b).

For a multigluon state we find in the same way

$$\langle n \rangle \sim \sum \ln(s_{i,i+1}/m_0^2) \approx \ln(s/m_0^2) + \sum \ln(k_{\perp i}^2/m_0^2) = \lambda \tag{6}$$

This expression, which we call $\lambda$, is an "effective rapidity range". It is given by the length of the baseline in fig. 2c.

It is possible to calculate the distribution $P(\lambda, s)$ in $\lambda$ for fixed $s$ [2]. We introduce the Laplace transform $\mathcal{P}$ and use the following notation

$$\begin{aligned} L &= \ln(s/\Lambda^2) \\ \mathcal{P}(\beta, L) &= \int d\lambda e^{-\beta\lambda} P(\lambda, L) \end{aligned} \tag{7}$$

It is then possible to derive the following differential equation

$$\begin{aligned} \frac{d^2}{dL^2}(\ln \mathcal{P}(\beta, L)) &= \frac{\alpha_0}{L}(\mathcal{P}(\beta, L) - 1) \\ where\ \alpha_0/L &= 3\alpha_s/2\pi \end{aligned} \tag{8}$$

together with the boundary conditions

$$\begin{cases} \mathcal{P}(\beta, 0) = 1 \\ \frac{d}{dL}\mathcal{P}(\beta, L = 0) = -\beta \end{cases} \tag{9}$$

This implies that it is possible to obtain the **full** distribution in $\lambda$ for **all** energies. (In the conventional approach to the cascade only asymptotic results are obtained [5].) Simple expressions are obtained for the first few moments, while the full distribution can be obtained by numerical calculations. Thus we find e.g.

$$\bar{\lambda} = \sqrt{\frac{L}{\alpha_0}} I_1(2\sqrt{\alpha_0 L}) \underset{s\ \text{large}}{\sim} (\ln s)^{1/4} \exp(2\sqrt{\alpha_0 \ln s}) \tag{10}$$

We note that in the conventional approach the same exponential expression is obtained for the average gluon multiplicity only in a very complicated way.

The quantity $\lambda$ can be generalized in an infrared stable way such that soft gluons give small but positive contributions [6]. This implies that no $k_\perp$-cutoff is needed. (With the expression in eq. (6) a negative contribution would be obtained for gluons with $k_\perp < m_0$.) This generalized measure $\lambda$ can be used in Monte Carlo simulations of the QCD cascade, where recoils and kinematical constraints are properly taken into account [6, 7] . The recoils are neglected in the strong ordering approximation which led to eqs. (8,9).

We see that for a fixed energy $W$ the perturbative cascade gives a parton state with definite parton momenta, and thus a definite value of $\lambda$. The soft hadronization mechanism then gives a certain hadronic state with a hadron multiplicity $n$. If the hadronization is described by the Lund string fragmentation model, we then find that the distribution in $n$ depends **only on** $\lambda$ and not on $W$ [6]. Thus for a high energy event with few gluons and a low energy event with many gluons, but with the same $\lambda$, we have the same effective string length and the same hadron multiplicity distribution.

It is then possible to separate the multiplicity fluctuations in two piece

$$\left.\frac{D^2(n)}{\langle n\rangle^2}\right|_{fixed\ W} = \left.\frac{D^2(\lambda)}{\langle\lambda\rangle^2}\right|_{fixed\ W} + \left.\frac{D^2(n)}{\langle n\rangle^2}\right|_{\lambda=\langle\lambda\rangle} \tag{11}$$

where the first term on the right hand side depends only on the perturbative cascade and the second term only on the soft hadronization. The results are shown in fig. 3. We note that the cascade contribution to the variance, $V_{casc}$ dominates at high energies while the fragmentation contribution, $V_{frag}$, is more important at lower energies.

For the momentum distribution of the hadrons it turns out that it is possible to generalize the hyperbolae in the $q\bar{q}$- and $q\bar{q}g$-cases (figs. 2a,b) and define a timelike curve in energy-momentum space [7]. This curve (which we call the x-curve) follows the (colour ordered) parton momenta in such a way that the corners are smoothed out with a resolution power given by a parameter $m_0$ (see fig. 2c). The length of the x-curve is given by $\lambda$ and we find that if we just cut it into equal pieces, then we obtain an average momentum distribution of the hadrons. The soft hadronization just adds limited fluctuations around this average.

These features give a quantitative meaning to the notion of local parton-hadron duality (LPHD) [8]. This is illustrated in fig. 4, which shows the distribution in rapidity and transverse momentum at a large energy. We note that if the cutoff is locally invariant, e.g. as a cutoff in dipole mass rather than parton mass, then $\lambda$ is proportional to the number of gluons and thus $n_{hadrons} \propto n_{gluons}$. The x-curve can also be constructed from experimental calorimeter data, as a direct check on the perturbative QCD shower.

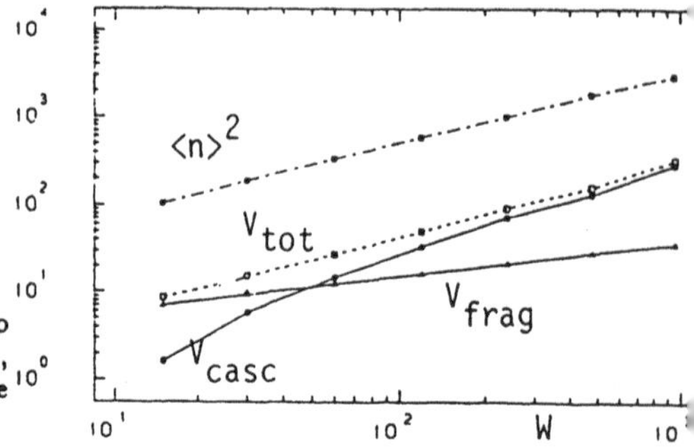


Fig. 3. The variance, $V$, separated into contributions from the QCD cascade, $V_{casc}$, and the soft fragmentation, $V_{frag}$. The dash-dotted curve shows $\langle n\rangle^2$.

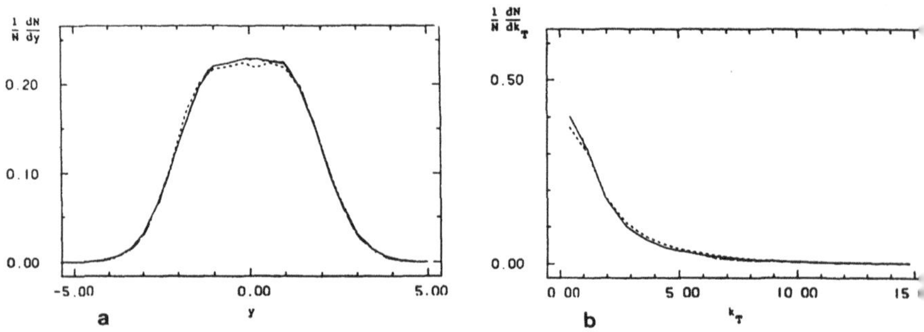


**Fig. 4. Distributions in rapidity (a) and $p_T$ (b) for $e^+e^-$-annihilation events at 200 GeV with sphericity > 0.1. The dashed curve corresponds to a partitioning of the x-curve and the solid line to standard Lund fragmentation.**

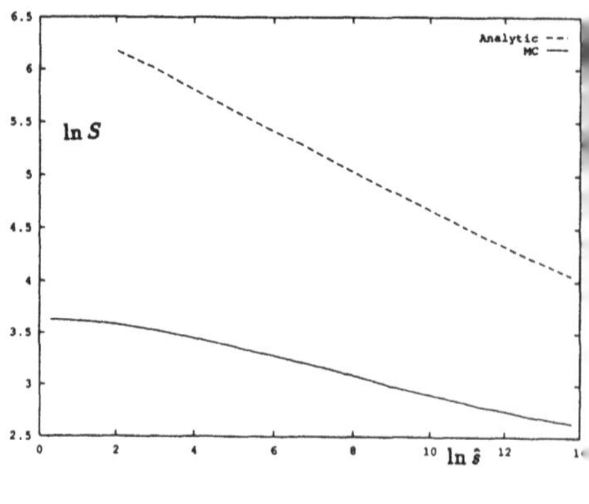


Fig. 5. The length $S$ of the x-curve, as defined in eq. (12), as a function of the resolution $\hat{s}$, in $e^+e^-$-annihilation at 1000 GeV. The dashed line is the analytic approximation and the solid line shows the MC results.

## 2. Fractal Structures and Intermittency in the Perturbative Phase [9]

*A. The x-curve has a fractal structure*

The x-curve is "longer" if it is studied with a higher resolution (cf Koch's snowflake curve). We divide the curve in pieces with a certain invariant length corresponding to a given resolution. If the length of the curve is determined by the number of pieces times the invariant length of each piece, then the total length will decrease with increasing resolution. This is therefore not a suitable way to measure the length of the more twisted curve. Instead we use as definition of the length of a piece with invariant length $\hat{s}$, the length of a hyperbola passing through the endpoints. This length is given by $\ln(\hat{s}/m_0^2)$ (if $\hat{s}$ is not too small). Thus we **define the length $S$ obtained with the resolution given by $\hat{s}$ in the following way**

$$S \sim (\text{no of pieces}) \cdot \ln(\hat{s}/m_0^2) \tag{12}$$

Within the analytic approximation discussed above it is possible to calculate $S$ for a given energy $s$ and resolution $\hat{s}$. The result can be expressed in terms of generalized Bessel functions. For asymptotic energies ($s >> \hat{s} >> m_0^2$) we obtain the result

$$S \sim (\ln s)^{1/4} \exp(2\sqrt{\alpha_0 \ln s})(\ln \hat{s})^{3/4} \exp(-2\sqrt{\alpha_0 \ln \hat{s}}) \tag{13}$$

Thus we see that the fractal dimension $D$ is given by the following expression

$$D = 1 - \frac{d(\ln S)}{d(\ln \hat{s})} \approx 1 + \sqrt{\frac{\alpha_0}{\ln \hat{s}}} = 1 + \sqrt{\frac{3\alpha_s(\hat{s})}{2\pi}} \tag{14}$$

In the square root we recognize the anomalous dimension of QCD. Thus the x-curve gives a geometrical interpretation of this anomalous dimension.

It is possible to calculate $S(s, \hat{s})$ also with the Monte Carlo simulation program, which takes recoils and kinematical constraints better into account. The result is shown in fig. 5 together with the analytic solution given in terms of Bessel functions. The curves are qualitatively similar, although they are quantitatively different. We note in particular that the derivative of $\ln S$ with respect to $\ln \hat{s}$ is decreasing with increasing $\hat{s}$, which reflects the running coupling constant in the expression for the dimension in eq. (14). Thus the fractal dimension is also varying and approaching one for very coarse resolutions.

*B. Distributions in y*

A lot of interest has recently focused on the notion of intermittency. It was suggested by Białas and Peschanski [10] to study the factorial moments of multiplicty distributions in a rapidity range $\delta y$.

$$F_q = \langle n(n-1)\cdots(n-q+1)\rangle / \langle n\rangle^q \tag{15}$$

They suggested that if $F_q$ increases with decreasing $\delta y$ it is a signal of an intermittent behaviour. For a Poissonic distribution $F_q$ would be independent of $\delta y$. In many experiments $\ln F_q$ are seen to rise linearly with $\ln(1/\delta y)$ down to $\delta y \sim 0.1$ and then level off [11].

There are standard methods developed to analyze fractal structures [12]. Let e.g. the total investigated region in $y$ and $\varphi$ be divided in intervals (in $y$ or $\varphi$) or squares (for the 2-dimensional distribution in $y$ and $\varphi$) with lengths or sides $\delta$. If the distribution has the scaling property

$$\sum_i n_i^q \sim \delta^\tau \tag{16}$$

where the sum goes over all intervals or squares, then this is interpreted in terms of a dimension $D_q$ given by

$$D_q = \frac{\tau}{q-1} \tag{17}$$

For $q = 0$ the dimension $D_0$ equals the Hausdorff dimension. (The use of these multifractal dimensions to analyze multiplicity distributions was suggested independently (as far as I know) in Lund, Vienna and Oregon [13].)

If we use normal moments as in eq. (16), or scaled normal moments

$$C_q \sim \frac{\langle n^q \rangle}{\langle n \rangle^q} \tag{18}$$

rather than the factorial moments in eq. (15), then it is possible to study also noninteger variables, e.g. the piece of the x-curve which has its tangent within an interval in $y$. This would represent the hadron distribution without the noise from the soft hadronization, and thus reveal the properties of the perturbative cascade.

The feature of QCD with jets within jets within jets, similar to a Cantor dust, gives fractal properties. If we look at $< \lambda^q >$ we see that for large $q$-values this is dominated by few events with large values of $\lambda$. These are events with a hard jet where the tip of the jet is inside the $y$-range $\delta$. We can calculate the angle between an emitted gluon in a gluon jet and the main jet axis, and see that for the situation shown in fig. 6 the piece indicated is contained within a $y$-region $\delta \approx e^{-\Delta}$, where $\Delta$ is defined in the figure. If we look at fig. 1c, we realize that the $\lambda$-value for this piece is just the same as for a full $e^+e^-$-annihilation event with an energy given by $\ln s = \ln k_\perp^2 - 2\Delta = \ln(k_\perp^2 \delta^2)$, where $k_\perp$ is the transverse momentum of the jet. To get the distribution $P_\delta(\lambda)$ of $\lambda$ within the $y$-region $\delta$ we note that the logarithm of the Laplace transform is obtained by adding the contribution from jets with different $k_\perp$. In this way we can show that the contribution to $< \lambda^q >$ is given by

$$\langle \lambda^q \rangle_{tip} \sim q\delta \sqrt{\frac{3\alpha_s(s\delta^2)}{2\pi}} (\bar{\lambda}_{e^+e^-}(s\delta^2))^q \tag{19}$$

Here $\bar{\lambda}_{e^+e^-}$ is the average $\lambda$ in an $e^+e^-$-annihilation event at an energy given by $s \cdot \delta^2$. Using eq. (10) we see that

$$\langle \lambda^q \rangle_{tip} \sim q\sqrt{\frac{3\alpha_s}{2\pi}} \bar{\lambda}^q_{e^+e^-}(s) \delta^{1+q\sqrt{\frac{3\alpha_s}{2\pi}}} \tag{20}$$

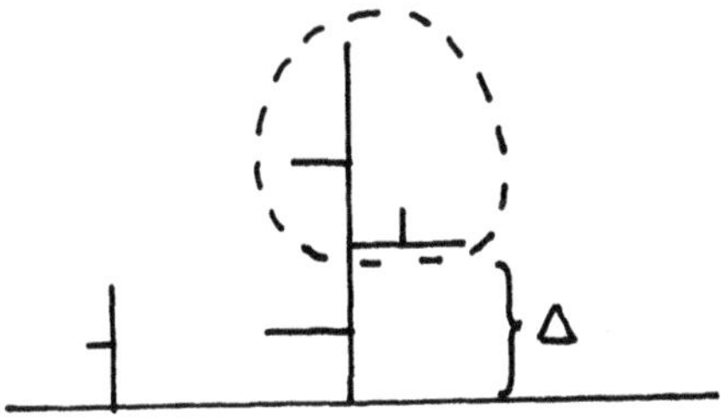


Fig. 6. The baseline of fig. 1c. The length of the curve corresponds to the number of hadrons. The branches and twigs show the QCD structure with jets within jets. The piece indicated is contained within a $y$-region $\delta \approx e^{-\Delta}$.

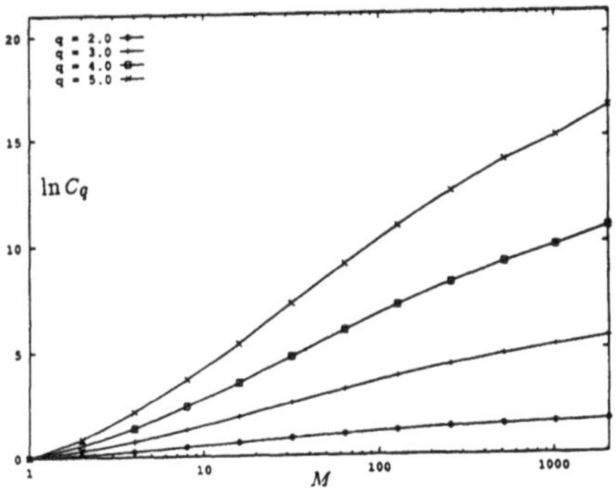


Fig. 7. Normalized moments $C_q$ for the x-curve pieces within varying rapidity regions $\delta y$. The energy is 200 GeV and the total y-range studied is $-2 < y < 2$. $M = 4/\delta y$ is the number of bins.

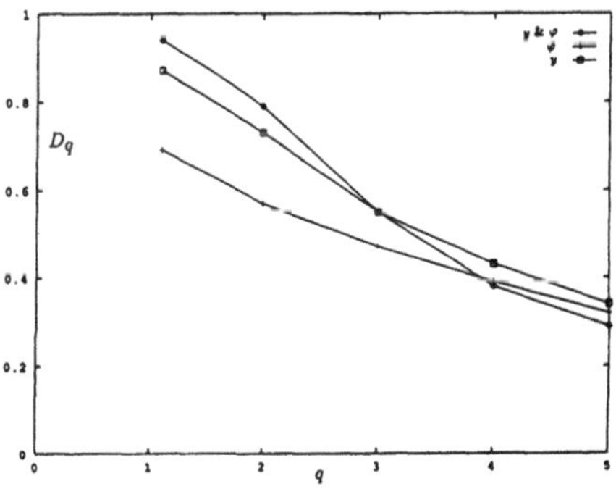


Fig. 8. Multifractal dimensions $D_q$ from moments of the x-curve when binned in $y$, in $\varphi$ or simultaneously in $y$ and $\varphi$. The energy is $200 GeV$.

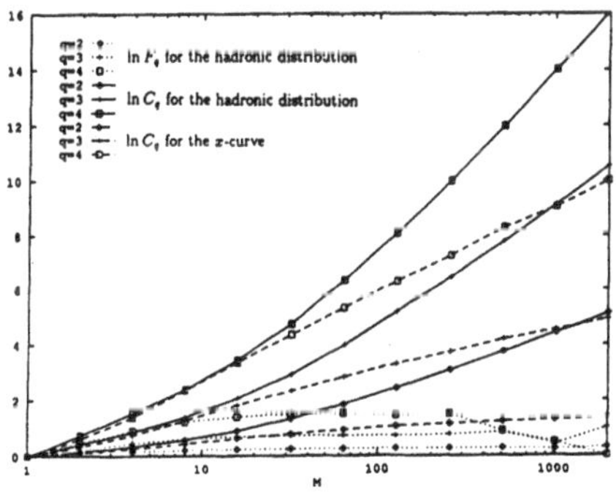


Fig. 9. Normal moments $C_q$ and factorial moments $F_q$ for the hadron distribution, compared with the moments of the x-curve from fig. 7.

If there are no jets we would obtain a background contribution $\lambda = \delta$ corresponding to a straight string. This would give

$$\langle \lambda^q \rangle_{BG} = \delta^q \tag{21}$$

We see that when $q$ and $s$ are large and $\delta$ is small, then the jet tips dominate and from the result in eq. (19) we get the following dimension

$$D_q \approx \sqrt{\frac{3\alpha_s(s\delta^2)}{2\pi}} \qquad q, s \text{ large, } \delta \text{ small} \tag{22}$$

When on the other hand $q$ and $s$ are small and $\delta$ large the background dominates. From eq. (21) we then get the dimension

$$D_q \sim 1 \qquad q, s \text{ small; } \delta \text{ large} \tag{23}$$

If we include the rest of the jets and not only the tips we obtain a sum of terms with dimensions between $\sqrt{\frac{3\alpha_s}{2\pi}}$ and 1. Therefore the result will be as seen in fig. 7 which shows MC results for the scaled moments of the $\lambda$-distribution at 200 GeV. Here the slope of the curve is given by $(q-1)(1-D_q)$. We note that for $\delta$ large the curve is rather flat because $D \sim 1$. For smaller $\delta$ it becomes steeper and $D$ is closer to $\sqrt{\frac{3\alpha_s}{2\pi}}$. For very small $\delta$ it flattens out again, because the running $\alpha_s$ becomes larger when $s\delta^2$ becomes small.

If we look at how the dimensions vary with $q$ we see in fig. 8 how for small $q$ the terms with higher dimensions dominate while for larger $q$ the jet tips dominate giving lower dimensions. We also see in fig. 8 that the dimensions are essentially the same if we divide the phase space in bins in $y$ or $\varphi$ or in twodimensional bins in $y$ and $\varphi$.

*C. Hadron distributions*

Let us assume that there are events of different types such that each type corresponds to a fixed average multiplicity $\bar{n} = \rho\,\delta y$ in one bin and that for this type of events there is a Poisson distribution around this average. If the different types correspond to a distribution $f(\bar{n})$ then the normal multiplicity moments blow up for small $\delta y$ while the factorial moments $F_q$ are given by

$$F_q \sim \int \bar{n}^q f(\bar{n}) d\bar{n} \tag{24}$$

Thus if the underlying dynamics determines $f(\bar{n})$, then one should look at the factorial moments, as was suggested by Białas and Peschanski [10].

This can be compared to the situation described above where the perturbative QCD cascade gives a partonic state corresponding to a certain x-curve, and the soft hadronization process gives a number of hadrons fluctuating around an average determined by the parton state. However, in string fragmentation the fluctuations are smaller than in a Poisson distribution, the variance $V$ is approximately equal to $\frac{2}{3}V_{Poisson}$. Furthermore, the production of particles in neighbouring bins are correlated.

In fig. 9 is shown both normal moments and factorial moments for the hadron distribution, compared with the moments from the x-curve. We see that indeed the normal moments shoot up above the x-curve for small $\delta y$ but that they nevertheless are closer to the x-curve than the factorial moments, which are much further below. It is at present not quite clear if this also implies that the normal moments are more related to the underlying dynamics of perturbative QCD, or if one could find some other quantity which interpolates between the normal and factorial moments, and which would be even better.

## 3. Intermittent Effects in Soft Hadronization [14]

I will here not discuss the question whether the rising factorial moments are indications for a basic intermittent structure. In many experiments on $e^+e^-$-annihilation [15, 16] and other reactions [11] $\ln F_q$ rises with $\ln(1/\delta y)$ with a steeper slope than Monte Carlo calculations based on Lund string fragmentation [17] or the Webber cluster model [18]. Thus one may wonder if these models are basically wrong.

These models are tuned to fit other variables, and are not retuned to reproduce the intermittency slopes. We have found that in the Lund model a large part of the effect is due to directly produced pions which are neighbours in rank (i.e. pions which are not decay products of $\rho$'s, $\omega$'s and other resonances). This implies that the result is very sensitive to the vector to pseudoscalar ratio. The default value in the MC is $P/(V+P) = 0.5$, and we note that e.g. the probability to have three directly produced pions in a row is proportional to $(P/(V+P))^3$. We also note that the vector meson production measured by NA22 [19] and EMC [20] is smaller than the corresponding MC results.

This observation also implies that the result is very sensitive to the fragmentation $p_\perp$ for direct pions, i.e. the $p_\perp$ (with respect to the string direction) given to the pions in the rest frame of the string. In string fragmentation the particles are ordered in rapidity. Thus if $p_\perp$ is smaller the particles are not only closer in $p_\perp$ but also closer in $p_L$.

The production of $q\bar{q}$ pairs in the string can be described as a tunnelling process [21, 17]. Classically the quark and the antiquark are produced at a certain distance $2\mu_\perp/\kappa$, and then pulled apart by the string force. Here $\mu_\perp$ is the transverse mass of the quark (and antiquark) and $\kappa \sim 0.2\ GeV^2$ is the string tension. Quantum mechanically the quark wavefunction has a tail out in the classically forbidden region, and the production probability is given by the overlap between the quark and the antiquark wavefunctions in this region.

The suppression of the wave functions from the classical boundaries to the overlap region can be estimated using the WKB method, and gives the following factor for the production probability

$$\text{Prob.} \sim \exp(-\frac{\pi}{\kappa}\mu_\perp^2) = \exp(-\frac{\pi}{\kappa}\mu^2)\exp(-\frac{\pi}{\kappa}k_\perp^2) \tag{25}$$

where $k_\perp$ is the transverse momentum of the quark and the antiquark and $\mu$ is their mass.

This would be the whole story for quarks pulled off to infinity by the string. In practice the quark must join an antiquark from a neighbouring pair and form a mesonic bound state. For different mesons we obtain from normalization different values for the wavefunction at the classical boundary, and therefore also at the production point in the classically forbidden region. This effect was estimated in ref. [17]. The result was that the probability for a quark to fit into a bound state meson wave function is approximately proportional to $(m_\perp)^{-1}$ where $m_\perp$ is the transverse mass of the meson.

This result was used to understand the suppression of vector mesons relative to pseudoscalar mesons. The ratio between $\rho$ and $\pi$ seems to be of the order 1 : 1 rather than 3 : 1 as obtained from spin counting. For heavier mesons, as e.g. $D$ and $D^*$, the mass difference is not so large and we therefore do not expect such a suppression, in good agreement with data.

In the MC program $q\bar{q}$ pairs are produced with a common Gaussian $k_\perp$ distribution, as obtained from eq. (25). For most mesons the $k_\perp$-dependence the factor $(m_\perp)^{-1}$ is also very small compared to this Gaussian distribution. For pions, which have a very small mass, this is however not the case, and we thus expect that the directly produced pions have smaller transverse momenta than other mesons. Because most pions are the decay products of $\rho$'s, $\omega$'s and other resonances, this effect is not noticeable in most observables, the intermittency being one exception.

We note that we get one factor $(m_\perp)^{-1}$ from each quark or antiquark. Thus we also have correlations between neighbouring hadrons. A $q\bar{q}$ pair with small $k_\perp$ has an increased probability to go into two neighbouring pions. The direct pions thus have a tendency to come in bunches, which increases the intermittency further.

In fig. 10 we show results obtained with $V/(V+P) = 0.35$ and where the $p_\perp$ of the direct pions is suppressed to $\langle p_\perp^2 \rangle = (0.07)^2$ (in the rest frame of the string). (The $p_\perp$ of the $\rho$-mesons is somewhat increased to give agreement with data for all final pions and for $\rho$-production [19].)

We note, however, that the TASSO data shown in fig. 10 are not corrected. When the MC generated events are processed through the detector simulation, the result is increased [15]. Thus the full difference between the old MC results and the data is not dynamical. We also note that the MC does not include the Bose-Einstein effect. This effect must increase the slopes of the factorial moments.[1] It can not be the dominant effect, however, because it would give a larger signal for same sign particles than for all charged particles, in contradiction with the data [15, 22]. Direct charged pions which are neighbours in rank must have opposite charges. Thus the effect discussed above is largest for particles of opposite charges.

We conclude that with about equal contributions from detector acceptance, Bose-Einstein interference and the effect from directly produced pions discussed here, it is likely that we can satisfactorily describe the difference between the MC simula-

[1] A MC version in which identical pairs are pulled together to reproduce the Bose-Einstein correlation is produced by T. Sjöstrand. This MC indeed gives an increased intermittency effect [24].

tions and the data in fig. 10.

Although not noticeable in most other variables, the effect discussed here should be observable also in the $K/\pi$ ratio at high $p_\perp$ in hadronic collisions. The ABCDHW collaboration at ISR [23] has observed a higher $K^+/\pi^+$ ratio than what is obtained in the MC calculations. Due to the trigger bias effect this ratio is most sensitive to the jet fragmentation for large values of $z$ and thus to the directly produced pions and kaons. As the experiment is measuring at an angle $\theta \sim 50°$ a smaller $p_\perp$ with respect to the jet direction will result in a smaller $p_\perp$ also with respect to the beam direction, and therefore to an increased $K/\pi$ ratio. As a smaller $p_\perp$ in the Lund model implies also a smaller $p_L$, we see that the $K/\pi$ ratio will be enhanced also at $\theta \sim 90°$ although to a somewhat lesser degree.

In fig. 11 we show results for the $K^+/\pi^+$ ratio as a function of $x_\perp = p_\perp/p_{\perp jet}$ for jets with $\theta = 50°$. We see that for larger $x_\perp$, relevant due to the trigger bias effect, the $K^+/\pi^+$ ratio is enhanced so that good agreement with data can be obtained.

We finally note that the effects discussed here should be noticeable not only in $e^+e^-$-annihilation, but also in DIS and hadronic collisions if the particles also here are produced from a stringlike colour field.

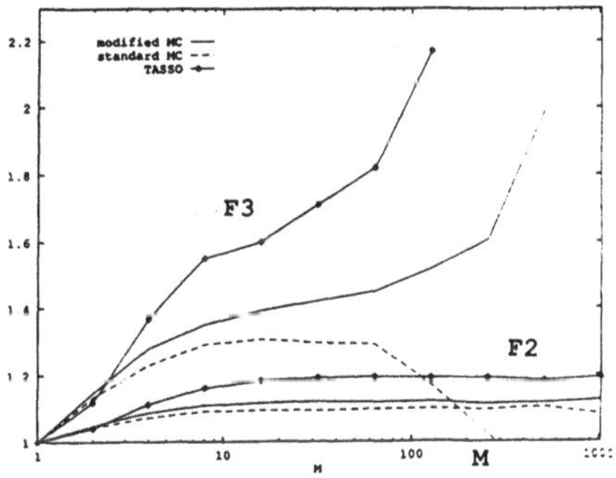


Fig. 10. Normalized factorial moments $F_2$ and $F_3$ for $e^+e^-$-annihilation at 36 GeV. The squares are the uncorrected data from TASSO [15]. The dashed lines show results from the ARIADNE MC with default values for the parameters and the solid lines show results where $V/(V+P) = 0.35$ and the directly produced pions have reduced $p_\perp$ with respect to the string.

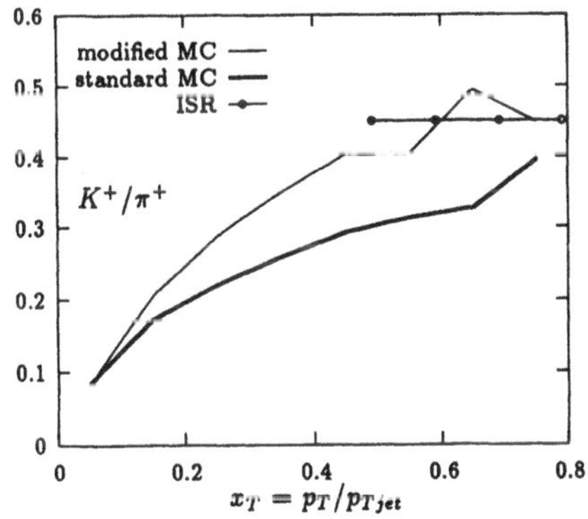


Fig. 11. The $K^+/\pi^+$ ratio as a function of $x_\perp = p_\perp/p_{\perp jet}$ for jets with $\theta = 50°$. The fat line shows results with default values for the parameters and the thin line when modified as in fig. 10. The ISR data from ref. [23] are also indicated.

### 4. Conclusions

Fractals and intermittency are new tools to study multiparticle dynamics. QCD is a nonlinear theory, and we therefore expect fractal structures. QCD is also not scaling, the coupling constant $\alpha_s$ is running, and thus we should expect the fractal dimensions to vary with energy and resolution. To understand these properties, it is very essential to have both a Monte Carlo simulation program and approximation methods suitable for analytic analyses.

**A.** For the **hard perturbative phase** of QCD we have found the following properties.

• The so-called **x-curve** is defined on a parton state and gives the average momentum distribution of the final hadrons. It is a **fractal curve**, embedded in fourdimensional energy-momentum space (cf Koch's snowflake curve). If suitably defined the length increases with the resolution according to a dimension $D = 1 + \sqrt{\frac{3\alpha_s}{2\pi}}$. In the square root we recognize the anomalous dimension of QCD, and thus the x-curve provides a geometrical interpretation of this anomalous dimension. We also note that the dimension varies with the resolution in accordance with the running coupling constant $\alpha_s$.

• For the **distributions in** $y$ we note that it is possible to study moments, not only for multiplicity but also for continuous variables like e.g. the piece of the x-curve with tangent within a certain angular region. The feature with jets within jets corresponds to a fractal structure (cf Cantor dust). For large energies and large moments the multifractal dimensions $D_q$ are given by $\sqrt{\frac{3\alpha_s}{2\pi}}$, i.e. we find again the anomalous dimension of QCD. We also note that the dimensions $D_q$ are the same whether we bin in $y$ or in $\varphi$ or do a twodimensional analysis in $y$ and $\varphi$.

• For the **hadronic multiplicity** we note that the normal moments, as expected, go above the moments of the continuous x-curve for small $\delta y$, but that they nevertheless are closer to the x-curve results than the factorial moments, which are much further below. Does this mean that the normal multiplicity moments also are closer to the dynamical structure of QCD, as the x-curve is defined directly from the parton state?

**B.** For the **soft hadronization phase** we have observed that in the Lund string fragmentation model the intermittency signal is most sensitive to directly produced pions which are neighbours in rank. From the $q\bar{q}$ tunnelling mechanism we expect that direct pions have smaller $p_\perp$ than other hadrons. This effect is neglected in the MC. It is unessential in most observables but noticeable in the intermittency signal and in the $K/\pi$ ratio at high $p_\perp$ in hadron collisions. We expect that this effect together with Bose-Einstein interference can account for the discrepancy between data and MC results for the intermittency signal.

# References

[1] G. Gustafson, *Phys. Lett.* **B175** (1986) 453; G. Gustafson, U. Pettersson, *Nucl. Phys.* **B306** (1988) 746

[2] B. Andersson, P. Dahlqvist, G. Gustafson, *Phys. Lett.* **B214** (1988) 604

[3] Ya.I. Azimov, Yu.L. Dokshitzer, V.A. Khoze, S.I. Troyan, Coherence effects in QCD jets, Leningrad preprint 1051 (1985); *Phys. Lett.* **B165** (1985) 147

[4] L. Lönnblad, U. Pettersson, ARIADNE 2 - A Monte Carlo for QCD Cascades in the Colour Dipole Formulation - an update, Lund preprint LU TP 88-15

[5] A.H. Mueller, *Nucl. Phys.* **B213** (1983) 85, **B228** (1983) 351; A. Bassetto, M. Ciafaloni, G. Marchesini, *Phys. Rep.* **100** (1983) 201

[6] B. Andersson, P. Dahlqvist, G. Gustafson, *Z. Physik* **C44** (1989) 455

[7] B. Andersson, P. Dahlqvist, G. Gustafson, *Z. Physik* **C44** (1989) 461

[8] Ya.I. Azimov, Yu.L. Dokshitzer, V.A. Khoze, S.I. Troyan, *Z. Physik* **C27** (1985) 65; L. van Hove, A. Giovannini, CERN preprint TH 4885/87; M. Garretto, A. Giovannini, T. Sjöstrand, L. van Hove, CERN preprint TH 5252/88

[9] G. Gustafson, A. Nilsson, LU TP 90-7 and to be published

[10] A. Białas, R. Peschanski, *Nucl. Phys.* **B273** (1986) 703, **B308** (1988) 857

[11] For a review, see e.g. W. Kittel, R. Peschanski, *Proceedings of the 1989 EPS Conference*, Madrid, 1989. (Saclay preprint SPhT/89-143)

[12] See e.g. T. Halsey et al., *Phys. Rev.* **A33** (1986) 1141

[13] P. Dahlqvist, G. Gustafson, B. Andersson, *Nucl. Phys.* **B328** (1989) 76, Proc. of XXIV Rencontre de Moriond, March 1989,; P. Lipa, B. Buschbeck, *Phys. Lett.* **B223** (1989) 465, Proc. of XXIV Rencontre de Moriond, March 1989; R.C. Hwa, Univ. of Oregon preprint OITS407 (March 1989)

[14] G. Gustafson, C. Sjögren, Lund preprint LU TP 90-11

[15] W. Braunschweig et al. (TASSO), *Phys. Lett.* **B231** (1989) 548

[16] B. Buschbeck, P. Lipa, R. Peschanski (HRS data), *Phys. Lett.* **B215** (1988) 788; K. Sugano (HRS), contr. to the workshop on Intermittency in high-energy collisions, Santa Fe, March 1990

[17] B. Andersson, G. Gustafson, G. Ingelman, T. Sjöstrand, *Phys. Rep.* **97** (1983) 31; T. Sjöstrand, *Computer Phys. Comm* **39** (1986) 347; T. Sjöstrand, M. Bengtsson, *Computer Phys. Comm* **43** (1987) 367

[18] B.R. Webber, *Nucl. Phys.* **B238** (1984) 492; G. Marchesini, B.R. Webber, *Nucl. Phys.* **B238** (1984) 1

[19] N.M. Agababyan et al. (NA22), *Z. Physik* **C46** (1990) 387

[20] M. Arneodo et al. (EMC), *Z. Physik* **C33** (1986) 167

[21] B. Andersson, G. Gustafson, T. Sjöstrand, *Z. Physik* C6 (1980) 235

[22] I.V. Ajinenko et al. (NA22), contrib. to the EPS conference, Madrid 1989

[23] A. Breakstone et al., *Phys. Lett.* **B135** (1984) 510

[24] W. Kittel, contr. to the workshop on Intermittency in high energy collisions, Santa Fe, March 1990

# MULTIFRACTAL STRUCTURES IN MULTIPARTICLE PRODUCTION IN $\bar{p}-p$ INTERACTIONS AT $\sqrt{s}=1800$ GeV.

CDF Collaboration*

Presented by Franco Rimondi

University of Bologna and INFN Bologna, ITALY.

## ABSTRACT

Fractal structure in multiparticle production of $\bar{p}-p$ "minimum-bias" interactions at $\sqrt{s}=1800$ GeV has been studied using the Collider Detector at Fermilab. Preliminary results are shown and compared with very simple Monte Carlo models.

## INTRODUCTION

In multiparticle production large fluctuations of the final state (pseudo)-rapidity density are observed[1-4]. In order to perform quantitative studies of the statistical significance of these fluctuations, Bialas and Peschanski[5-6] have developed a method of analysing the data in terms of scaled (normalised) factorial moments as functions of decreasing rapidity bin width. In the case of non statistical, self-similar density fluctuations those moments are expected to increase according to a power law with decreasing bin width, down to the experimental resolution. Evidence for such behaviour, usually referred to as intermittency, has been shown to exist in ll[7], hh[8], lh[9] and NN[10] interactions.

Intermittency is closely connected with fractal structures. Theoretical approaches to the study of the multifractal structures in multibody production in particle interactions have been proposed[11-13]. The analysis procedure developed in references[12-13] has been applied to minimum bias $\bar{p}-p$ data at $\sqrt{s}=1800$ GeV. In the following, preliminary results from this analysis are presented. The data have been recorded with the Collider Detector at the Fermilab Tevatron Collider.

* Argonne National Laboratory,Brandeis University,University of Chicago, Fermi National Accelerator Laboratory,Laboratori Nazionali di Frascati, Harvard University, University of Illinois,National Laboratory for High Energy Physics(KEK),Lawrence Berkeley Laboratory,University of Pennsylvania,Istituto Nazionale di Fisica Nucleare Pisa,Purdue University,Rockfeller University,Rutgers University,Texas A&M University, University of Tsukuba,Tufts University,University of Wisconsin.

## THE FRACTAL MOMENTS

In order to allow a quantitative study of the multifractal structure in multiparticle production a set of moments, the fractal moments $G_q$, are defined. A (pseudo)-rapidity interval $\Delta\eta$ is divided into M bins of width $\delta = \Delta\eta/M$. If $k_i$ denotes the number of particles in the $i^{th}$ bin, the multifractal moments are defined as follows:

$$G_q = \sum_{i=1}^{M} p_i^q \quad = \sum_{i=1}^{M} (k_i/n)^q \tag{1}$$

where $p_i = k_i/n$, n is the event multiplicity in $\Delta\eta$ and q is a real number. The summation in (1) is carried over the non-empty bins only.

$G_q$ can be determined as a function of $\delta$. The particle production process exhibits self-similar behaviour when there is a range of $\delta$ in which:

$$G_q \quad \propto \quad \delta^{\tau(q)} \tag{2}$$

It is the relation (2) which has to be studied and the slope $\tau(q)$ has to be extracted from the data.

It should be noticed that for finite resolution and limited multiplicity n, since in (1) only $k_i > 0$ are taken into account, the behaviour (2) does not necessarly appear for $\delta \rightarrow 0$ [13].

In practice the average of $G_q$ over many events has to be considered. This is done by the following procedure. First the average of $Ln(G_q)$, over all the events of the sample, is computed as a function of $\nu$, where $\nu = \log(M/2)$.

$$< Ln(G_q) >= 1/N_{ev}(\sum_{j=1}^{Nev} Ln(G_q)_j) \tag{3}$$

In the plot $< Ln(G_q) >$ versus $\nu$, a region in $\nu$ of approximate linearity should be identified for each q. From those intervals, the mean value of $\tau$ (q) can be computed as:

$$< \tau(q) >= (-\Delta < Ln(G_q) > /\Delta\nu)(1/Ln(2)) \tag{4}$$

Once $< \tau$ (q)$>$ is determined, the two quantities:

$$\alpha_q = d < \tau(q) > /dq \quad \text{and} \quad f(\alpha_q) = q\alpha_q - \tau(q) \tag{5}$$

can be evaluated, as they come from the theory of multifractals. The function $f(\alpha_q)$ describes the fractal structure of the pseudorapidity distribution. For a discussion of the function $f(\alpha_q)$ and of its properties see references 12-14 and references therein.

## THE DATA SAMPLE

Detailed descriptions of the experimental apparatus, as well as of the minimum bias trigger and of the event selection, have been reported elsewhere[15,16]. Here only a few things relevant to the present analysis will be described. A sample of about 140,000 fully reconstructed minimum bias events from the 1988/1989 run has been used. The trajectories of the charged particles were measured by a set of eight time-projection chambers (VTPC) surrounding the beam pipe at the interaction point. The chambers cover a region of about $\pm$ 3 units in $\eta$ and $2\pi$ in azimuth. They provide good $\eta$ determination, with precision better than 0.07 units of $\eta$ for $\eta > 3$ and around 0.005 near $\eta = 0$. For about the 25% of the tracks only a poor $\phi$ measurement is provided and only for about 20% of the tracks it is possible to measure the momenta. The VTPC are segmented azimuthally into octants; track pairs going through the same octant are resolved if they are separated by at least 0.06 units of $\eta$, they are unambiguously resolved if they pass through two different azimuthal segments. Charged tracks within the $\Delta\eta$ interval: [ -2.9 - -0.1 ] $\cup$ [ 0.1 - 2.9 ] have been selected and only events with multiplicity greater than 5 in the above interval have been kept. The preliminary results presented here are not efficiency corrected. The tracking inefficiency is less than 5%, while the charged particle background from photon conversion, particle decay and secondary hadronic interactions is estimated to be about 4-5% in central region, rising to about 10% at the edge of the interval[16].

## RESULTS

In fig.1 the mean values of Ln($G_q$) are plotted as a function of $\nu$ for some values of q. Actually $<$ Ln($G_q$) $>$ has been computed for q ranging from -6 to 6 in steps of 0.5, except in the region -1 $<$ q $<$ 1 where the q step is 0.1. The range of $\nu$ corresponding to the first three points at the lower values of $\nu$ have been chosen to extract the mean slope $< \tau$ (q) $>$, assuming it is given by the mean slope among those three points as determined by the least squares method. From $< \tau$ (q) $>$, $\alpha_q$ and f($\alpha$) have been computed according to formulas (2) and (3). The f($\alpha$) spectrum is shown in fig. 2. It has a smooth behaviour with a maximum at $\alpha_o \approx 0.89$, corresponding to q$\approx$0 as expected.

In fig. 2 results from two very simple Monte Carlo models are also shown. In the first model, tracks are generated with a uniform distribution and the event multiplicity is distributed according to the observed multiplicity distribution. In the second one, tracks are generated in clusters of gaussian shape with the following parameters: $\bar{\eta}$ uniformly distributed in $\Delta\eta$, $\sigma_\eta = 0.5$, cluster multiplicity between

1 and 10. The number of clusters per event is determined by the total event multiplicity which is again distributed according to the observed distribution.

The same analysis has been performed dividing the events in classes, containing events with multiplicities in different restricted intervals. In fig. 3 the f($\alpha$) spectra for each multiplicity interval are shown. The data in fig. 3 correspond to a range of q from -1 to 1 in steps of 0.1. A clear dependence of the f($\alpha$) spectrum on multiplicity is observed: as multiplicity increases, the f($\alpha$) spectrum becomes wider.

Comparisons with the two previously discussed models are also shown.

## CONCLUSIONS

We are looking at a new analysis method of multiparticle final state processes. It can be complementary to intermittency, but it may be too early to draw physical conclusions. The only other experimental results are the $e^+ e^-$ data presented by K. Sugano at this conference. The ability and reliability of the method to extract useful physical information needs to be tested comparing results from different kind of data from the same experiment and from different experiments as well as from Monte Carlo. The preliminary results presented here encourage further studies.

## REFERENCES


1) T.H. Burnett et al.(JACEE): Phys.Rev.Lett.,50,2062,(1983).

2) G.J.Alner et al. (UA5): Phys.Rep.,154,247,(1987).

3) M.Adamus et al. (NA22): Phys.Lett.,B185,200,(1987).

4) M.I. Adamovich et al. (EMU-01): Phys.Lett.,B201,397,(1988).

5) A.Bialas,R.Peschanski: Nucl.Phys.,B273,703,(1988); Nucl.Phys.,B308,857,(1988).

6) A.Bialas, K.Fialkowski, R.Peschanski: Europhys. Lett., 7, 125, (1988); A.Bialas, R.Peschanski: Phys. Lett., B207, 59, (1988); B.Buschbeck, P.Lipa, R.Peschanski; Phys. Lett., B215, 788, (1988).

7) W.Braunschweig et al. (TASSO): Phys.Lett., B231,548,(1989).

8) I.V.Ajinenko et al. (NA22): Phys. Lett., B222, 306, (1989); Phys. Lett., B235, 373, (1990); W.Kittel (EHS/NA22) Collaboration: Santa Fe Workshop on Intermittency in High-Energy Collisions, March 18-21,(1990); C.Albajar

et al. (UA1 Collaboration): CERN-EP/90-56; B.Buschbeck (UA5): Proc. Meeting on Multiparticle Dynamics, La Thuile, Italy, (1989).

9) I.Derado et al.: Preprint, MPI-PAE/Exp.EL.221,(1990).

10) R.Holynski et al. (EMU-07): Phys. Rev. Lett., 62, 733, (1989); R.Albrecht et al. (WA80): Phys. Lett., B221, 427, (1989); I.Derado (NA35): Proc. Meeting on Multiparticle Dynamics, La Thuile, Italy, (1989); E.Gladysz-Dziadus: Mod. Phys. Lett., A4, 2553, (1989).

11) P.Lipa, B.Buschbeck: Phys. Lett., B223, 465, (1989); Ph.Brax, R.Peschanski: Preprint, Saclay SPhT/89-203, (1989).

12) R.Hwa: Preprint, OITS-404,(1989); Preprint, OITS-407, (1989); preprint, OITS-418, (1989).

13) C.B.Chiu,R.Hwa: Preprint, OITS-431 DOE-40200-203,(1990).

14) T.C.Halsey et al.: Phys. Rev.,A33,1141,(1986).

15) F.Abe et al. (CDF Collaboration): Nucl. Instrum. Methods Phys. Res. Sect., A271, 387, (1988).

16) F.Abe et al. (CDF Collaboration): Phys. Rev.,D41,2330,(1990).

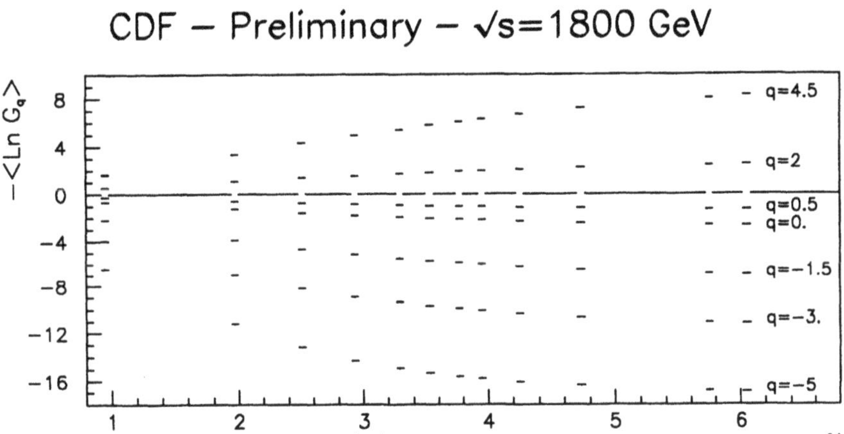


FIG. 1 – Mean value of $Ln(G_q)$ versus $\nu$ for different values of q.

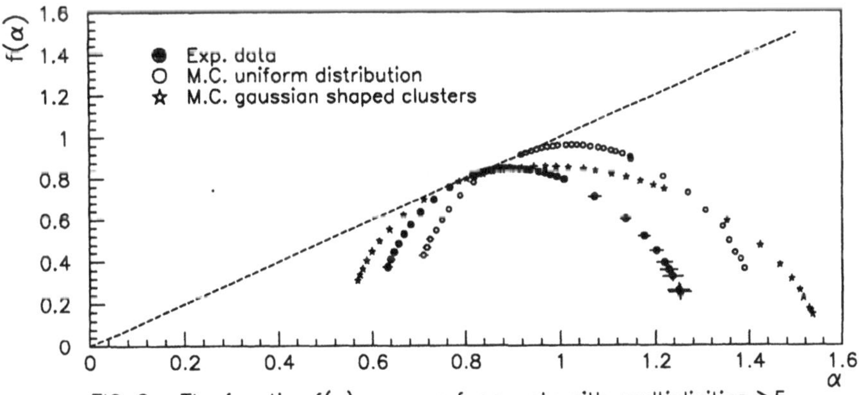


FIG. 2 – The function $f(\alpha)$ versus $\alpha$ for events with multiplicities >5 in $\Delta\eta$ [ −2.9 – −0.1 ]& [ 0.1 – 2.9 ]. Results from the two simple Monte Carlo models described in the text are also shown.

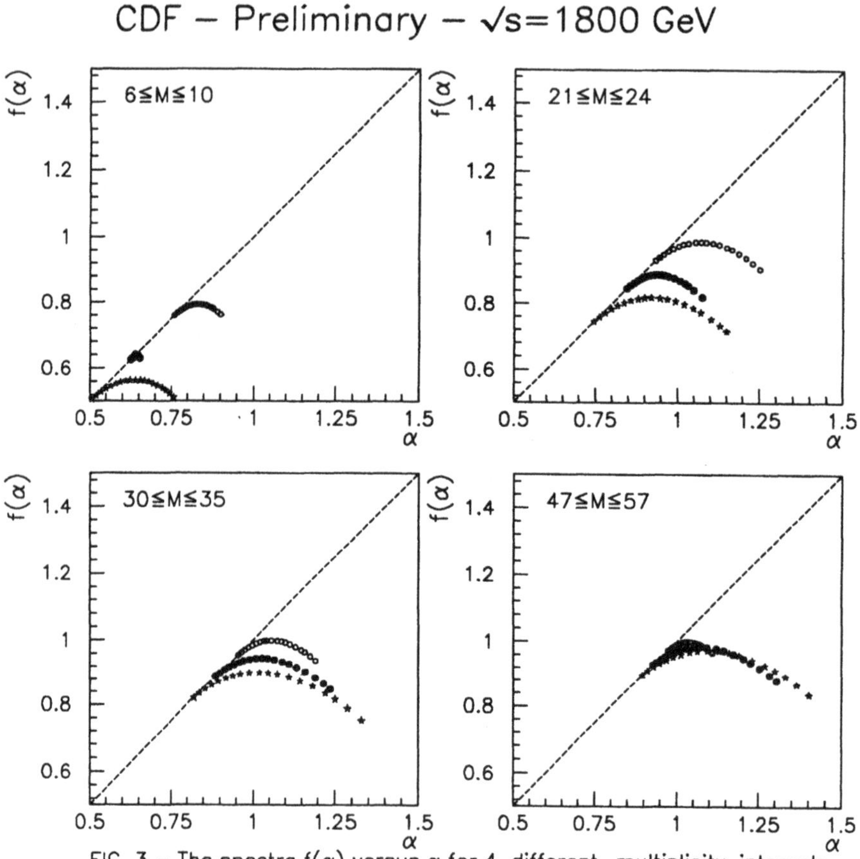


FIG. 3 – The spectra f(α) versus α for 4 different multiplicity intervals. Also shown are the results from the uniform distribution model ( O ) and the cluster model ( ☆ ).

# MULTIPLICITY DISTRIBUTIONS FROM CALORIMETRIC MEASUREMENTS OF TRANSVERSE ENERGY*)

S.Raha[1], R.M.Weiner[1], G.Wilk[1,2] and Z.Włodarczyk[1,3]

[1] Dept. of Physics, University of Marburg, Marburg, FRG
[2] Soltan Institute for Nuclear Studies, Warsaw, Poland
[3] Inst. of Physics, Pedagogical University, Kielce, Poland

Using the recent UA1 data and the Interacting Gluon Model as a theoretical tool we demonstrate how to get multiplicity distributions $P(n)$ from the transverse energy distributions $P(E_T)$.

## 1. Introduction

Multiplicity distributions $P(n)$ constitute one of the most characteristic property of strong interaction dynamics. Of the late, however, measurement of transverse energy distributions $P(E_T)$ have become more and more popular. On the one hand, such measurements are essential to identify the class of events where the Quark-Gluon Plasma may be produced. More importantly, however, with increasing beam sizes ( heavy ions ) as well as increasing beam energies, the multiplicity of particles is expected to be so large that detection of individual particles may tax the technology beyond limits. Calorimetric measurements may then provide the only avenue to pursue.

It is therefore important to understand the relation between the probability of finding a certain transverse energy $E_T$, $P(E_T)$ and the probability of producing a given number of particles $n$, $P(n)$ [1]:

$$P(E_T) = \sum_n \omega(E_T|n)\cdot P(n) \qquad (1)$$

where $\omega(E_T|n)$ is the conditional probability of obtaining a transverse energy $E_T$ in a certain event with a given multiplicity $n$. Equivalently, one can invert this relation to obtain $P(n)$ from a knowledge of $P(E_T)$:

$$P(n) = \sum_{E_T} \bar{\omega}(n|E_T)\cdot P(E_T). \qquad (2)$$

*) presented by Z.Włodarczyk

These quantities may be calculated within presently available phenomenological models of multiparticle production. Furthermore, at currently accessible energies, where both $P(n)$ and $P(E_T)$ are measurable, these conditional probabilities are also measurable. We therefore suggest a new strategy in the investigation of the relationship between $P(E_T)$ and $P(n)$ which might ultimately lead to a method for determining $P(n)$ from the measurement of $P(E_T)$. The idea is to compare at present energies the experimental $\bar{\omega}(n|E_T)$ with that calculated from theoretical models and then to tune these models so that they can be expected to work satisfactorily at higher energies where measurements of $P(n)$ ( and $\bar{\omega}(n|E_T)$ ) would not be possible. We would like to present here some progress along this line based on the Interacting Gluon Model (IGM)[2,3].

## 2. P(n) and P($E_T$) in the Interacting Gluon Model

The IGM was originally intended to provide inelasticity distributions and leading particle spectra for both pp and pA collisions. It has also proved to be successful in describing energy density distributions in heavy ion collisions [3]. Actually the model describes only the first stage of the multiparticle production process, namely the formation of the lump of gluonic matter in the central region of reaction ( which we shall call Central Fireball - CF ). The picture proposed ( cf. ref [4] for more details ) is that the valence quarks of colliding hadrons pass through without losing their energy whereas the gluonic contents of these hadrons ( among which we count also the $q\bar{q}$ sea pairs ) interact violently forming a CF. The gluons which did not interact join their corresponding valence partners and together they form what we identify as leading particles [2-4].

Actually in order to obtain $P(E_T)$ and $P(n)$ one has to supplement the IGM as presented here with some prescription for the hadronization properties of the CF. This will be done by making use of the highly model independent method of maximization of the information entropy. [5]. In actual calculations we have used a Monte Carlo event generator implementation of the IGM as decribed in ref.[4]. It provides us, in a given run, among other things with a CF of mass $M$. We assume now that this CF of mass $M$ will decay on average into $N = \langle N(M) \rangle = a \cdot M^b$ number of secondaries of mass $\mu$ ( which we take to be just pions ). The maximization of information entropy principle [5] tells us then that the single particle distribution $f(y,p_T)$ <u>has to be</u> of the form

$$f(y,p_T) = \frac{1}{Z}\cdot \exp[\, -\alpha\mu_T - \beta\mu_T\cdot \cosh y \,] \tag{3}$$

where $\mu_T=[\mu^2+p_T^2]^{1/2}$ and where $\alpha=\alpha(M,N,\mu)$ and $\beta=\beta(M,N,\mu)$ are to be obtained from the following constraint equations:

$$\begin{aligned} &\int d^2p_T dy\cdot\mu_T [\, \cosh y - \frac{M}{N} \,]\cdot \exp[\, -(\alpha + \beta\cosh y\,)\mu_T \,] = 0 \\ &\int d^2p_T dy\, [\, \mu_T - \frac{\langle\mu_T\rangle}{N} \,]\cdot \exp[\, -(\alpha + \beta\cosh y\,)\mu_T \,] = 0 \end{aligned} \tag{4}$$

and

$$Z = \left[ \int d^2p_T dy\cdot \exp\{-(\alpha+\beta\cosh y)\mu_T\} \right]^{-1}. \tag{5}$$

We can now continue the algorithm presented in ref.[4]. For a given $M$ we calculate $N=N(M)$ and solve eqs.(4) and (5) for $\alpha$ and $\beta$ ( actually in present calculations $\alpha$ was chosen to fit experimental $P(E_T)$ spectra ). With fixed $\alpha$ and $\beta$ we continue to sample consecutive $\{y,p_T\}$'s (i.e. secondaries) according to the distribution $f(y,p_T)$ as long as energy is available.

In this way we can finally get any distribution we wish, among them $P(E_T)$, $P(n)$ or $\omega(E_T|n)$ and $\bar{\omega}(n|E_T)$ which are of immediate interest to us.

## 3. Numerical results

As an example we have used $P(E_T)$ data of UA1 [6], cf. Fig.1, as our input. The parameters of the IGM were the same as those quoted in ref.[4] for the realistic case of average inelasticity decreasing with energy. For $\langle N\rangle$ we have used $a=5.2$ and $b=0.42$ which reproduce the observed energy dependence of mean multiplicity in a broad range of energies [7]. The $P(E_T)$ spectra then allows us to fix $\langle\mu_T\rangle$ in eq.(4). The resulting fit to UA1 data is shown in Fig.1. Figs.2a and 2b represent our prediction for $P(n)$. Notice that although it agrees well with the UA5 data on $P(n_{ch})$ for the <u>full</u> rapidity range [8], it does not show the scaling property seen in the UA1 data on $P(n_{ch})$ obtained from smaller rapidity windows [9]. In Figs.3a and 3b the examples of the actual distributions $\omega(E_T|n)$ and $\bar{\omega}(n|E_T)$ for a c.m. energy of 500 GeV are displayed. Notice, that with the exceptions of $n<5$ one can represent $\omega(E_T|n)$ via the following distribution

$$\omega(E_T|n) = 12.5\cdot\varepsilon\cdot\exp(-5\cdot\varepsilon), \qquad \varepsilon = (E_T|n)/n. \tag{6}$$

For $\bar{\omega}(n|E_T)$ one could also write similar distribution, $\bar{\omega}(n|E_T)=c_1\cdot\bar{\varepsilon}\cdot\exp(-c_2\bar{\varepsilon})$, $\bar{\varepsilon}=(n|E_T)/E_T$, but with $(c_1,c_2)$ changing from (2.3, 1.0) for $E_T=0\div2$ GeV to (5.3, 0.43) for $E_T\geq15$ GeV. Figs.4a,4b and Figs.5a,5b represent other details of the $\omega(E_T|n)$ and $\bar{\omega}(n|E_T)$ distributions for energies of 200, 500 and 900 GeV, namely the behaviour of the corresponding mean values and dispersions as functions of $n$ or $E_T$ respectively. One observes linear behaviour of $\omega(E_T|n)$ and $\bar{\omega}(n|E_T)$ with practically energy independent slopes ( for example: $\langle E_T|n\rangle=0.4\cdot n+1$ ) and constant relative widths ( for example: $\sigma_\varepsilon/\surd\langle\varepsilon\rangle = [0.2+0.5/n]^{1/2} \longrightarrow 0.45$ for $n\longrightarrow\infty$ ).

The above example shows the feasibility of the program proposed at the beginning. It can be performed with any model which correctly reproduces the main features of the observed multiparticle data [10]. Because the presented results are highly insensitive to the details of the model ( once $P(E_T)$ is correctly reproduced ) the eventual differences are likely to show up only in higher order correlation functions.

This work was supported in part by the Heraeus-Stiftung, Hanau, the Federal Minister for Research and Technology (BMFT) under the contract number 06 MR 713 and the Gesellschaft für Schwerionenforschung, Darmstadt.

**References**

[1] M.Plümer and R.M.Weiner, Phys.Rev. D37, 3136 (1988).
[2] G.N.Fowler et al., Phys.Rev.Lett. 55, 173 (1985); Phys.Rev. D35, 870 (1987) and Phys.Rev. C40, 1219 (1989).
[3] G.Wilk, in:*Multiparticle Production*, eds.R.Hwa, G.Pancheri and Y.Srivastava, World Scientific 1989. p.455.
[4] Cf. talk by G.Wilk (R.Weiner, G.Wilk and Z.Włodarczyk, these proceedings).
[5] F.Takagi and T.Tsukamoto, Phys.Rev. D38, 2288 (1988) and references therein; cf. also: G.Wilk and Z.Włodarczyk, *Information theory and phenomenology of multiple hadronic production*, SINS/Zd-PVIII preprint 7/90
[6] C.Albajar et al.,UA1 Coll., Nucl.Phys.B335, 261 (1990)
[7] G.Wilk, *Interacting Gluon Model for hadronic and nuclear collisions*, SINS Report 2075/P-VIII/PH/A, March 1989.
[8] R.E.Ansorge et al., UA5 Coll. Z.Phys. C43, 357 (1985).
[9] This apparent incosistency between the two different types of data obtained by UA1 may be caused either by some internal experimental problems ( transverse energy distributions were obtained by calorimeters whereas multiplicity distributions by means of counters ) or by our neglect of more detailed information provided by UA1, namely the correlation between the density of particles in rapidity and mean transverse momentum. We plan to address this problem elsewhere.
[10] Wu Yuanfang and Liu Lianshou, Int.J.Mod.Phys. A4, 4505 (1989).

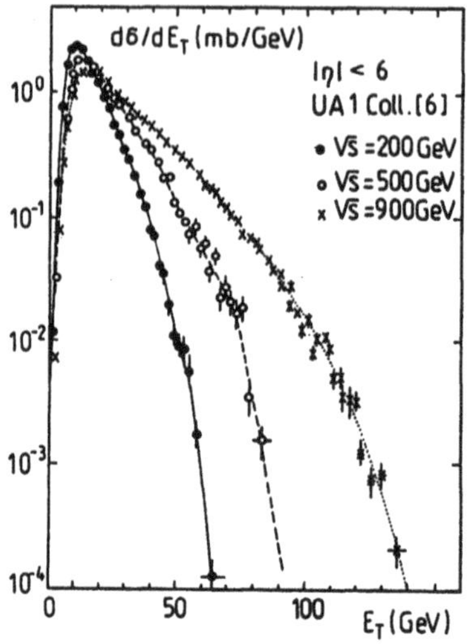


Fig.1 Fit to the data of UA1 [6]

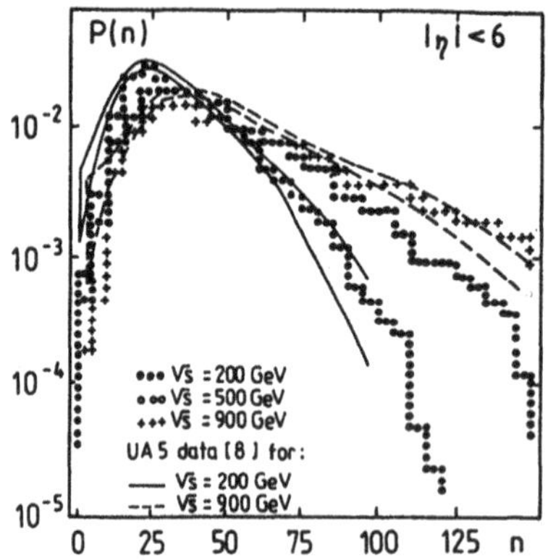


Fig. 2a

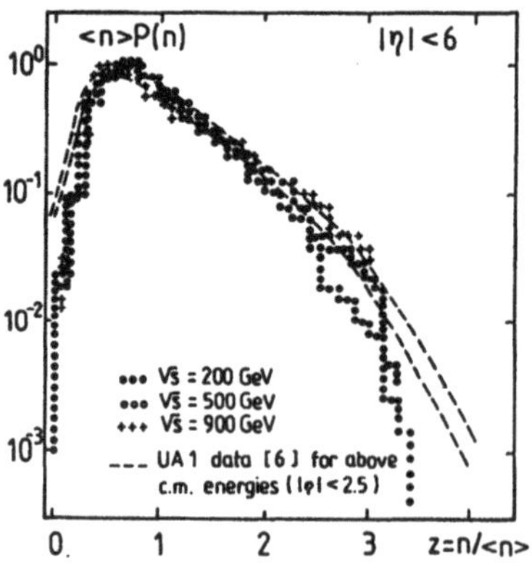


Fig. 2b

Fig.2 Our predictions for $P(n)$ ( compared with UA5 data for full rapidity range [8] ) – (a), and for $\langle n \rangle P(n)$ ( compared with UA1 data for $|\eta|<2.5$ ) – (b).

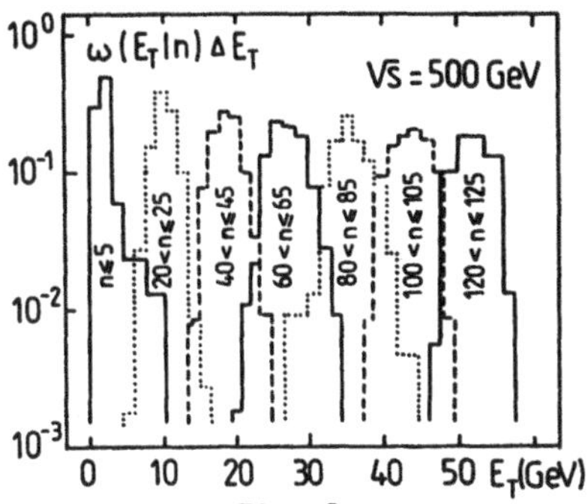


Fig. 3a

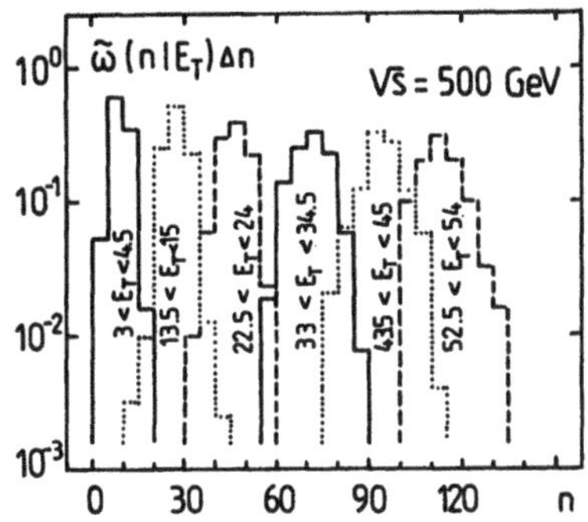


Fig. 3b

Examples of $\omega(E_T|n)$ and $\bar{\omega}(n|E_T')$ distributions for energies 200, 500 and 900 GeV.

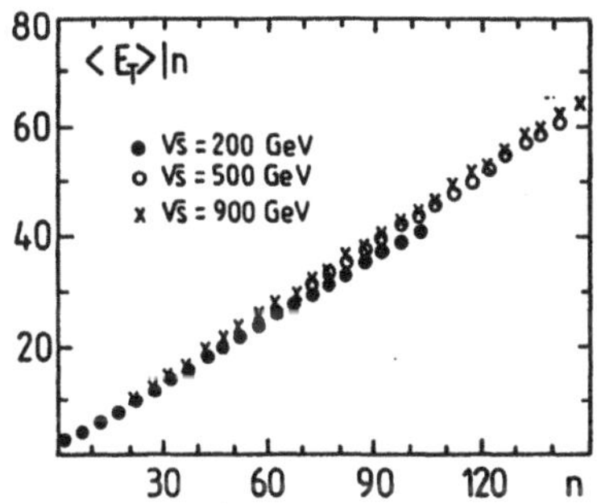


Fig. 4a

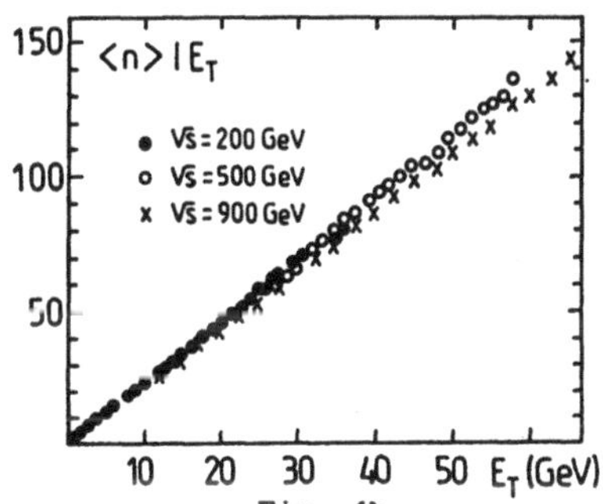


Fig. 4b

$\langle E_T\rangle$ for fixed $n$ as a function of $n$ - (a) and $\langle n\rangle$ for fixed $E_T$ - (b) as a function of $E_T$.

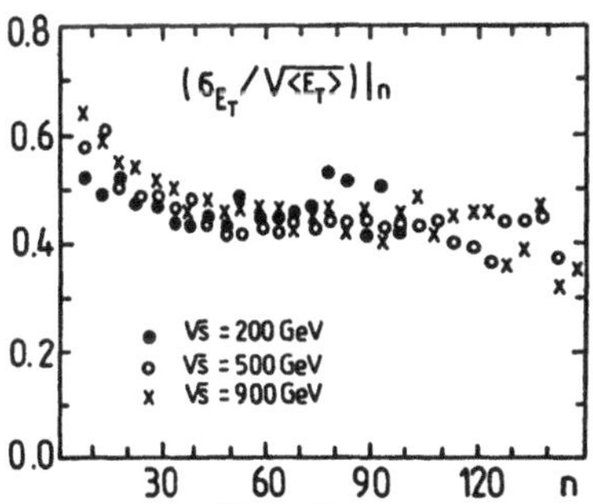


Fig. 5a

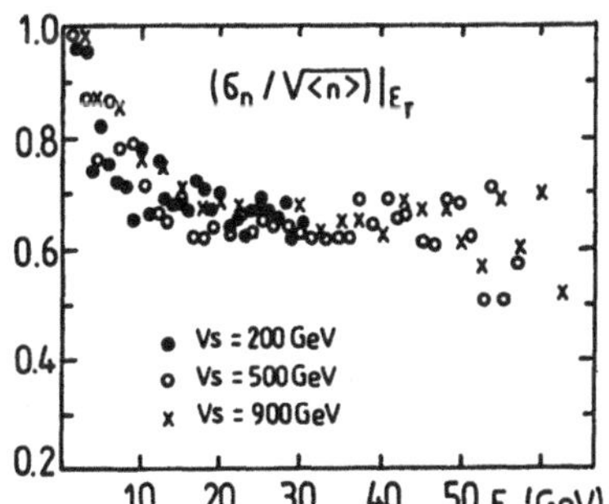


Fig. 5b

Dispersions of $\omega(E_T|n)$ and $\bar{\omega}(n|E_T)$ distributions.

# RAPIDITY and TRANSVERSE MOMENTUM DISTRIBUTIONS FROM FLUID DYNAMICS and FROM A STRING MONTE-CARLO REACTION MODEL

N. S. Amelin, and L. P. Csernai

Department of Physics, University of Bergen,
Allégt. 55, 5007 Bergen,
Norway

## INTRODUCTION

The rapidity and transverse momentum distribution of different emitted particles in a high energy nuclear collision are among the most important observables. The most commonly measured quantities are the average transverse momentum $< p_\perp >$, or transverse mass distributions. The transverse momentum spectrum $dN/(p_\perp dp_\perp)$ is a very important quantity, especially if it is measured for several particle species at the same time. These quantities reveal important features of the collective dynamical features of the collision, and so, they serve as the best possible signals of a possible phase transition to Quark Gluon Plasma (QGP).

Viewed from a continuum aspect the source of transverse momentum is at least threefold in nuclear collisions: i) Random thermal motion, ii) Collective flow, iii) Preequilibrium direct emission from underlying microscopic processes. In a cascade or molecular dynamics picture the transverse momentum of particles is generated in microscopic collisions and these collisions are not classified into different classes. However, when the number of collisions is large in a given volume (e.g. $>$ 3-5 per particle) the system is usually close to local thermal equilibrium. In such cases it is in principle possible to extract local collective macroscopic parameters, like flow velocity or temperature, from a microscopic cascade type model.

Some of the microscopic, cascade type, models, used for ultra-relativistic energies, are very detailed, including not only elastic collisions, but also particle creation and annihilation [1]. Furthermore, the space-time picture of particle creation is also simulated via string formation and subsequent decay [2, 3]. One can imagine that the EOS represented by a microscopic model like this is not simple and it can be far from that of an ideal gas. The temporary string formation with about 1 GeV/fm string tension, may absorb an essential part of the thermal (kinetic) energy in the high density high temperature matter. Unfortunately to evaluate the underlying EOS in a microscopic model is not easy and could be a quite expensive computational task.

Collective dynamical reaction models are different in how much thermalization they assume. Two extremes are the one-fluid (or Landau) fluid dynamical model assuming rapid (immediate) local equilibration and the Bjorken fluid dynamical model assuming complete transparency, with the exception of a color charge exchange between the projectile and the target. We discuss first how one can evaluate measurables related to the transverse momentum in the one fluid model, and will discuss some characteristic results based on Refs. [4].

## FLUID DYNAMICS

The fluid dynamical model is uniquely defined (one-fluid) and the only input is the nuclear EOS. If the system is close to local thermalization one fluid dynamics might be a good approximation to the collision dynamics, and it can reveal some features of the EOS.

Recent SPS and AGS experiments lead to the conclusion that in central heavy ion collisions hadronic matter is strongly stopped and a significant portion of the incident energy can be observed in transverse directions. The normalized pseudorapidity distributions of charged secondaries are strongly peaked at mid rapidities for 14.6, 60, and 200 GeV/nucleon O+Ag/Br collisions and do not show the development of a central plateau as shown by the KLM collaboration [5, 6]. The present data from WA-80 and NA-35 also indicate that the stopping of baryons is strong even at energies near to 200 GeV/nucleon [7, 8, 9, 10], although in S + S reactions the baryon rapidity distribution seems to indicate already a double peaked structure. At lower energies (14.5 and 60 GeV) and for heavier nuclei there is a stong indication for complete stopping in central collisions where no energy is detected around zero degrees in calorimeters.

Thus we attempted the description of these latter two reactions in terms of a one-fluid hydrodynamical model in which there is complete stopping [4, 11]. The selected EOS had a first order phase transition [12], with a critical temperature at $n = 0$ is $T_c = 216$ MeV, which is consistent with recent lattice QCD results: $T_c = 177 - 276$ MeV [13]. The energy density increases monotonically with $T$ until $T_c$ is reached where the critical energy density is $e_{cH}(n = 0) = 280$ MeV/fm$^3$. The details of the EOS are discussed in Ref. [12].

For comparison we performed two sets of fluid dynamical calculations with EOS's with and without a phase transition to QGP. The equations of relativistic perfect fluid dynamics were solved numericly by the PIC method [14] on CRAY-XMP computers in Trondheim and Los Alamos. We can illustrate the results of the one-fluid model by using numerical calculations for $^{32}S+^{208}Pb$ at 14.5 and 60 GeV/nucleon beam energy and at one impact parameter $b = 0.3(R_T + R_P)$ [4]. The phase transition to QGP contributes to an appreciable increase of the maximum density and to a longer expansion time. In case of the purely hadronic EOS at 14.5 (60) GeV/nucleon the volume average of density reached a maximum of 2.3 (7) $n_0$ at 2.5 (1.2) fm/c, while this average density dropped below $2n_0$ at 4.5 (4.5) fm/c.

In order to account for both the collective fluid-dynamical motion and for the random thermal motion of the nucleons, we have to consider the thermal baryon distribution locally in every fluid cell and boost it by the velocity of the cell.

The shape of the transverse momentum distribution is a very important observable. It was shown by Siemens and Rasmussen that deviations from thermal emission are clearly observable on $p_\perp$ spectra. The collective flow causes a characteristic deviation from exponential type thermal spectra: heavy emitted particles $(p, K)$ show a concave shape, while light particles like pions show a convex shape on logarithmic plots of the spectra.

## PION MEASURABLES

We can calculate the pion distributions, assuming a thermal volume emission at the breakup inside each cell. This is consistent with our EOS where thermal pions were also considered. We also assume that the pion chemical potential vanishes, in accordance with the one fluid model where local thermodynamical equilibrium is assumed for all particles.

At the breakup the local pion momentum distribution at a space-time point $x$ is assumed to be a relativistic Bose distribution: $f_\pi(x, p; U^\mu, T)$. The pions are emitted in their own local rest frame (LR), and their number is proportional to this volume $V_{LR}$ as well as to $T^3$. The pion multiplicity is extremely sensitive to the breakup condition. The number of fluid cells filled with matter is important: a dilute system corresponds to a larger source volume. Furthermore, if the dilute system is also expånding faster the local rest (LR) volume of the pion source, $V_{(LR)} = \gamma V_{cell}$, is further enhanced. Due to these competing mechanisms and to the long lifetime of the mixed phase at higher energies, it is not possible to draw a unique conclusion regarding how the phase transition affects the pion multiplicities quantitatively. Breakup conditions play an essential role in determining pion multiplicities, and these conditions are outside the predictive power of fluid dynamics.

The pion transverse momentum distribution $dN_\pi/(p_\perp dp_\perp)$ from the relativistic three - dimensional one - fluid dynamical model using the PIC method for the reaction S+Pb at $E_{Lab} = 14.5$ GeV/nucleon, at impact parameter $b = 0.3(R_p + R_t)$ is shown in ref. [4]. The breakup times are 8.4 and 10.7 fm/c. With increasing breakup time the pion multiplicity decreases. Two different EOS's were used without (dashed lines) and with (full lines) a phase transition into QGP. The formation of QGP decreases both the temperature and the transverse flow so the $p_\perp$ spectra are steeper. The observable effective temperature or slope parameter is higher than the average temperature of the fluid cells at the same time.

The spectrum shows a concave shape which is a consequence of the collective expansion. This shape, (also seen in several experiments [15]), is connected to transverse radial flow as pointed out by Atwater, Freier, and Kapusta using a model with idealized cylindrical geometry [15]. The fact that the same shape can be obtained from a one-fluid dynamical calculation shows that the concave shaped $p_\perp$ spectrum is a consequence of the collective flow but it is not necessarily related to complete transparency and the Bjorken scaling expansion [16]. In our calculations the spectra with hadronic EOS are more concave indicating a larger transverse flow.

The strongest effect of the EOS can be seen in the pion rapidity disributions where the QGP leads to a shift of the peak position to one unit higher rapidities [4]. The Gaussian rapidity distribution of pions peaks at a small rapidity for hadronic EOS, closer to the nucleus nucleus center of mass than to the nucleon nucleon center of mass. This is because the hadronic EOS is stiffer and it leads to a large target participation. At the same time the QGP EOS results in almost straight clean cut collision geometry with minimal target participation.

## MICROSCOPIC STRING DYNAMICS and CASCADE

In a microscopic picture the nuclei are described as a set of nucleons initially. We consider their quark-parton composition and describe the separate nucleon nucleon collisions in a string picture. At a hadron-hadron impact one or more strings are formed. The leading partons of the string (quarks or diquarks) carry a given fraction of the initial momentum of the nucleon, which is given by random selection following appropriately chosen structure functions [3]. The interaction of the strings is described approximately by the interaction of the leading partons. The main parameters of the model were taken from the standard Reggeon phenomenology, leaving very little fredom to influence the predictions of the model. The nuclear stopping power measured in p+A collisions [17] was fitted by the same parameter set. This stopping power has a crutial role in heavy ion collisions [18]. The most important parameter in this regard is the string tension which was chosen to be 1 GeV/fm.

After about 1fm/c proper time the string rehadronizes. This hadronization follows essentially the Field-Feynmann algorithm [19], but conservation laws were strictly enforced and the longitidual momentum and the energy of the produced hadron were determined through the variable $z = \frac{(E+P_\parallel)_h}{(E+P_\parallel)_q}$, which was generated by using the distribution

$$f_q^h \sim (1-z)^{\alpha_q^h(p_t)}, \tag{1}$$

the so called sharing function. As shown in ref. [20], the index $\alpha$ depends on the flavour of the constituent, on the type of the hadron it is transformed into, and on transverse momentum of the hadron. The values of $\alpha$ were taken in each case from ref. [20].

After hadronizarion the formed secondary hadrons are allowed to rescatter. In the space time the formation of a hadron can be chosen according to different algorithms: The *constituent* formation time, which happens earlier when the string breakes and the constituents of the hadron are created as a pair, or the *yo-yo* formation time, which happens somewhat later, when the constituent partons reach each other first after their

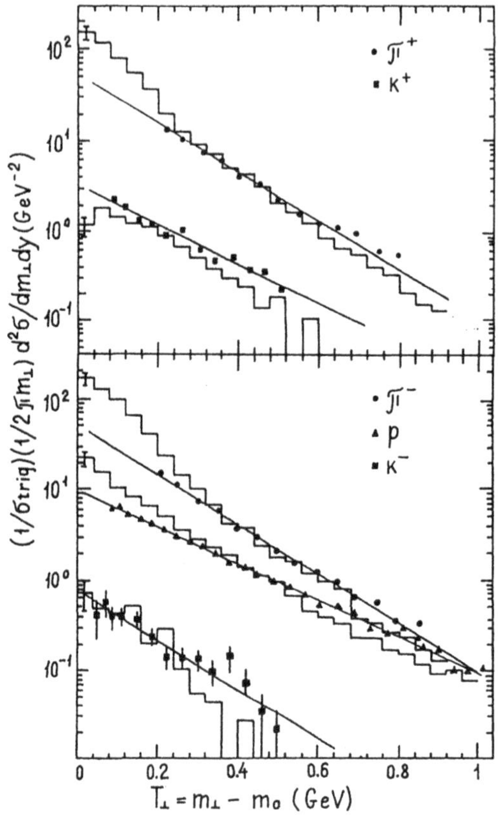


Figure 1: Invariant cross section for $\pi^\pm$, $K^\pm$, and protons measured in the rapidity interval $1.2 \leq y \leq 1.4$ in the reaction $Si + Au$ at 14.6 GeV/nucleon energy. The solid lines show the exponential fits to the data [22]. The solid histograms show results of computation.

formation. It was pointed out by Bialas and Gyulassy [21] the imporance of this choice. We have found that the constituent formation time leads to a more compact hadron formation while the yo-yo formation time leads to a hadron formation on a 20-30% longer longitudinal range. It turned out that most experimental data are fit better with the constituent formation time prescription [3].

The model yields a generally good overall fit to most experimental data. It is interesting to mention that both the proton and pion rapidity distributions are reproduced by the model at 200 GeV A energy, while if we neglect the rescattering of secondaries on each other the model fails to reproduce the data especially the rapidity distribution of protons.

One of the interesting results is that the model fits satisfactorily the measured pion and kaon transverse energy distribution data from the reaction $Si + Au$ at 14.6 GeV/nucleon, see Fig. 1. The calculation, however, shows an excess in the pion spectra at low transverse energies above the exponential. This increase (or peak) is mainly due to resonance decays according to the model. The kaons do not show such an increase according to our calculations. This leads to an interesting conclusion: The calculated $K^+/\pi^+$ ratio is about 0.05 which is similar to the value observed in p+p collisions. At the same time the experiment yields roughly 0.2 if one assumes an exponential or Boltzmann shaped spectrum.

Thus it would be important to know the low energy behaviour of the transverse energy spectra experimentally, if one want to reach a final conclusion on strangeness enhancement in heavy ion collisions.

## CONCLUSION

In conclusion, it is expected that the heavy particle (Proton and Kaon) spectra are most sensitive to plasma formation and provide the structure in the $< p_\perp >$ versus $< dN_c/dy >$ correlation. Viewed from the fluid dynamical model both the momentum spectra and the average momenta support the existence of a collective flow which increases with increasing multiplicity. On the other hand in microscopic models an enhancement is obtained for pion spectra at low $p_\perp$ from resonance decays. This finding causes an uncertainty in the estimated amount of strangeness enhancement, and also in the accurate method to estimate qualitatively the collective flow or expansion. Further theoretical and experimental investigations are needed to clarify these problems.

## References


[1] N.S. Amelin, K.K. Gudima and V.D. Toneev, University of Bergen, Dept. of Phys., Scientific/Technical Report no. 223/1990.

[2] N.S. Amelin, K.K. Gudima, S.U. Sivoklokov and V.D. Toneev, Sov. Jad. Fiz. **52** (1990) 272.

[3] N.S. Amelin, K.K. Gudima, V.D. Toneev and S.U. Sivoklokov, University of Bergen, Dept. of Phys., Scientific/Technical Report no. 224/1990.

[4] E.F. Staubo, A.K. Holme, L.P. Csernai, M. Gong and D. Strottman, Phys. Lett. **B229** (1989) 351.

[5] H. von Gersdorff, J. Amundson, P.S. Freier, J. Kapusta, C. J. Waddington, L.M. Barbier, W. v. Jones, K. H. Moon, I. G. Park, O. E. Pruet, J. P. Wefel, R. Holynski, A. Jurak, A. Olszevski, M. Szarska, B. Wilczynska, H. Wilczynski, W. Wolter, B. Wosiek and K. Wozniak, Phys. Rev. **C39** (1989) 1385.

[6] L. M. Barbier *et al.*, Phys. Rev. Lett. **60** (1988) 405.

[7] H. R. Schmidt, *et al.* (WA-80 collaboration), Proc. of the 6th Int. Conf. on Ultra-relativistic Nucleus-Nucleus Collisions, 1987, Z. Phys. **C38** (1988) 109.

[8] F. Plasil, *et al.* (WA-80 collaboration), Proc. of the Winter Workshop on Nuclear Dynamics V, Feb. 21-27, 1988, Sun Valley, Idaho (unpublished)

[9] B. Jacak, private communication (1989).

[10] K. van Dijk, *et al.* (E802 collaboration), Proc. of the Winter Workshop on Nuclear Dynamics V, Feb. 21-27, 1988, Sun Valley, Idaho (unpublished)

[11] T. Rentsch, G. Graebner, J. A. Maruhn, H. Stöcker and W. Greiner, Mod. Phys. Lett. **2** (1987) 193.

[12] A.K. Holme, E.F. Staubo, L.P. Csernai, E. Osnes and D. Strottman, Phys. Rev. **D40** (1989) 3735.

[13] F. Karsch, Z. Phys. **C38** (1988) 147.

[14] H. Stöcker and W. Greiner, Phys. Rep. **137** (1986) 277.
R. B. Clare and D. D. Strottman, Phys. Rep. **141** (1986) 177.

[15] T. Atwater, P. S. Freier, and J. I. Kapusta, Phys. Lett. **B199** (1987) 30.

[16] J. D. Bjorken, Phys. Rev. **D27** (1983) 140.

[17] C. De Marzo, *et al.*, Phys. Rev. **D27** (1982) 1019.

[18] L.P. Csernai and J.I. Kapusta, Phys. Rev. **D29** (1984) 2664, and Phys. Rev. **D31** (1985) 2795.

[19] R. D. Field and R. P. Feynman, Nucl. Phys. **B136** (1978) 1.

[20] A. B. Kaidalov, 11-th Physics School of the ITEP, **4** (Moscow, Energoatomizdat, 1983), 3; Yad. Fiz. **45** (1987) 1452.

[21] A. B. Bialas and M. Gyulassy, Nucl. Phys. **B291** (1987) 793.

[22] T. Abbott, *et al.*, Phys. Rew. Lett. **64**, (1990) 847.

# FACTORIAL MOMENTS AND CORRELATIONS IN MESON–NUCLEUS INTERACTIONS AT 250 GeV/c

EHS–NA22 Collaboration

Antwerp/Brussels – Berlin(Zeuthen) – Helsinki – Krakow – Moscow
Nijmegen – Rio de Janeiro – Serpukhov – Tbilisi – Warsaw – Yerevan

**F. VERBEURE**
*Department of Physics, Universitaire Instelling Antwerpen,*
*Universiteitsplein 1, B-2610 Wilrijk, Belgium*

ABSTRACT

A sample of 8000 interactions of 250 GeV/c $\pi^+$ and $K^+$ mesons on $Al$ and $Au$ nuclei is used to search for intermittency effects by calculating the normalized factorial moments of order two to four. No significant effect is observed. Bose-Einstein correlations in these interactions are presented, both in terms of the Kopylov-Podgoretskii and the Weiner parametrizations.

## 1 Introduction

The possibility that particle production in multihadron final states may exhibit so-called "intermittent" behaviour was first put forward by Białas and Peschanski[1]. They suggested to study the normalized factorial moments $F_i$ (i=2. 3. ...) of the multiplicity distributions, and in particular the dependence of these moments on $\delta y$, a rapidity interval of decreasing size. Intermittency is observed when the factorial moments exhibit a power law behaviour

$$< F_i > \sim \left(\frac{Y}{\delta y}\right)^{f_i} \qquad \text{with} \qquad f_i > 0 \tag{1}$$

Intermittent behaviour of particle production is now claimed in all studied types of interactions, such as $e^+e^-$, $h-h$, $h-A$ and $A-A$ interactions[2]. Moreover, it is remarked that the effect seems to decrease with increasing complexity of the collision partners and with increasing complexity of the final state.

The results presented here, are obtained with the european hybrid spectrometer (EHS), irradiated by a beam of $\pi^+$ and $K^+$ mesons of 250 GeV/c momentum. An $Al$ and a $Au$ foil are inserted in the rapid cycling bubble chamber (RCBC), which serves as track detector with $4\pi$ angular acceptance for all interactions in the foils. The experimental set-up, the minimum bias trigger and the selection criteria for the nuclear interactions, are described in more detail in[3,4]. The selection criteria aim at isolating a set of well measured and reconstructed inelastic interactions in either the $Al$ or the $Au$ target, thereby eliminating both quasi-elastic and coherent interactions. The events are selected on the basis of the following criteria: the incident particle is well measured in the bubble chamber and its track matches with hits in the upstream wire chamber; the reconstructed vertex position is within one of the nuclear targets; the outgoing tracks are satisfactorily measured with an error on the momentum $\Delta p/p < 0.25$; the accepted loss of tracks, due to measurement or reconstruction problems, is at most one for charge multiplicity up to 10, and at most 20% for higher multiplicities.

Electrons and positrons are identified up to $p_{\rm lab} \simeq 200$ MeV/c and protons up to $p_{\rm lab} \simeq 1.2$ GeV/c, by visual inspection of their ionization in the $H_2$ of RCBC. These particles are not used in the subsequent analysis. Among the accepted charged tracks remains a small admixture of unidentified electrons and positrons from $\gamma$-conversions in the foils. This admixture is estimated to be smaller than 2% in the $Al$ and smaller than 7% in the $Au$ sample. Since no statistically significant difference is observed for $\pi^+$ and $K^+$ induced interactions, we use the combined sample of $\pi^+$ and $K^+$ collisions with 4200 interactions on $Al$ and 3700 on $Au$. The combined samples are denoted as $M^+Al$ and $M^+Au$.

# 2 Factorial moments

The scaled factorial moment of order $i$ is defined as

$$< F_i > = \frac{1}{< \overline{n}_m >^i} \left\langle \frac{1}{M} \sum_{m=1}^{M} n_m(n_m - 1)\ldots(n_m - i + 1) \right\rangle \tag{2}$$

with

$$< \overline{n}_m > = \left\langle \frac{1}{M} \sum_{m=1}^{M} n_m \right\rangle \tag{3}$$

whereby the considered rapidity interval $Y$ is subdivided in $M$ equal subintervals, each of size $\delta y = Y/M$. The multiplicity is $n_m$ in bin $m$ ($m = 1 \ldots M$) and the averages $<>$ are taken over all events in the sample. The definition Eq. 2 is often called "horizontal averaging". If non-statistical, self similar fluctuations of many different sizes exist, the $\delta y$ dependence of the factorial moments is expected to exhibit the power law behaviour Eq. 1. in the limit $\delta y \rightarrow 0$. Such behavior shows up as a linear rise of $\ln < F_i >$ versus $(-\ln \delta y)$. Fig. 1 shows the factorial moments of order 2, 3 and 4 for the $M^+Al$ and $M^+Au$ events. A rapidity interval $Y$ of 3 units $(-2 < y < 1)$ in the meson–nucleon c.m.s. is used, where the density of particles is almost constant.

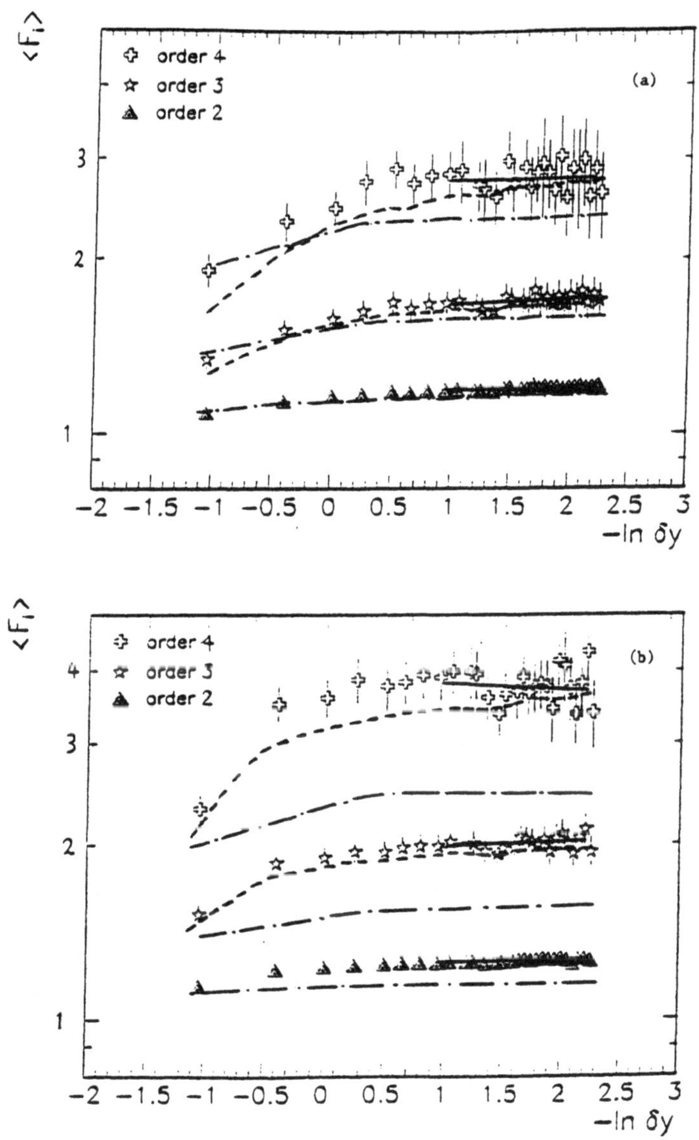


**Fig. 1.** Moments $F_i$ ($i$=2, 3 and 4) for **a** $M^+Al$ and **b** $M^+Au$. Fits to Eq. (1) are shown as straight lines, predictions of the Fritiof model as dash-dotted lines and calculated values of $F_3$ and $F_4$ as dashed lines.

From Fig. 1 it is clear that the slope of $\ln F_i$ versus $(-\ln \delta y)$ varies continously with the latter quantity, until $-\ln \delta y \approx 0.7$. We therefore fit the slopes in the interval $0.9 < -\ln \delta y < 2.2$ ($0.4 > \delta y > 0.1$). The results of the fits are collected in Table 1 and given as full lines on Fig. 1. The fitted slopes are small in all cases and compatible with zero within one standard deviation. For comparison, we also show in Table 1 the fitted slopes for 20000 Monte Carlo events per channel, generated according to the FRITIOF model[5] (version 3.0) and subjected to the same cuts as the experimental data. The slopes $f_i$ found in $M^+Al$ and $M^+Au$ collisions are considerably smaller than the ones in $M^+p$ interactions from the same experimental set-up[6].

**Table 1.** Slopes $f_i$ of the fits to Eq. 2 of the experimental factorial moments $F_i$ and of FRITIOF Monte Carlo events, in the interval $-2 < y < 1$, with the fits performed in the range $0.4 \geq \delta y > 0.1$.

| | $M^+Al$ | | $M^+Au$ | |
|---|---|---|---|---|
| | Data | Fritiof | Data | Fritiof |
| $f_2$ | 0.0132±0.0138 | -0.0004±0.0067 | 0.0092±0.0141 | 0.0047±0.0052 |
| $f_3$ | 0.0301±0.0325 | 0.0114±0.0157 | 0.0186±0.0313 | 0.0085±0.0108 |
| $f_4$ | 0.0088±0.0825 | 0.0101±0.0401 | -0.0247±0.0648 | 0.0095±0.084 |

We are thus led to the conclusion that intermittency effects, if at all present, are weak in positive meson induced interactions on $Al$ and $Au$ nuclei.

Recently, several authors have pointed to the relationship between the factorial moments of order $i$, and the parametrization of the multiplicity distribution (MD) in terms of a negative binomial distribution (NBD). If the MD is described by the NBD with parameters $\bar{n}$ and $1/k$, then the factorial moments can be expressed in the form

$$F_i = (1 + \frac{1}{k})(1 + \frac{2}{k}) \dots (1 + \frac{(i-1)}{k}) \tag{4}$$

which implies that the higher order moments $F_3$, $F_4$,... can be expressed in terms of the moment $F_2$, with $\frac{1}{k} = F_2 - 1$. Fig. 1 also shows the factorial moments $F_3$ and $F_4$, calculated from $F_2$ according to Eq. 4. The calculated values follow the trend of the experimental data but fall below the latter ones. The deviation is amplified with increasing order $i$ of the moments.

# 3 Bose–Einstein correlations

Results were published[7] on Bose–Einstein correlations in $K^+p$ and $\pi^+p$ interactions at 250 GeV/c, derived from the same experiment with the same set-up as used here. One considers the ratio of the distribution of a suitable variable for like-sign pions, to this distribution for a reference sample. The reference sample used is the

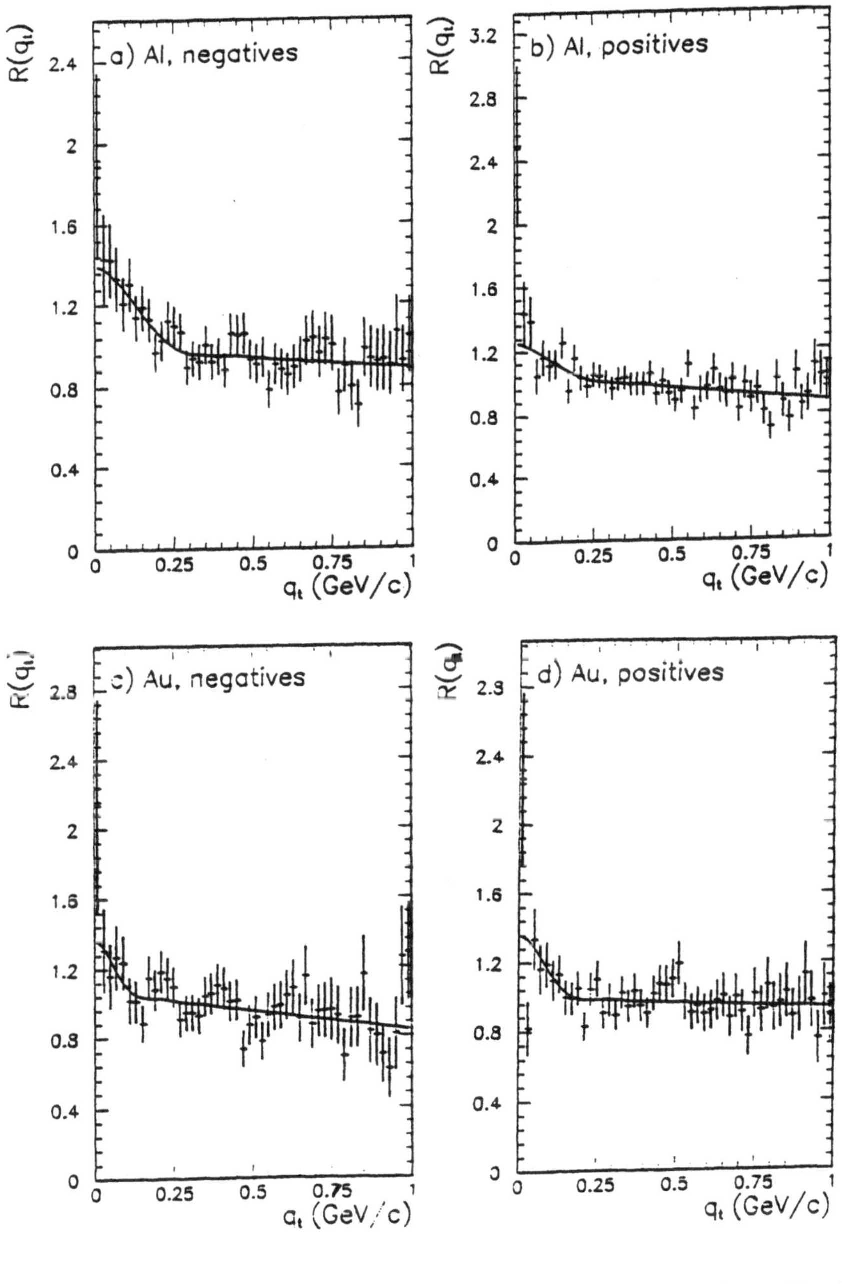


**Fig. 2** $R(q_t)$ for $--$ and $++$ combinations, for $Al$ and $Au$, with fits to Eq. (5) superposed. Events with $6 \leq n_{ch} < 20$ are used.

"mixed" sample, constructed by combining pions from one event with pions of the same charge, randomly chosen in other events with approximately the same charge multiplicity.

The first parametrization which we used is the Kopylov-Podgoretskii one[8]:

$$R(q_T) = \gamma \left[1 + 4\lambda J_1^2(\beta q_T)/(\beta q_T)^2\right] (1 + \delta q_T) \quad (5)$$

at small values of $q_0 = |E_1 - E_2|$. The fit parameters are the normalization $\gamma$, the "correlation strength" $\lambda$, the spatial dimension of the pion source $\beta$ (with $r = 0.197\beta$ fm) and $\delta$, introduced to account for the slow variation of the background.

Fig. 2 shows the ratio $R(q_T)$, obtained with the mixed reference sample, for positive and negative like-pion combinations and both targets for the sample with charge multiplicity $6 \leq n_{ch} < 20$. In all four plots of Fig. 2, a significant enhancement is observed at small $q_T$ values. Fits according to Eq. 5 for $q_0 \leq 0.2$ GeV/c are superimposed. The fit parameters are collected in Table 2. Taking into account that a radius $r = 1.4 \pm 0.1$ fm was found in $M^+p$ collisions[7], we observe an increase of $r$ with $A$.

**Table 2.** Fit results for the Kopylov-Podgoretskii parametrization (Eq. 5) to the ratio $R(q_t)$ for like-pion distributions for events with $6 \leq n_{ch} < 20$. The results for $M^+p$ interactions are reproduced from ref. 7.

| Sample | Charge | $\lambda$ | $\beta$ (GeV/c)$^{-1}$ | $r$ (fm) | $\gamma$ |
|---|---|---|---|---|---|
| $M^+p$ | $--$ | $0.30 \pm 0.03$ | $7.0 \pm 0.5$ | $1.4 \pm 0.1$ | $0.91 \pm 0.03$ |
| $M^+Al$ | $--$ | $0.41 \pm 0.11$ | $12 \pm 2$ | $2.3 \pm 0.3$ | $0.99 \pm 0.05$ |
| $6 \leq n_{ch} < 20$ | $++$ | $0.19 \pm 0.09$ | $13 \pm 4$ | $2.5 \pm 0.7$ | $1.05 \pm 0.04$ |
| $M^+Au$ | $--$ | $0.28 \pm 0.17$ | $24 \pm 6$ | $4.8 \pm 1.1$ | $1.07 \pm 0.04$ |
| $6 \leq n_{ch} < 20$ | $++$ | $0.37 \pm 0.13$ | $18 \pm 4$ | $3.5 \pm 0.8$ | $1.00 \pm 0.22$ |

Weiner[9] advocated a different approach to Bose–Einstein correlations , based on quantum statistical considerations and assuming the particle fields to be a superposition of a chaotic and a coherent part. The correlation function versus $y_{\text{diff}} = |y_1 - y_2|$ for negative pion pairs in the interval $-1.8 < y < 1$ with a mixed reference sample, is given in Fig. 3 for $M^+Al$ and $M^+Au$ collisions. Fits to the form

$$R(y_{\text{diff}}) = \gamma \left(1 + 2p(1-p)\exp[-(y_{\text{diff}})^2 r^2] + p^2 \exp[-2(y_{\text{diff}})^2 r^2]\right) \quad (6)$$

yeild correlation lengths $r$ is 0.92±0.36 (0.97±0.21) in $M^+Al$ ($M^+Au$) interactions, to be compared to 1.6 in $pp$ and $p\bar{p}$ interactions[10]. The chaoticity parameter $p$ is only 0.06±0.02. implying that the last term in (6) essentially does not contribute to $R$.

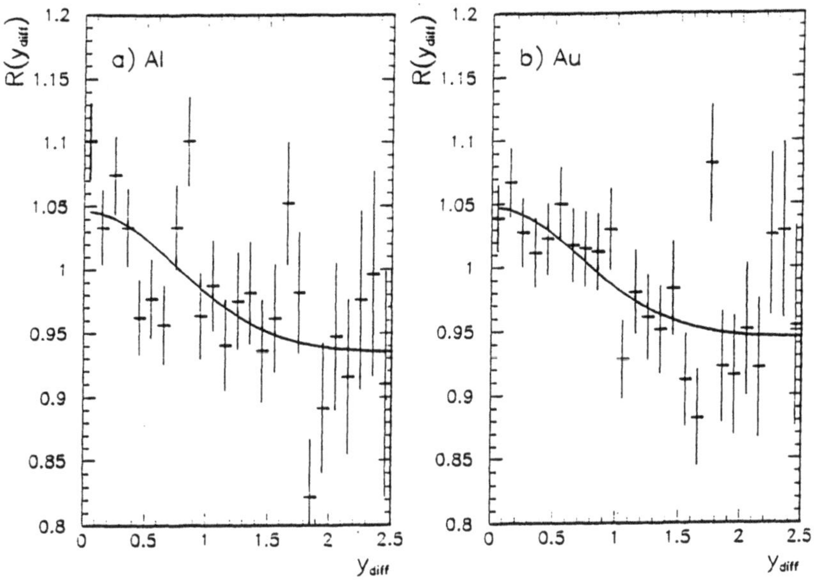


**Fig. 3.** The ratio $R(y_{diff})$ for −− combinations in the $Al$ and $Au$ samples, with fits to Eq. (6) superposed.

# 4 References


1. A. Białas, R. Peschanski: Nucl. Phys. **B273** (1986) 703; ibid. **B308** (1988) 857 and Phys. Lett. **B207** (1988) 59
2. W. Kittel, R Peschanski: *Review on intermittency in particle production*, Nijmegen preprint HEN-325 and Saclay preprint SphT/89-143
3. M. Adamus et al.: Z. Phys. C – Particles and Fields, **C32** (1986) 475
4. I.V. Ajinenko et al.: Z. Phys. C – Particles and Fields, **42** (1989) 377
5. B. Andersson, G. Gustafson, B. Nilsson-Almqvist: Lund preprint LU TP 86-3
6. I.V. Ajinenko et al.: Phys. Lett. **B235** (1990) 373
7. M. Adamus et al.: Z. Phys. C – Part. and Fields, **37** (1988) 347
8. G.I. Kopylov, M.I. Podgoretskii: Sov. J. Nucl. Phys. **15** (1972) 219; **18** (1974) 336; G.I. Kopylov: Phys. Lett. **B50** (1974) 472
9. R.M. Weiner: Phys. Lett. **B232** (1989) 278 and references therein
10. K. Kulka, B. Lörstad: Z. Phys. C – Particles and Fields **45** (1990) 581

# FLUCTUATIONS OF FLUCTUATIONS AND SPIKY SPIKES

*I.M.Dremin*[1], *P.E.Lasaeva*[2], *A.A.Loktionov*[2], *M.T.Nazirov*[2], *E.K.Shabalina*[3], *L.A.Tikhonova*[3], *S.A.Zotkin*[3]

**Abstract**

It is shown that the factorial moments widely used as a measure of fluctuations strongly fluctuate themselves from event to event and their distribution is very non–monotonous discrete function. If the fluctuations (spikes) are chosen as dense isolated homogeneous groups of particles on the pseudorapidity scale, the distribution of their centers reveals two peaks at cms angles about $75^0$ and $105^0$ for pp–interactions at 205 and 360 GeV. These two features should be taken into account when conclusions about intermittent behaviour are made.

1/ P.N.Lebedev Physical Institute, Moscow, USSR

2/ Institute of High Energy Physics, Alma–Ata, USSR

3/ Nuclear Physics Institute, Moscow University, USSR

Strong fluctuations of the pseudorapidity distribution of particles in an individual inelastic event have been observed at high energies. The dynamical origin of such spikes is widely discussed.

One of the proposals is to ascribe them to a stochastical dynamics similar to turbulence where the scaling regime is reached. A manifestation of a corresponding intermittent behaviour is provided by the power–like dependence of factorial moments on the (pseudo)rapidity bin width (or by the linear rise on the double–logarithmic scale). The factorial moments are defined as

$$F_q(\delta y) = \frac{M^{q-1}}{<n>^q} \sum_{k=1}^{M} n_k(n_k-1)\ldots(n_k-q+1) \qquad (1)$$

where M is the number of bins of the width $\delta y$ in a rapidity interval Y so that $Y = M\delta y$, $<n>$ is an average multiplicity of a sample, $n_k$ is a number of particles inside of the k–th interval. The intermittency indices $\varphi(q)$ are defined according to a scaling dependence

$$<F_q(\delta y)> \sim (\delta y)^{-\varphi(q)} \qquad (2)$$

at small values of the bin width $\delta y$ after averaging over all events in a sample.

However such a procedure which is usually called the vertical averaging (in contrast to the horizontal averaging within a given event provided by the sum in the formula (1)) should be treated very carefully. We would like to point out one of the specific features of the factorial moments which reveals strong event by event fluctuations of their values. These fluctuations are inherent to the way how the factorial moments are defined by (1). To show it, let us start with a simplest case of rather small bin width (say, $\delta y = 0.1$ for hadron–hadron reactions at accelerator energies) for a definite moment. Dealing, for example, with $F_3$ one has to calculate the sum $\sum_{k=1}^{M} n_k(n_k-1)(n_k-2)$ which differs from zero*) only for the bins with three particles at least. For small enough width, the probability decreases strongly with increase of the number of particles so that events with 3 particles in a bin provide the main peak while 4, 5 particles give rise to satellite peaks shifted to much higher values of $F_3$ (equal to 4, 10,...). Each of these peaks has its own hypersatellites due to events with two or more dense groups containing 3, 4, 5,... particles. If one denotes the main peak of the q–th moment $F_q$ distribution by indices s=0, h=0, its satellites by s = 1, 2,..., h=0 and hypersatellites by h = 1, 2,... then the positions of all of them may be described according to (1) as

$$\frac{F_{qsh}}{F_{q00}} = \frac{(q+s)!}{q!\,s!} + h, \tag{3}$$

$$\frac{F_{(q+1)00}}{F_{q00}} = \frac{M}{<n>}(q+1) \tag{4}$$

There are q–1 hypersatellites between the main peak at $F_3 \neq 0$ and its first satellite. The heights of the satellites should decrease exponentially with their numbers as may be guessed from multiplicity distribution in narrow bins or from experimental data of NA22–collaboration[1]:

$$\frac{H_{qs0}}{H_{q00}} \sim \exp[-\gamma_q(\delta y)\cdot s] \tag{5}$$

Concerning the heights of the hypersatellites of the main peak, they should satisfy an inequality

---

*) The events with $F_3=0$ do not contribute to average value of $F_3$ and we do not take them into account. Still, they are important for dispersion.

$$H_{q02}H_{q00} < H_{q01}^{\ 2} \tag{6}$$

It may be guessed by considering the limiting case of the infinite multiplicity when the probability to find n groups is given by the product of n probabilities to find a single group. The inequality appears due to the decrease of the subsequent probabilities for finite multiplicities.

Thus, the general structure of the factorial moments distribution at small $\delta y$ and high enough q is determined by well separated main and satellite peaks which are accompanied by much lower hypersatellites. We may say that the measure of fluctuations strongly fluctuates itself so that its distribution should be described by a large set of its moments and may not be characterized by its mean value and dispersion alone.

The whole picture changes if one enlarges the bin width or deals with processes with higher particle densities (instead of lepton or hadron induced reactions one may treat nuclei collisions). Surely, the main peak shifts to higher values of the multiplicity inside a bin so that the wide bump could appear. The above fluctuations would be still present at the tail of the distribution or at higher moments.

All these features are nicely confirmed by experiment. In Fig.1, we show the events distribution over the third factorial moment for the pp–interactions at 400 GeV and $\pi^- p$ interactions at 360 GeV. The full pp statistics contains 480000 events measured in the forward hemisphere in cms and $\pi^- p$ statistics contains 150000 events. The distributions, shown in Fig.1 are obtained for the rapidity interval $\Delta Y = [0, 2.0]$ which is divided into M=10 bins with a width $\delta y = 0.2$ each. For comparison, the distribution for $\pi^- p$ sample is normalized to the total number of events of pp–sample and to the same average multiplicity in the rapidity interval $\Delta Y$. Very non–monotonous behavior of both distributions is clearly seen.

It shows that higher moments are surely very important. The peak at $F_3=0$, while unimportant for average value of $F_3$, contributes strongly to dispersion and higher moments. They increase both with the increase of the order of factorial moments and with the decrease of the bin width. As we have checked, the similar features are typical for other factorial moments as well.

Above, we have described one of the problems in deriving and interpreting the average values of factorial moments. One could also add some others. For example, the overall normalization of a factorial moment to an average multiplicity $<n>$ is often questioned. The use of $\ell n<F_q>$ instead of $<\ell nF_q>$, as is required to get $<\varphi(q)>$, may be misleading (especially, for small values of $\ell n<F_q>$). The sensitivity to the binning procedure and to the sample choice should be checked as well. Only after the careful study of all the technicalities involved is done, one can be sure in quantitative conclusions about intermittency.

Thus, there still exists the general problem of the dynamics at the origin of the dense groups of particles which determine the qualitative features of factorial moments. In some parts, they may be ascribed either to the chaotic (intermittent) dynamics or to conventional (but special) dynamical processes, in particular, to gluon radiation at the finite length of the colour deconfinement.

These two hypothesis lead to quite different predictions about the angular distribution of the dense groups. There is no preferable emission angle of a group for a stochastic dynamics and one may await for a smooth bell–like pseudorapidity distribution of their centers. Vice a versa, the coherent gluon radiation at a finite (and fixed) length provides a peak at a definite polar angle.

Having these typical features in mind, we tried[2] to get an answer from experimental data about pp–interactions at 205 and 360 GeV. To proceed, one has to define the notion of the dense group of particles on the pseudorapidity scale. The previously exploited methods suffer from some deficiencies. When using them, one has to omit either the dense groups with a small number of particles or those particles which appear very close (but outside) to the interval chosen. Our method allows for variation both of the content and of the width of the group. According to it, the group is called a dense, isolated and homogeneous one if it satisfies the following criteria:

1) its density (DN) exceeds the threshold value (DN>TDN);

2) the density gradient (GDN) becomes high enough if any of the neighbouring particles is added to the group (GDN>TGDN);

3) there are no wide gaps (GAP) inside of the group; that is, it can not be split up into two or more separate groups (GAP<TGAP).

The density is defined as a number of particles in a group divided by its pseudorapidity extension. The gradient equals to the density change when adding the nearest neighbour. The gap is a largest interval within the group.

The particular values of the thresholds TDN, TGDN are shown in Table I. They have been chosen from the experimental distributions. The three–particle combinations of nearest neighbours in any event have been plotted at their DN and GDN values. The histogram has a huge tower at low values of density and its gradient with a long tail extending to large values of them. The borderlines between the tower and tail parts of the distribution define the values of TDN and TGDN in Table I. The tail of the three–particle distribution over DN and GDN in pp–events at 360 GeV is shown in Fig.2. The value of TGAP = 0.291 has been chosen from the analysis of the distributions of the maximal internal gap inside the groups and of the next one (which appeared to be twice smaller, on the average).

After defining the dense group, one finds out its center as an average of the pseudorapidities of its constituent particles and gets the distribution of the group centers $\eta_c$ on the pseudorapidity scale. It is shown in Fig.3 for 33228 events of pp interactions at 360 GeV. (The similar distribution has been obtained for 5024 pp–events at 205 GeV but we do not reproduce it here.) It is remarkable that the centers of groups are not monotonously distributed on the pseudorapidity axis but prefer some special values exhibiting the peaks (with more than four standard deviations from the background) of their distribution at the polar angle near 75° (and, correspondingly, 105°) in the center of mass frame, that is, at $\eta_c \approx \pm 0.3$. This effect is not reproduced by Lund model Fritiof 1.6

as is seen in Fig.4. Surprisingly enough, these peaks are very narrow ($\Delta\eta_c \sim 0.1$). While the widths of the groups are rather large ($\delta\eta \sim 0.3$), their centers are placed in a tiny interval. One could speculate about some peaks at large values of $|\eta_c|$ but their statistical significance is much lower.

Among other features of the groups, discussed in ref.[2], let us mention the well established ring–like structure in the plane perpendicular to the collision axis and their low transverse momentum.

The only explanation of the spiky distribution of spikes, we are aware of, is the coherent gluon radiation at a finite length caused by the motion of a free colour charge within the confinement range. The share of such a mechanism is not very high as may be estimated from small integral contribution of peaks compared to background but it is quite noticeable. Besides providing peaks at large center–of–mass frame polar angles (the confinement length is assumed to be about 1 Fm) it predicts some other features, in particular, the absence of such peaks for $\pi$p and p$\bar{\text{p}}$ interactions due to the different interference of the radiation of quarks and antiquarks.

Such a feature could be the only predicted difference for inelastic pp and p$\bar{\text{p}}$ processes at ISR energies. It needs experimental verification.

The higher experimental statistics is needed also to reduce the error bars in the above distributions. However, one should note that the $4\pi$ geometry is preferred or, at least, there should be no bias within the acceptance range of about two units of pseudorapidity around the center ($|\eta|\leq 2$). We are not aware of any experimental data besides the ones treated above which satisfy these requirements.

In conclusion, we have shown that, first, the shape of the factorial moment distribution is very non–monotonous so that its average value and dispersion are not enough to describe it and, second, the pseudorapidity positions of the centers of dense groups of particles (blamed for intermittency patterns) prefer some definite values about $\eta_c = \pm 0.3$, what is hard to interpret for stochastical processes and may imply coherent gluons. These findings ask for more careful analysis of conclusions about the relative roles of stochastical and conventional dynamical schemes.

We are indebted to NA23 and NA27–collaborations and to FNAL bubble chamber group whose experimental data has been used.

*References*

1. Adamus M. et al. Phys. Lett. 1987, B185, 200.
2. Dremin I.M. et al. Sov. J. Nucl. Phys. 1990, 52, N8.

Table I. The threshold values for k–particle group densities (TDN) and their gradient (TGDN)

| K | 3 | 4 | 5 | 6 | 7 | 8 | 9 | 10 | 11 | 12 | 13 | 14 |
|---|---|---|---|---|---|---|---|---|---|---|---|---|
| TDN | 5.5 | 5.75 | 6.0 | 7.0 | 8.0 | 9.0 | 10.0 | 11.0 | 12.0 | 13.0 | 14.0 | 15.0 |
| TGDN | 3.00 | 2.75 | 2.50 | 2.25 | 2.25 | 2.25 | 2.25 | 2.25 | 2.25 | 2.25 | 2.25 | 2.25 |

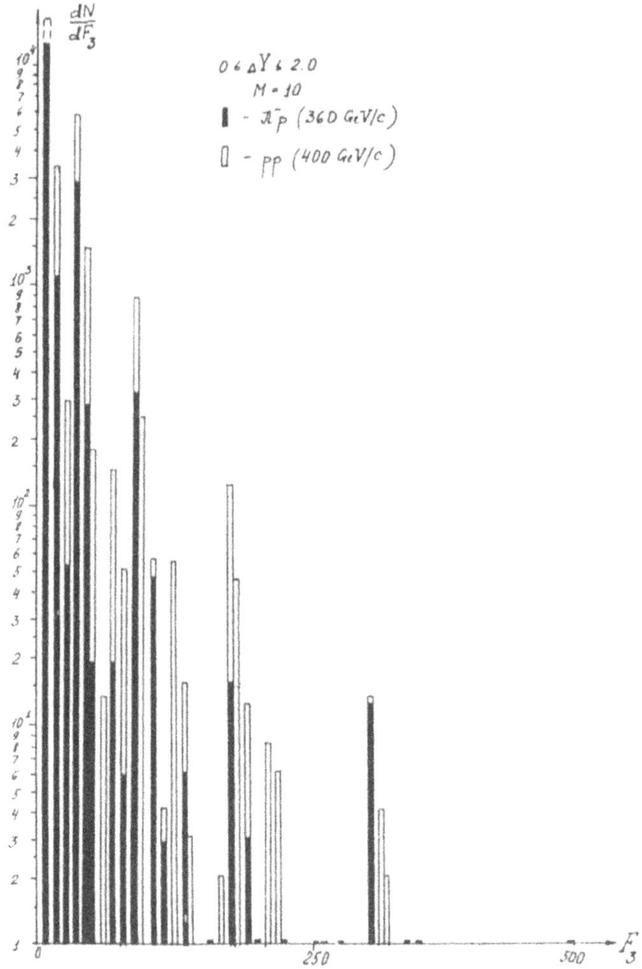


Fig.1. The distributions of inelastic pp–events at 400 GeV (empty histogram) and $\pi^-$p–events at 360 GeV (shaded histogram) over the values of the third factorial moment are very non–monotonous with pronounced and well separated peaks.

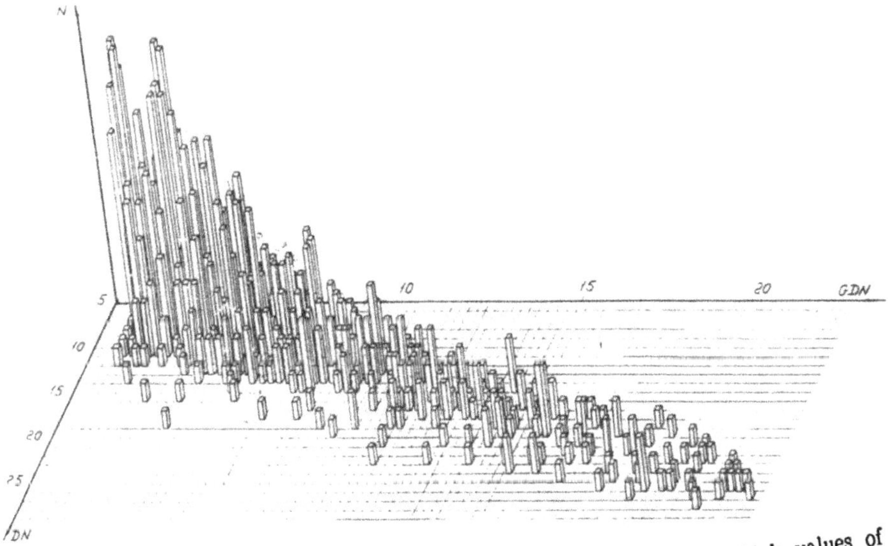


Fig.2. The tail of the distribution of three-particle groups at high values of density and of density gradient for pp-interactions at 360 GeV.

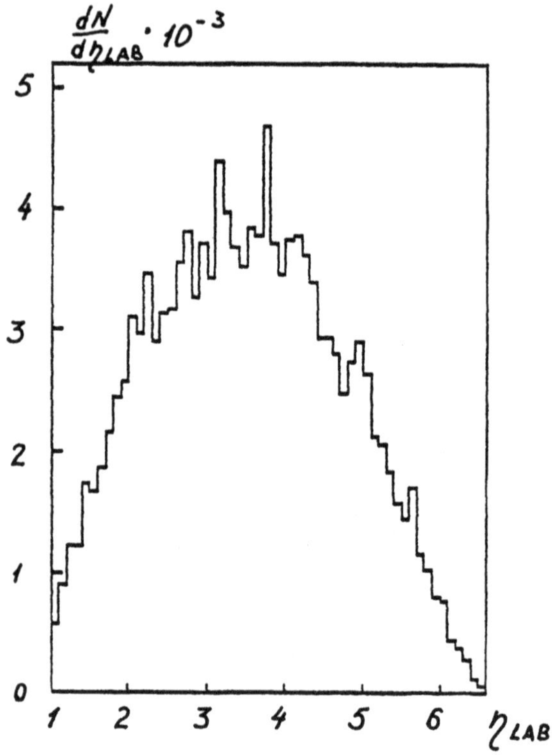


Fig.3. The distribution of the centers of dense isolated homogeneous groups in pp–interactions at 360 GeV possesses pronounced peaks at pseudorapidity $|\eta_c| = 0.3$ what favors the noticeable contribution from the process of the coherent gluon radiation.

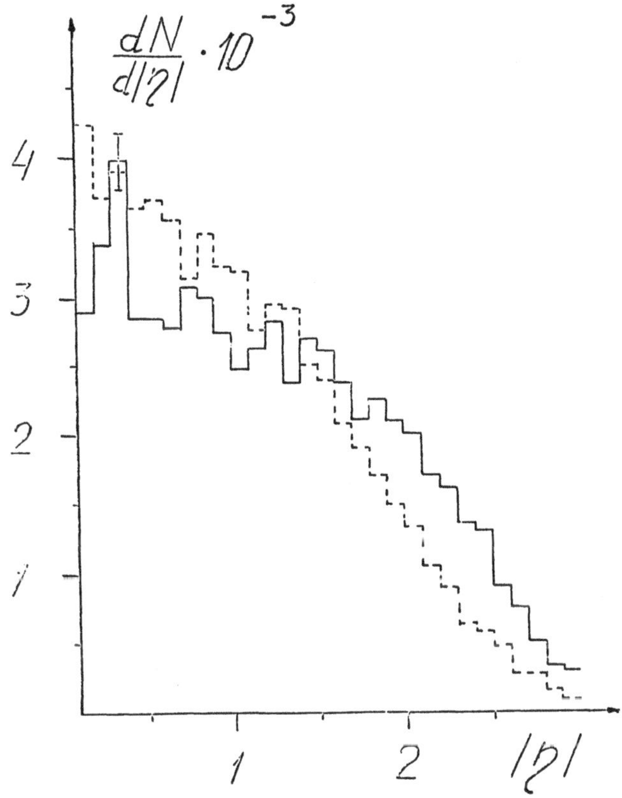


Fig.4. The Lund–model (version Fritiof 1.6) predicts maximum at $\eta=0$ for the positions of the centers of groups while experimental data reveal narrow maximum at $|\eta_c| = 0.3$ and minimum at $\eta=0$. The normalization differs from Fig.3 since we show the data at somewhat different criteria of the group choice. It is done intentionally to demonstrate the persistence of qualitative features with peaks at $|\eta_c| = 0.3$.

# INTERMITTENCY IN MULTIPARICLE PRODUCTION ANALYZED BY MEANS OF STOCHASTIC THEORIES*

A. BARTL
Institut für Theoretische Physik, Universität Wien, Vienna, Austria

M. BIYAJIMA**
Department of Physics, Faculty of Liberal Arts, Shinshu University, Matsumoto 390, Japan

T. MIZOGUCHI
Toba National College of Maritime Technology, Toba 517, Japan

N. SUZUKI
Matsusho Gakuen Junior College, Matsumoto 390-12, Japan

**Abstract**

Intermittency in multiparticle production is described by means of probability distributions derived from pure birth stochastic equations. The UA1, TASSO, NA22 and cosmic ray data are analyzed.

In the present contribution we report some new results of our studies of intermittency in multiparticle production at high energies by means of stochastic theories[1]. We investigate under which conditions a behaviour like[2]

$$\log\langle F_i\rangle \overset{\delta y\to 0}{\longrightarrow} f_i \cdot \ln\frac{1}{\delta y} \tag{1}$$

can be obtained for the normalized factorial moments

$$\langle F_i\rangle = \frac{\langle n(n-1)\dots(n-i+1)\rangle}{\langle n\rangle^i} \tag{2}$$

in certain stochastic distributions. $\delta y$ is the rapidity interval considered and $f_i$ is the slope parameter. (For a recent review of intermittency in multiparticle production, see e.g. Ref. 3.) We thereby rely on the quantum statistical approach in describing hadronic multiplicity distributions[4–6]. It is well known that the negative binomial distribution (NB) and the pure birth distribution (PB) give good descriptions of the multiplicities in different $y$ windows[7–10]. Furthermore, the solutions of the underlying stochastic differential equations show the property of self-similarity, the essential ingredient leading to intermittent behaviour[2,11].

* Presented by A. Bartl.

** Partially supported by the Federal Minister of Science and Research of Austria, and the Japan Society for the Promotion of Science (JSPS).

We start with the PB stochastic differential equation[1,9,10,12]

$$\partial P(n,t)/\partial t = \lambda(n-1)P(n-1;t) - \lambda n P(n;t) \tag{3}$$

with the initial condition at $t = 0$

$$P(n;t=0) = \langle m\rangle^n e^{-\langle m\rangle}/n!. \tag{4}$$

In this case the normalized factorial moments of the PB distribution read

$$\langle F_i\rangle = \Gamma(i)\xi^{1-i}L_{i-1}^{(1)}(-\xi) \tag{5}$$

where $L_n^{(1)}(x)$ is the associated Laguerre polynomial and

$$\xi(t) = \langle m\rangle(p+1)/p, \qquad p = e^{\lambda t} - 1. \tag{6}$$

This means that the $\langle F_i\rangle$ can be expressed completely in terms of the single variable $\xi$. For $i = 2$ we obtain

$$\langle F_2\rangle = \langle n(n-1)\rangle/\langle n\rangle^2 = 1 + 2/\xi. \tag{7}$$

Choosing as initial condition

$$P(n;t=0) = \delta_{nk} \tag{8}$$

one obtains the Furry distribution[9] as solution of eq. (3). In this case the normalized factorial moments are

$$\langle F_i\rangle = k\sum_{j=0}^{i}\binom{i}{j}\frac{(k+i-j-1)!}{(k-j)!}p^{i-j}/\langle n\rangle^i. \tag{9}$$

The Furry distribution was also obtained in Ref. 12 for a certain class of branching models.

We shall also use the NB distribution[7–10] in our analysis of the $\delta y$ dependence of the moments. It follows from a more complicated differential equation than eq. (3) which also contains migration terms. The normalized factorial moments are

$$\langle F_2\rangle = 1 + 1/k \tag{10}$$

$$\langle F_{i+1}\rangle = \langle F_i\rangle\cdot(1 + i/k). \tag{11}$$

Here the $\langle F_i\rangle$ can also be expressed in terms of a single variable, $k$.

In order to compare the normalized factorial moments of the PB distribution, eq. (5), and of the NB distribution, eq. (11), with the data, we make the ansatz $\xi^{-1} = \alpha + \beta\log(1/\delta y)$, and $k^{-1} = \alpha' + \beta'\ln(1/\delta y)$, and determine the parameters $\alpha$, $\beta$, $\alpha'$, $\beta'$ using the data for $\langle F_2\rangle$ as input. In the following we present some of our numerical results of the comparison with the experimental data of Refs. 13–16. Further results are also in Ref. 1, and will be given elsewhere.

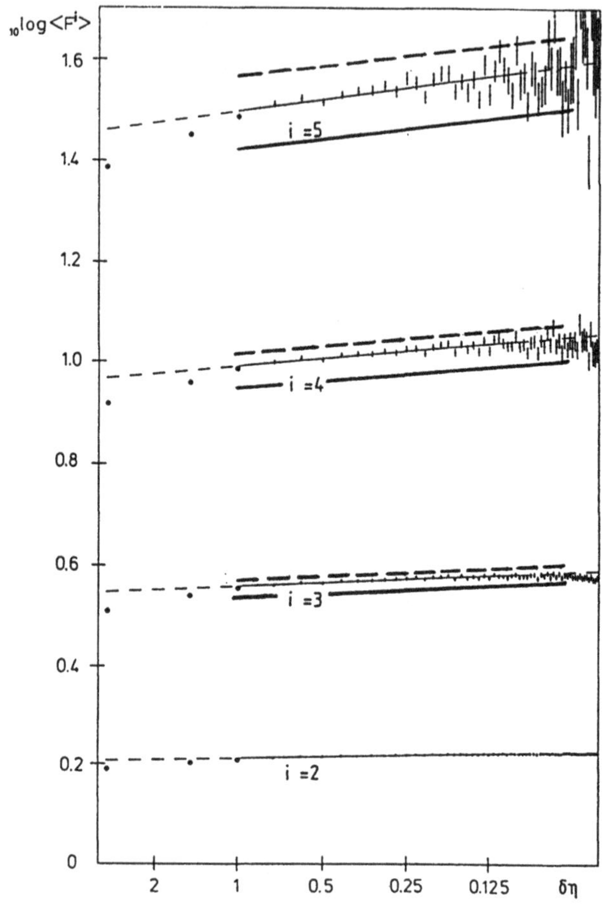


**Fig. 1:** Comparison of the UA1 $\bar{p}p$ data at $\sqrt{s} = 630$ GeV for $\log\langle F_i\rangle$ vs. $\log(1/\delta\eta)$ with the predictions of the PB distribution (full line) and the NB distribution (dashed line), eqs. (5) and (11), taking $\log\langle F_2\rangle$ as input

The UA1 data[13] for $\log\langle F_i\rangle$, $i = 2-5$, at $\sqrt{s} = 630$ GeV have very good statistics and show a significant intermittency signal. They are shown in Fig. 1 for $\log\langle F_i\rangle$ vs. $\delta\eta$, together with our predictions for the PB distribution and the NB distribution. Both distributions reproduce the slopes very well, whereas the intercepts of the higher $\langle F_i\rangle$ show some deviation from the data. The numerical results for the slopes $f_i$ are given in Table 1.

In Table 1 we also compare the $e^+e^-$ data of TASSO[14] and the cosmic ray data of Ref. 15, for the slopes $f_i$, with the predictions of the PB distribution and the NB distribution. For determining the slopes $f_i$ of the TASSO data we used the rapidity intervals between 0.7 and 0.12, as was also done in Ref. 14. Furthermore, the normalization of the factorial moments was taken as $1/\langle n\rangle^i$, like in eq. (2), with $\langle n\rangle = 8.6$, instead of the normalization used in Ref. 14. Quite generally, multiplicity data in $e^+e^-$ annihilation and in cosmic rays have the highest values for the slopes $f_i$[17]. As can be seen in Table 1, the experimental values of Refs. 14 and 15 are well reproduced by the PB and NB distributions.

Table 1 also shows that in the case of the NA22 data[16] neither the PB nor the NB distribution can reproduce the slope $f_i$ in a satisfactory manner. A possible explanation could be that at lower energies such as in Ref. 15 a stochastic equation more complicated than eq. (3) has to be used[18].

Apart from the NA22 data, in most of the examples studied we found that the slopes $f_i$ are better reproduced by the PB and NB distributions than the intercepts of $\log\langle F_i\rangle$. The capability of the NB distribution to reproduce intermittency has also been studied in Refs. 19–21. In the case of the Furry distribution, eq. (9), the normalized factorial moments do not depend on a single variable, but on $\langle n\rangle$ and the variable $p$ (see eq. (6)) separately. Consequently, for performing a similar analysis of the $\log\langle F_i\rangle$ with the Furry distribution, additional information on the dependence of $\langle n\rangle$ on $\delta y$ is necessary.

Finally, we want to comment on the problem of expressing all moments $\langle F_i\rangle$, $i > 2$, via a linked-pair ansatz for the normalized factorial cumulants, by the lowest moment $\langle F_2\rangle$ which itself is related to the two-particle correlation function. This was proposed in Refs. 22 and 23 and investigated in the case of the NB distribution in Ref. 21 (see also Ref. 24). The multiplicative constants $a_i$, $i = 3, 4, 5, \ldots$ introduced in Ref. 22 and obtained by a fit to the data on $\langle F_i\rangle$, $i = 3, 4, 5$, are completely determined in the PB distribution. They follow from eq. (5) as $a_3 = 1.2$, $a_4 = 1.4$, $a_5 = 1.7, \ldots$, to be compared with the values $a_3 = 1.3$, $a_4 = 1.6$, $a_5 = 2.8$ of Ref. 22 (for the NB distribution they were obtained in Ref. 21 as $a_3 = 1.4$, $a_4 = 1.8$, $a_5 = 2.2, \ldots$). On the other hand, we obtain for the Furry distribution from eq. (9) the normalized factorial cumulants: $k_2 = \rho$, $\rho = (p-1)/\langle n\rangle$, $k_3 = 2\rho^2 + 2\rho/\langle n\rangle + 2/\langle n\rangle^2$ etc. While $k_2$ is given by the single variable $\rho$, the higher normalized factorial cumulants $k_i$, $i > 3$, are determined by two variables, $\rho$ and $\langle n\rangle$. The Furry distribution, therefore, is a counterexample to the linked pair approximation.

**Table 1:** Comparison of the data of Refs. 13–16 for the slope parameters $f_i$ with our results from the PB and the NB distributions, eqs. (5) and (11)

**UA1** $\bar{p}p$ at $\sqrt{s} = 630$ GeV[13]

| $f_i$ | Exp | PB | NB |
|---|---|---|---|
| $i = 2$ | $0.011 \pm 0.001$ | input | input |
| 3 | $0.025 \pm 0.003$ | 0.026 | 0.027 |
| 4 | $0.050 \pm 0.005$ | 0.042 | 0.047 |
| 5 | $0.077 \pm 0.011$ | 0.060 | 0.068 |

**TASSO** $e^+e^-$ at $\langle\sqrt{s}\rangle = 35$ GeV[14]

| $f_i$ | Exp | PB | NB |
|---|---|---|---|
| $i = 2$ | $0.023 \pm 0.003$ | input | input |
| 3 | $0.080 \pm 0.014$ | 0.067 | 0.068 |
| 4 | $0.134 \pm 0.052$ | 0.124 | 0.130 |

**Cosmic rays, hadrons**[15]

| $f_i$ | Exp | PB | NB |
|---|---|---|---|
| 2 | $0.22 \pm 0.08$ | input | input |
| 3 | $0.61 \pm 0.16$ | 0.49 | 0.51 |
| 4 | $1.04 \pm 0.14$ | 0.71 | 0.80 |
| 5 | $1.60 \pm 0.24$ | 1.08 | 1.20 |
| 6 | $2.03 \pm 0.22$ | 1.42 | 1.59 |

**NA22** $(n_{ch} < 7 - 8)$ at $\sqrt{s} = 22$ GeV[16]

| $f_i$ | Exp | PB | NB |
|---|---|---|---|
| 2 | $0.0127 \pm 0.0008$ | input | input |
| 3 | $0.0499 \pm 0.0022$ | 0.0312 | 0.0330 |
| 4 | $0.148 \pm 0.007$ | 0.0534 | 0.0583 |
| 5 | $0.328 \pm 0.019$ | 0.0783 | 0.0872 |

**Acknowledgements:** We are very grateful to Prof. E. Lohrmann for reanalyzing the TASSO data and communicating his results. We are also very grateful to Dr. B. Buschbeck for informing us about the UA1 data prior to publication, and to her and Prof. W. Kittel for useful discussions. One of the authors (M.B.) would like to thank the Federal Minister of Science and Research of Austria and the Japan Society for the Promotion of Science (JSPS) for financial support.

# References

1. M. Biyajima, A. Bartl, T. Mizoguchi, N. Suzuki, Phys. Lett. B237 (1990) 563.
2. A. Bialas and R. Peschanski, Nucl. Phys. B273 (1986) 703; Nucl. Phys. B308 (1988) 857.
3. R.C. Hwa, "Self-Similarity in Multiplicity Fluctuations", University of Oregon preprint OITS 430, 1989.
4. P.A. Carruthers and C.C. Shih, Int. J. Mod. Phys.A2 (1987) 1447.
5. G.N. Fowler and R.M. Weiner, Phys.Rev. D17 (1978) 3118.
6. E.M. Friedlander, these Proceedings.
7. UA5 Coll., G.J. Alner et al., Phys. Lett. B160 (1985) 193; Phys. Lett. B160 (1985) 199; Phys. Lett. B167 (1986) 47; Phys. Rep. 154 (1987) 247; R.E. Ansorge et al., Z. Phys. C37 (1988) 191.
8. A. Giovannini and L. Van Hove, Z. Phys. C30 (1986) 391; Act. Phys. Polonica B19 (1988) 495; Act. Phys. Polonica B19 (1988) 931.
9. M. Biyajima, T. Kawabe and N. Suzuki, Phys. Lett. B189 (1987) 466.
10. M. Biyajima, K. Shirane and N. Suzuki, Phys. Rev. D37 (1988) 1824.
11. P. Carruthers and M. Duong-Van, preprint LA-UR-83-2419.
12. R.C. Hwa, Nucl. Phys. B328 (1989) 59.
13. UA1 Coll., C. Albajar et al., preprint CERN-EP/90-56, and HEPHY-PUB 531/90, to be publ. in Nucl. Phys. B.
14. TASSO Coll. W. Braunschweig et al., Phys. Lett. B231 (1989) 548.
15. C.Gładysz-Dziaduś, Mod. Phys. Lett. A4 (1989) 2553.
16. NA22 Coll., J.V. Ajinenko et al., Phys. Lett. B222 (1989) 306.

17. W. Kittel, Proc. XXIV Int. Conf. on High Energy Physics, Munich 1988, ed. R. Kotthaus and J.H. Kühn, Springer-Verlag, p. 625.

18. I. Šarčevič, private communication.

19. W. Ochs and J. Wosiek, Phys. Lett. B232 (1989) 271.

20. L. Van Hove, Phys. Lett. B232 (1989) 509.

21. E.A. De Wolf, "Intermittency, Negative Binomials and Two-Particle Correlations", submitted to Acta Phys. Polonica.

22. P.A. Carruthers and I. Šarčevič, Phys. Rev. Lett. 63 (1989) 1562.

23. A. Capella, K. Fiałkowski and A. Krzywicki, Phys. Lett. B230 (1989) 149.

24. C.C. Shih, these Proceedings.

# THE SELF-SIMILAR CASCADE PATTERN AT HIGH ENERGIES

Ina Sarcevic*

*University of Arizona*
*Physics Department*
*Tucson, AZ 85721*

ABSTRACT: We find that in very high energy hadronic collisions the self-similar cascade represents a new mechanism for multiparticle production. We show that in the large rapidity region, this pattern leads to a universal power-law behavior of the multiplicity moments as a function of rapidity bin size, in agreement with the UA5 data. We propose further test of this new pattern of multiparticle production at Tevatron energies.

## 1. INTRODUCTION

In recent years, the study of particle density fluctuations in different rapidity intervals in high energy collisions has created a new excitement in the field of multiparticle production. A proposal by Bialas and Peschanski [1] to detect intermittent behavior in high energy collisions by measuring the factorial moments has attracted a lot of attention. These bin-averaged factorial moments, $F_i$, are defined as

$$F_i = \frac{< \frac{1}{M} \sum n_m(n_m - 1)\ldots(n_m - i + 1) >}{< \frac{1}{M} \sum n_m >^i}, \tag{1}$$

where $n_m$ is the number of particles in a rapidity bin $m$ and M is a total number of bins in the interval $\Delta Y$ ($\delta y = \Delta Y/M$). In Ref. 1 it was claimed that, in the small $\delta y$ region, a power law behavior of the factorial moments, i.e. $F_i(\delta y) \sim (\delta y)^{-\nu_i}$, as $\delta y \to 0$, signals an onset of intermittency in analogy with turbulence in hydrodynamics.

* Humboldt Fellow

Several experimental groups have recently reported the observation of this new phenomenon in hadronic, leptonic and nuclear collisions [2]. In our previous work we have shown that the increase of the factorial moments with decreasing rapidity bin size in the small $\delta y$ region is due to the conventional short range correlations [4]. However, here we consider the possibility to have a signal for intermittency or self-similar structure for *large* $\delta y$, which is outside the usual resonance formation region ($\delta y \geq 1-2$) [3].

## 2. THE SELF-SIMILAR PATTERN AT VERY HIGH ENERGIES

At very high energies the phase space available for particle production becomes large enough so the self-similar cascade with many branches can develop. Clearly, this new mechanism for multiparticle production has some treshold energy. For example, at energy $\sqrt{s} = 20GeV$ the cascade has only few branches because the maximum rapidity available ($Y \equiv \ln s$) is only slightly above the resonance formation treshold. In contrast, at CERN Collider energies $Y \geq 10$ and the self-similar cascade with many branches can develop. We find that the treshold energy for this self-similar cascade mechanism is of the order of a few hundred GeV. Since we expect that at these energies the application of perturbative QCD is well justified, this self-similar cascade should be related to the low $p_T$ jet production. In support of this conjecture, we mention recent observation of "minijets" [5], which indicates that the fraction of "semi-hard" events responsible for the low $p_T$ jet production increases very fast with energy. At $\sqrt{s} \sim 20-50GeV$, for example, it is only few percent, while at CERN Collider energies it is about $15-17\%$ [5]. Keeping this in mind, we construct a simple self-similar cascade model [3], in which collision takes place in several steps. First, "heavy mass particle" is created (this could be a jet, for example). Then, this particle decays into two lighter particles of mass $m_1$, which by conservation of energy is related to the initial particle of mass $m_0$ ($m_0^2 = 4(m_1^2 + p_1^2)$). This decay pattern continues until the initial mass is reduced to the mass of the resonance ($m_{\pi\pi} \sim 0.5GeV$). This happens after $r$ steps and therefore $m_{\pi\pi} \equiv m_r = (1/\alpha)m_{r-1} = (1/\alpha)^r m_0$, where $\alpha$ is the fraction of the mass which goes into each of the produced particles ($\alpha \leq 2$). For $\alpha = 2$, for example, there would be no kinetic energy in the products of the decay. The schematic drawing of the self-similar cascade is shown on Fig. 1. We note that after $r$ steps there will be $n(r)$ particle produced, each one with mass $m_r$. These particles will not cover the full mass length, namely,

$$n(r)m_r = 2^r(\frac{1}{\alpha})^r m_0 \leq m_0. \tag{2}$$

This will be true even in the $r \to \infty$ limit. The larger the value of $r$, the less space is covered. This situation is well known in the theory of fractal structure [6]. Consequently, we can define the generalized fractal dimension [7]

$$d_F = \lim_{r\to\infty} |\frac{\ln n(r)}{\ln(m_0/m_r)}| = \frac{\ln 2}{\ln \alpha}. \tag{3}$$

The number of particles produced after $r$ steps is then given by

$$n(r) = (m_0/m_r)^{d_F} \tag{4}$$

Clearly, for $\alpha = 2$ the whole space is covered and $d_F = 1$ (or $n(r) = m_0/m_r$). We note that $r \to \infty$ limit could be taken by either fixing $m_0$ and letting $m_r \to 0$ or by fixing $m_r$ and letting $m_0 \to \infty$. The first option corresponds to an infinite decay of the fixed mass $m_0$, while the second one corresponds to having the final state mass fixed and increasing the initial mass $m_0$. The latter will turn out to be the relevant case for the multiparticle production at very high energies.

Now we apply these ideas to multiparticle production in high energy collisions. Since the multiplicity of a jet is proportional to the rapidity, we identify mass with rapidity, so that $m_0 = Y = \ln s$ and $m_r = \delta y$. The Eq. (4) can then be written as

$$n(r) = (Y/\delta y)^{d_F}. \tag{5}$$

The normalized multiplicity moments are defined by

$$C_l(\delta y) =< \frac{1}{M}\sum_{m=1}^{M} n_m^l/(\frac{1}{M}\sum_{m=1}^{M} n_m)^l >, \tag{6}$$

where $n_m$ is the number of particles in the $m$th bin and $< ... >$ denotes the average over the events. Since $\sum n_m = N$ and $M = Y/\delta y$, we can rewrite Eq. (6) in the following form:

$$C_l(\delta y) = (Y/\delta y)^{l-1} < \sum_{m=1}^{M} P_m^l > . \tag{7}$$

Here $P_m$ is the probability to find particle in a bin $m$ ($P_m \equiv n_m/N$). In a high energy collision the probability $P_m$ fluctuates greatly from bin to bin and it is difficult to extract much meaningful information on this quantity. However, we can map this one-dimensional self-similar cascade onto a $d_F$-dimensional ($d_F \leq 1$) fractal space where probability distribution is "smooth" and is proportional to the inverse of the produced particles in a given rapidity bin ($P_m = 1/n_m$). Thus, using Eq. (5), the multiplicity moments given by Eq. (7) can be cast in the following form

$$C_l = (Y/\delta y)^{(l-1)(1-d_F)}. \tag{8}$$

This implies that the plot of $\ln C_l$ versus $\ln(Y/\delta y)$ is a straight line with a slope given by

$$\lambda_l \equiv (l-1)(1-d_F), \tag{9}$$

where $d_F$ is a fractal dimension. Strictly speaking, the derived power-law should hold in the $Y/\delta y \to \infty$ limit. In the discussion of the cascade of Fig. 1, we have seen that this limit could be taken either by fixing $Y$ and letting $\delta y \to 0$, or by fixing $\delta y$ and letting $Y \to \infty$. Since the self-similar cascade exists only in the region outside the resonance formation, we consider the latter limit. This implies that the straight line behavior of $\ln C_l$ versus $\ln(Y/\delta y)$ will flatten out at the point where the short-range correlations take over. By increasing the energy, the cascade will be "longer" and the flattening will be moved to higher values of $\ln(Y/\delta y)$. In the limit of infinite energy ($Y \to \infty$), there should be no deviation from the straight line. This is schematically presented in Fig. 2.

Now we apply our theory of multiparticle production to the experimental data. We consider the UA5 measurements of the normalized multiplicity moments $f_l(\delta y)$, defined as

$$f_l(\delta y) \equiv F_l(\delta y)/F_l(Y) \tag{10}$$

where

$$F_l = \frac{< n_0(n_0-1)\ldots(n_0-l+1) >}{< n_0 >^i}, \tag{11}$$

and $n_0$ is the number of particle in the central rapidity bin. We find that UA5 data at energies $\sqrt{s} = 200, 546$ and $900 GeV$ follow the predicted universal power-law

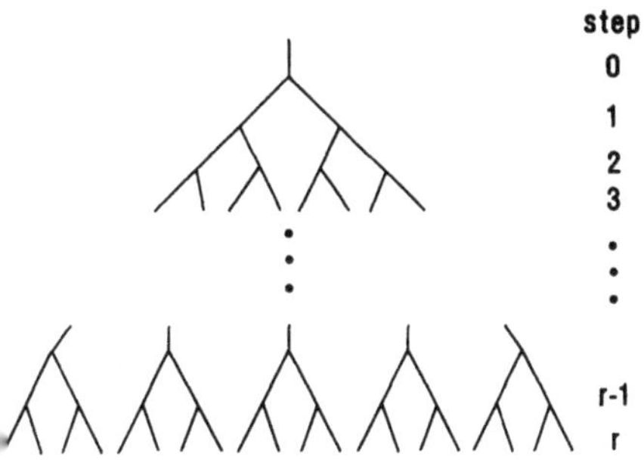


Figure 1: The self-similar cascade.

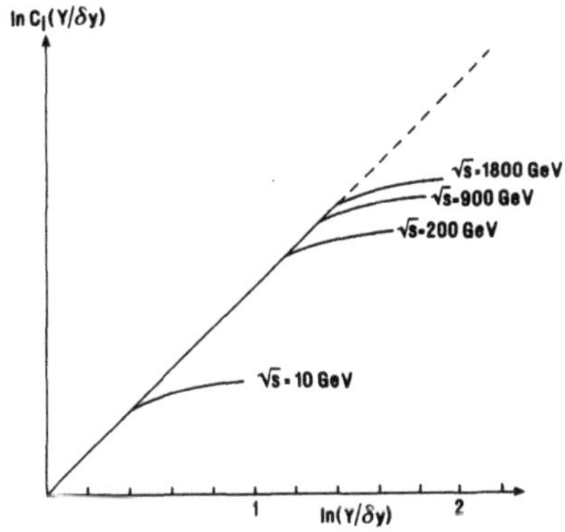


Figure 2: Schematic illustration of the multiplicity moments $C_i(Y/\delta y)$ for fixed $i$ and $\delta y_0 = 1.8$ for the energies $\sqrt{s} = 22GeV$, $200GeV$, $900GeV$ and $1800GeV$. The dashed line represents moments $C_i$ in the limit $\sqrt{s} \to \infty$.

behavior (Figs. 3-6), with the slopes having the following values [3]:

$$\lambda_2 = 0.140 \pm 0.008 \tag{12a}$$

$$\lambda_3 = 0.172 \pm 0.009 \tag{12b}$$

$$\lambda_4 = 0.203 \pm 0.013 \tag{12c}$$

$$\lambda_5 = 0.228 \pm 0.016 \tag{12d}$$

In a simple fractal picture, the ratio of the slopes should be constant ($\lambda_l/\lambda_{l+1} = (l-1)/l$). From Eqs. (12) we note that the data requires additional $l$ dependence of the fractal dimension $d_F$. This is the well known case of "multifractals" [7]. Furthermore, from Figs. 3-6 it is apparent that the data shows the expected deviation from the straight line at the point where short range correlations come in ($\delta y_0 \sim 1.8$), which corresponds to $\ln(Y/\delta y) = 1.2$ at $\sqrt{s} = 200GeV$, $\ln(Y/\delta y) = 1.3$ at $\sqrt{s} = 546GeV$ and $\ln(Y/\delta y) = 1.4$ at $\sqrt{s} = 900GeV$.

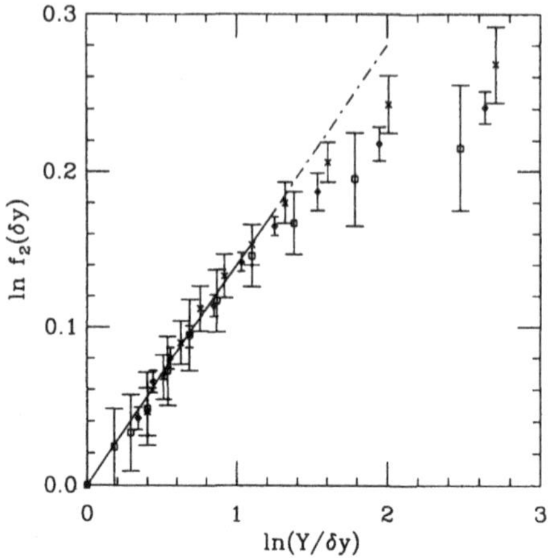


Figure 3: Experimental data on multiplicity moments $\ln f_2(\delta y)$ ($f_2(\delta y) \equiv F_2(\delta y)/F_2(Y)$) as a function of $\ln(Y/\delta y)$ for energies $\sqrt{s} = 200$GeV (squares), $\sqrt{s} = 546$GeV (diamonds) and $\sqrt{s} = 900$GeV (crosses) and the straight-line fit to the data at large $\delta y$, from Ref. 3.

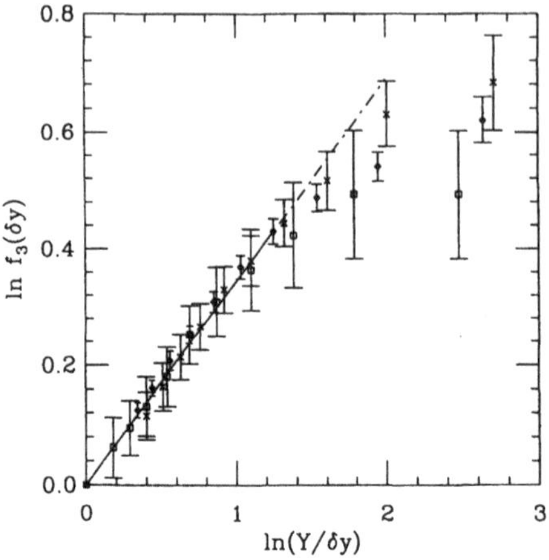


Figure 4: Same as Figure 3, but for $\ln f_3(\delta y)$ ($f_3(\delta y) \equiv F_3(\delta y)/F_3(Y)$).

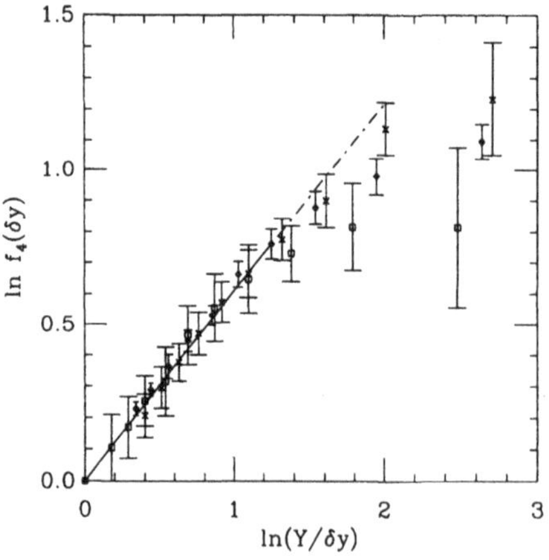


Figure 5: Same as Figure 3, but for $\ln f_4(\delta y)$ ($f_4(\delta y) \equiv F_4(\delta y)/F_4(Y)$).

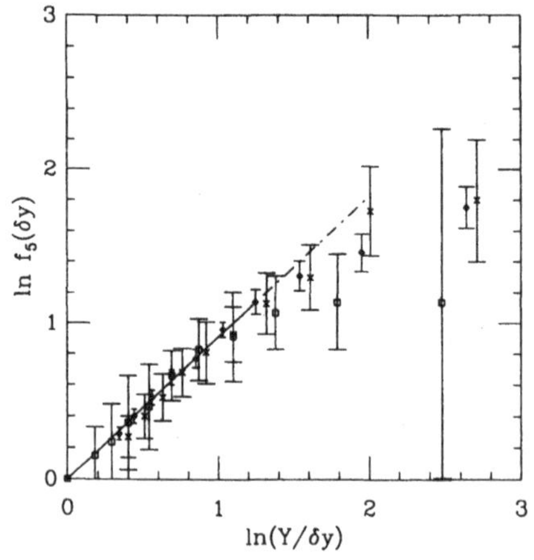


Figure 6: Same as Figure 3, but for $\ln f_5(\delta y)$ ($f_5(\delta y) \equiv F_5(\delta y)/F_5(Y)$).

From our considerations it follows that the multiplicity moments $C_l(\delta y)$ at Tevatron energies will obey the predicted universal power-law behavior, with the same slopes as the UA5 data. The deviation from the straight-line behavior will start at slightly higher value of $\ln(Y/\delta y)$. The remaining theoretical challenge is to construct a QCD-based cascade. Such a microscopic model should in principle be able to predict the values of the slopes. Work along these lines has been initiated in Ref. 8.

## ACKNOWLEDGEMENTS

The work presented here was done in collaboration with H. Satz whom I also thank for many useful discussions. I am also indebted to Freie University in Berlin for the hospitality while this manuscript was completed. This work was supported in part by the United States Department of Energy, Division of High Energy and Nuclear Physics, NATO Research Collaboration Grant and the Humboldt-Stiftung.

## REFERENCES

1. For a review, see W. Kittel and R. Peschanski, talk presented at the *EPS Conference*, Madrid, Sep. 6-13, 1989, Saclay preprint Saclay PhT/89-143 and these proceedings.

2. A. Bialas and R. Peschanski, *Nucl. Phys.* **B273** (1986) 703;ibid. **B308** (1988) 857; *Phys. Lett.* **207B** (1988) 59.

3) I. Sarcevic and H. Satz, *Phys. Lett.* **233B** (1989) 251.

4) P. Carruthers and I. Sarcevic, *Phys. Rev. Lett.* **63** (1989) 1562; P. Carruthers, H. Eggers, Q. Gao and I. Sarcevic, University of Arizona preprint AZPH-TH/90-9.

5) UA1 Collaboration, C. Albajar et al., *Nucl. Phys.* **B309** (1988) 405.

6) B. Mandelbrot, The fractal geometry of nature (Freeman, San Francisco, 1982).

7) F. Hausdorff, *Math. Ann.* **79** (1919) 157.

8) W. Ochs and J. Wosiek, *Phys. Lett.* **232B** (1989) 271.

# Scaling of correlation function moments for small fluctuations

David Seibert
School of Physics and Astronomy
University of Minnesota, Minneapolis, MN 55455 USA

**Abstract**

We propose a scaling law that relates all moments of a distribution to the second moment. The scaling law is derived for the scaled multiplicity moments, $C_i$, and the inclusive scaled factorial moments. The scaling law is independent of the source of the correlations, as it holds whenever quantities can be expanded about their mean values in a power series. We demonstrate that the multifractal moments obey a second scaling law. Moments that obey the second scaling law are dominated by single-particle effects and thus are not good correlation functions.

Weak two-particle correlations have been measured for many quantities in high-energy collisions. Often, higher moments of the same quantities have been measured,[1-5] but little attention has been paid to the extraction of higher-order effects from these higher moments. We describe here a theory-independent method for the separation of higher-order effects from two-particle correlations. This method may also be generalized to remove all lower-order correlations from any given correlation function, although the usefulness of this procedure decreases as the order of the correlations to be removed increases. We also give an experimental test to see whether moments are good correlation functions, or whether they are dominated by single-particle effects.

We first derive a scaling law for the scaled multiplicity moments, using a power-series expansion of the moments. This scaling law is identical to the scaling law that we have derived for scaled factorial moments.[6] Using the two scaling laws, we show that the same moment scaling law is observed by the inclusive scaled factorial moments.[7] This scaling law should hold, in the small fluctuation limit, for almost all sets of moments that are good correlation functions.

Finally, we show a set of moments that does not obey our scaling law, the multifractal moments,[8] $G_q$. These moments obey a second scaling law. Any set of moments that obeys this second scaling law in the small fluctuation limit is dominated by single-particle effects. As a result, moments that obey the second scaling law are not good correlation functions *for the system in question.*

The simplest correlation functions are the scaled multiplicity moments,

$$C_i = \frac{\langle N^i \rangle}{\langle N \rangle^i} , \tag{1}$$

where $N$ is the particle number, and the averages are taken over all events. Knowing the set of all moments is equivalent to knowing the probability, $P(N)$, of any given multiplicity $N$. Therefore, we assume that we know $P(N)$ and calculate the scaled multiplicity moments.

We first rewrite the moments in terms of fluctuations from the mean multiplicity:

$$C_i = \sum_{j=0}^{i} \frac{i!}{j!(i-j)!} \frac{\langle \delta N^j \rangle}{\langle N \rangle^j} , \tag{2}$$

where

$$\delta N = n - \langle N \rangle . \tag{3}$$

We treat Eq. (2) as a power series, because typically

$$\frac{\langle \delta N^i \rangle}{\langle N \rangle^i} \sim \left[ \frac{\langle \delta N^2 \rangle}{\langle N \rangle^2} \right]^{i/2} = (C_2 - 1)^{i/2} . \tag{4}$$

We then expand $C_i$ in powers of $C_2 - 1$, obtaining

$$C_i = 1 + \frac{i(i-1)}{2}(C_2 - 1) + \mathcal{O}\left((C_2 - 1)^{3/2}\right) . \tag{5}$$

Thus, we find that

$$\frac{C_i - 1}{i(i-1)} = \frac{C_2 - 1}{2} + \mathcal{O}\left((C_2 - 1)^{3/2}\right) , \tag{6}$$

assuming only that Eq. (4) holds. Higher-order corrections are discussed in Refs. 9 and 10.

We can also expand $\ln(C_i)$ as a function of $\ln(C_2)$, rather than $C_i - 1$ as a function of $C_2 - 1$. Because we only expanded to first order, the scaling law (5) remains the same, except that $\ln(C_i)$ and $\ln(C_2)$ replace $C_i - 1$ and $C_2 - 1$ everywhere. We then obtain

$$C_i \approx C_2^{i(i-1)/2} . \tag{7}$$

To lowest order, Eq. (5) and (7) are identical, and so they can be used interchangeably as scaling equations for moments in the weak correlation limit. Both equations give successfull predictions of higher moments in the small fluctuation limit.[10]

Equation (6) is the same scaling law that we derived earlier for scaled factorial moments, or SFMs:[6]

$$F_i(\delta y; \Delta Y) = \frac{\left[\prod_{j=1}^{i} \int_{y}^{y+\delta y} \frac{dy_j}{\delta y}\right] \rho^{(i)}(y_1, \ldots, y_i)}{\left[\prod_{j=1}^{i} \int_{y_0}^{y_0+\Delta Y} \frac{dy_j}{\Delta Y}\right] \rho^{(i)}(y_1, \ldots, y_i)} , \tag{8}$$

where $y$ is rapidity and $\rho^{(i)}$ is the $i$-particle rapidity density. These rapidity correlation functions satisfy the scaling law

$$F_i = 1 + \frac{i(i-1)}{2}(F_2 - 1) \tag{9}$$

in the weak-correlation limit. This is a somewhat more demanding condition than Eq. (5), as Eq. (9) holds for all values of $\delta y$ and $\Delta Y$. The entire functional form of $F_i$ is thus determined by $F_2$.

It is interesting that the same scaling law holds for both the scaled multiplicity moments and the SFMs. The inclusive SFMs,[7]

$$F_i(\delta y;\Delta Y)=\frac{\left[\prod_{j=1}^{i}\int_{y}^{y+\delta y}\frac{dy_j}{\delta y}\right]\rho^{(i)}(y_1,\ldots,y_i)}{\left[\int_{y_0}^{y_0+\Delta Y}\frac{dy}{\Delta Y}\,\rho(y)\right]^i}\ , \tag{10}$$

are another commonly used set of correlation functions. These moments are the product of multiplicity and rapidity correlation functions:[6]

$$F_i^{(\text{inclusive})}(\delta y;\Delta Y)=\frac{\langle N(N-1)\cdots(N-i+1)\rangle}{\langle N\rangle^i}\,F_i(\delta y;\Delta Y)\ , \tag{11}$$

where $N$ now is the number of particles in a rapidity window of width $\Delta Y$.

We can rewrite the inclusive SFMs in terms of the $C_i$'s and the SFMs, obtaining

$$F_i^{(\text{inclusive})}(\delta y;\Delta Y)=\left[C_i-\left(\frac{i(i-1)}{2\langle N\rangle}\right)C_{i-1}+\mathcal{O}\left(\langle N\rangle^{-2}\right)\right]F_i(\delta y;\Delta Y)\ . \tag{12}$$

Combining Eq. (12) and our two scaling laws, Eqs. (5) and (9), we find that the inclusive SFMs should also exhibit scaling behavior in the weak-correlation limit. We have shown that this is true for data from O-Em collisions[4] at 200 GeV/A in Ref. 10.

The scaling law will be obeyed, in the small fluctuation limit, by almost any good correlation function, as it depends only on the fact that we can expand the correlation function in a power series, using the deviation of some quantity from its mean value as an expansion parameter. It is possible that, in some cases, both the first-order and second-order terms will vanish. In this case, $C_0=C_1=C_2=1$, and we obtain the scaling law

$$C_i=1+\frac{i(i-1)(i-2)}{6}\,(C_3-1)\ , \tag{13}$$

as three-particle effects will dominate in the weak-correlation limit. Higher-order scaling laws are also possible, but these are progressively less likely, as they only hold if all of the lower moments are equal to unity.

Finally, there are sets of moments that obey a linear scaling law. One example of this is the set of multifractal moments,[8]

$$G_q = \sum_{m=1}^{M} \frac{\langle \theta(n_m)\, n_m^q \rangle}{\langle n_m \rangle^q} , \tag{14}$$

where each event is divided into $M$ bins, $n_m$ is the number of particles in the $m$-th bin, and

$$\theta(n) = \begin{cases} 0 & n = 0 , \\ 1 & n > 0 . \end{cases} \tag{15}$$

The function $\theta$ is necessary for $G_q$ to be well-defined for $q < 0$. We show the scaling behavior of the multifractal moments in Ref. 10, using preliminary data from $\pi$-p and K-p collisions[5] at $\sqrt{s} = 23$ GeV. We find that the moments are largely determined by first-order (single-particle) effects:

$$\ln G_q = (1 - q)\, \ln G_0 . \tag{16}$$

The scaling law (16) tells us that the first-order term in our expansion of $G_q$ does not vanish. Thus, $G_q$ is dominated by single-particle effects, and these moments are not good correlation functions. The physics of these moments is readily determined from the fact that all of the moments depend strongly on $G_0$. $G_0$ is always equal to the number of occupied bins, and so *all* of these moments are essentially functions of the mean number of occupied bins.

In summary, we have derived a scaling law that holds, in the weak-correlation limit, for almost any set of moments that are good correlation functions. This scaling law has been demonstrated to hold for scaled multiplicity moments, scaled factorial moments, and inclusive scaled factorial moments. We have shown how to extract higher-order contributions to a set of moments, and discussed possible violations of the scaling law. Finally, we have shown that some moments obey a linear scaling law, but that these sets of moments are not good correlation functions as they are dominated by single-particle effects.

## Acknowledgements

We thank Dominique Vautherin and the Institute for Nuclear Physics of the University of Paris – XI for their hospitality while these proceedings were written. This research was supported by the U. S. Department of Energy under Contract No. DOE/DE-FG02-87ER-40328. Computer facilities were provided by the Minnesota Supercomputer Institute.

## References

1. M. Derrick *et al.* (HRS Collaboration), *Phys. Rev.* **D34** (1986) 3304.
2. G. J. Alner *et al.* (UA5 Collaboration), *Phys. Lett.* **138B** (1984) 304.
3. K. Sengupta, G. Singh and P. L. Jain, State University of New York at Buffalo preprint (November 1989), to be published in *Modern Physics Letters* **A**.
4. R. Holynski *et al.* (KLM collaboration), *Phys. Rev. Lett.* **62** (1989) 733.
5. W. Kittel (EHS/NA22 Collaboration), preliminary results, to be published in the proceedings of the Santa Fe Workshop on Intermittency, Santa Fe, New Mexico, March 1990.
6. D. Seibert, University of Minnesota preprint (November 1989), to be published in *Phys. Rev.* **D**, June 1, 1990; University of Minnesota preprint (March 1990), to be published in the proceedings of the Santa Fe Workshop on Intermittency, Santa Fe, New Mexico, March 1990.
7. A. Białas and R. Peschanski, *Nucl. Phys.* **B273** (1986) 703; *ibid,* **B308** (1988) 857.

8. R. Hwa, University of Oregon preprint OITS-430 (November 1989), to appear in the review monograph *Quark-Gluon Plasma*, edited by R. C. Hwa, in the Advanced Series on Directions in High Energy Physics (World Scientific, Singapore, 1990).

9. E. Stenlund, material presented at this workshop.

10. D. Seibert, University of Minnesota preprint (May 1990), submitted to *Phys. Rev. Lett.*.

# NON-STATISTICAL FLUCTUATIONS IN RELATIVISTIC HEAVY-ION COLLISIONS

Evert Stenlund
Cosmic and Subatomic Physics, University of Lund, Sölvegatan 14,
S - 223 62 Lund, Sweden

and

M I Adamovich, M M Aggarwal, Y A Alexandrov, Z U Ameeva, N P Andreeva, Z V Anzon, R Arora, S K Badyal, K B Bhalla, A Bhasin, V S Bhatia, V I Bubnov, T H Burnett, X Cai, I Y Chasnikov, L P Chernova, M M Chernyavski, B Dressel, G Z Eligbaeva, L E Eremenko, E M Friedlander, S I Gadzhieva, A S Gaitinov, E R Ganssauge, S Garpman, S G Gerassimov, J Grote, K G Gulamov, S K Gupta, V Gupta, H H Heckman, H Huang, B Jakobsson, B Judek, S Kachroo, F G Kadyrov, G S Kalyachkina, E K Kanygina, G L Kaul, M Kaur, S P Kharlamov, T Koss, V Kumar, P Lal, V G Larionova, P J Lindstrom, L S Liu, S Lokanathan, J Lord, N S Lukicheva, S B Luo, L K Mangotra, N V Maslennikova, I S Mittra, S Mookerjee, C Mueller, S H Nasyrov, V S Navotny, G I Orlova, I Otterlund, H S Palsania, N G Peresadko, S Persson, N V Petrov, W Y Qian, R Raniwala, S Raniwala, N K Rao, J T Rhee, N A Salmanova, W Schulz, C I Shakhova, B Singh, D Skelding, K Söderstrom, S C Strausz, J F Sun, L N Svechnikova, M I Tretyakova, H Q Wang, Z Q Weng, R J Wilkes, G F Xu, D H Zhang, P Y Zheng, D C Zhou, J C Zhou

EMU01-collaboration

ABSTRACT

In this talk I will summarize the recent results concerning non-statistical fluctuations obtained by the EMU01 collaboration. We have observed that, in $^{32}S$ + Au interactions, scaled factorial moments increase with decreasing rapidity window, $\delta\eta$, down to 0.1 rapidity units, but for $\delta\eta < 0.1$ the increase stops. The slopes, $\phi_q$, in the region $2 \geq \delta\eta \geq 0.1$ are observed to be inversely proportional to the average particle density, $\langle\rho\rangle$.

## 1. INTRODUCTION

The topic of non-statistical fluctuations in the particle densities is a topic of a lot of recent interest[1]. A lot of effort has been devoted to the description of these phenomena, but more theoretical work is still needed for the understanding of the effects. In the meantime the experiments produce a lot of data showing different aspects of these fluctuations[2]. In this talk I will summarize the experimental achievements performed by the EMU01-collaboration. This experiment utilizes two different emulsion techniques; ordinary emulsion stacks, exposed parallel to the emulsion plates, and emulsion chambers, irradiated perpendicular to the plates. Some of the chambers are equipped with thin target foils of gold. The measurements in the chambers are performed with semi-automatic systems. In the conventional stacks the resolution in pseudo-rapidity is about 0.1 units in the central region, whereas in the chambers a resolution of 0.013 units can be obtained. More details on the experiment are published elsewhere[3].

In the second chapter I will discuss the data under the assumption that correlations between the particles in the central region are absent. In the next chapter I will introduce the scaled factorial moments[4], and in chapter 4 I will discuss their slopes. In chapter 4 I will also comment on error estimates, the influence of events having largely differing multiplicities and furthermore discuss some proposed scaling rules. Chapter 4 ends with a section on the uncertainties introduced by electron pairs produced either by the Dalitz decay or by $\gamma$-conversion. Chapter 5 will contain the conclusions.

## 2. STOCHASTIC EMISSION

Due to the many sources expected in heavy-ion collisions, one can expect that most of the correlations between the produced particles will be washed out. Let us consider a pseudo-rapidity interval $\Delta\eta$, within which we look at the smaller window $\delta\eta$. If we assume that correlations between particles can be neglected and that the shape of the pseudo-rapidity distribution is independent of the multiplicity (or centrality), we obtain that $F_2(\eta,\delta\eta) \equiv \langle n\cdot(n-1)\rangle/\langle n\rangle^2$ is independent of the size and location of the window. $F_2$ is generally larger than unity due to the width of the global multiplicity distribution. If $\langle n\rangle\cdot(F_2-1)^{½}$ is large enough we find

$$\sigma \simeq \langle n\rangle\cdot\sqrt{F_2-1} + \frac{1}{2\cdot\sqrt{F_2-1}} \qquad (1)$$

i e we have a linear relationship between $\sigma$ and $\langle n\rangle$ for different windows, where both intercept and slope are determined by the value of $F_2$. In fig 1 this is tested for minimun bias oxygen-induced interactions at three different energies, for $\eta$-windows centered around mid-rapidity[5]. For all three energies we find a nice linear behaviour as long as the windows are

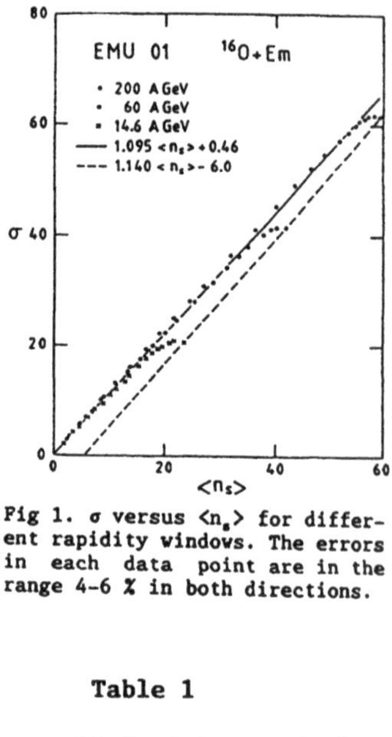


**Fig 1.** $\sigma$ versus $\langle n_s\rangle$ for different rapidity windows. The errors in each data point are in the range 4-6 % in both directions.

**Table 1**

$F_2$-values obtained for central $\delta\eta$-windows of size 2.

| Type | $E_{inc}$ (A GeV) | $F_2$ |
|---|---|---|
| Data | 200 | 2.28±0.08 |
| Data | 60 | 2.20±0.09 |
| Data | 14.6 | 2.04±0.06 |
| FRITIOF | 200 | 2.28(2.25)±0.02 |
| FRITIOF | 60 | 2.23(2.19)±0.02 |

narrow enough to exclude the fragmentation regions. Only the line for the 60 A GeV data is indicated in the plot. The $F_2$-values are determined for $\delta\eta = 2$, and given in table 1, where also Fritiof[4] has been utilized. The values within brackets are values obtained when real rapidity is used instead of pseudo-rapidity. We see that the $F_2$-values as well as the slopes increase with energy, and the asymptotic value of $F_2$ can be found in the same plot, given by the slope of the line connecting the three points representing the global distributions. This gives the asymptotic value of $F_2$ to be 2.30, in good agreement with the general trend in table 1.

The results clearly indicate that this kind of stochastic emission works well in the central region and that the correlation between the produced particles are well hidden in the combinatorial background. I will return to this idea in section 4.1. when the error estimates on the intermittency indices are discussed.

## 3. SCALED FACTORIAL MOMENTS

The most commonly used methods for studying non-statistical fluctuations are based on two papers by Bialas and Peschanski[4]. These methods are based on the notion of scaled factorial moments and differ in the procedures of normalization and how to average over events and over different regions of phase space. One method, horizontal averaging, calculates the moments for each event and then averages over the events. A second method, vertical averaging, first calculates the moments for a given $\eta$-bin in all events and then averages over all bins. Horizontal averaging is normally followed by a correction for the variation of particle densities as a function of $\eta$, here referred to as Fialkowski correction[6]. The moments can either be normalized to the whole event sample or to the individual events (individual bins for the vertical averaging). The moments are studied as a function of the size of the chosen $\eta$-bin and the variation of the moments with varying $\delta\eta$ may indicate an intermittent behaviour.

With the resolution of our chambers we have the possibility to study a region which is normally not accessible for other fixed target experiments. In fig 2 I show our results for central $^{32}S$+Au events[7]. The moments are calculated by both vertical averaging and corrected horizontal averaging and give basically the same results. Also shown is a model calculation with built in intermittency, based on the $\alpha$-model by Bialas and Peschanski. The Fritiof model[8] shows no variation with $\delta\eta$ (not shown in the figure). Our conclusions are that for $\delta\eta \geq 0.1$ the moments rise with decreasing $\delta\eta$, but for smaller

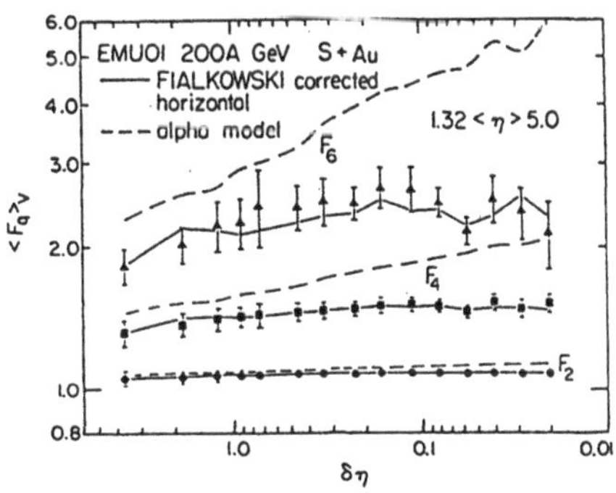


Fig 2. Scaled factorial moments calculated with two different methods and an M C simulation using a model with built in intermittency.

values of $\delta\eta$ the moments stay essentially flat, indicating a smallest inherent scale for the particle production. The rise for $\delta\eta \gtrsim 0.1$ is consistent with other experimental findings[9].

In the following chapter I will deal with five different data samples; $^{16}O$+Em at 14.6, 60 and 200 A GeV, $^{32}S$+Em and $^{32}S$+Au at 200 A GeV. Each of the samples have been divided into one central and one semi-central subsample, based on the forward charge-flow, $Q_{ZD}$. All samples contains data from the two different experimental techniques. The method used for calculating the moments is

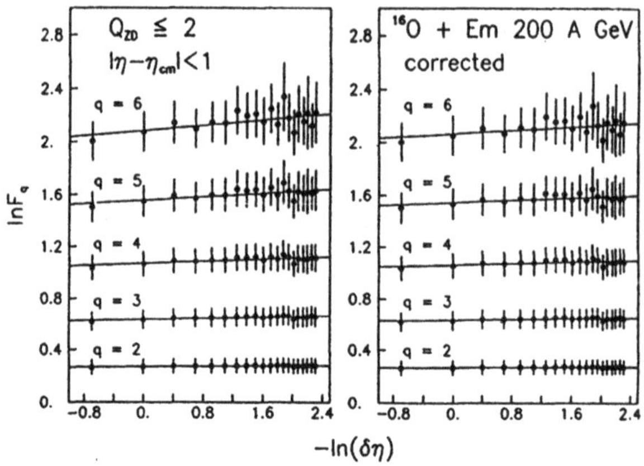


Fig 3. Scaled factorial moments for central $^{16}O$+Em interactions at 200 A GeV. Left uncorrected and right Fialkowski corrected.

$$F_q = M^{q-1} \cdot \sum_{m=1}^{M} \frac{k_m \cdot (k_m-1) \cdot \cdots \cdot (k_m-q+1)}{\langle N \rangle^q} \tag{2}$$

which is then averaged over all events and corrected for the shape of the pseudo-rapidity distribution[6]. Here M is the number of $\delta\eta$-bins and $k_m$ is the multiplicity in bin m. $\langle N \rangle$ is the average multiplicity in the whole window under consideration. I will furthermore concentrate on the central region here defined as $|\eta - \eta_{cm}| < 1$, and I will only study the region $0.1 < \delta\eta < 2$. In fig 3 I show $\ln F_q$ versus $-\ln(\delta\eta)$ for the central $^{16}O$+Em sample before and after the shape correction. For all moments a linear behaviour is observed. As can be seen in the plot, for this sample the correction is marginal.

## 4. SLOPES OF SCALED FACTORIAL MOMENTS

### 4.1. Error Estimates

When lines are fitted to data as in fig 3, one has to keep a few facts in mind. For a given moment the points are strongly correlated since the same sample is used, although with a different binning, for each of the points. This makes it very hard to estimate the error of the intermittency index, i e the slope of the line. Furthermore for the same reason the different moments are strongly correlated, which I will come back to in section 4.3. In order to study these correlations and to estimate the errors of the slopes we have devised a simple Monte Carlo method based on the stochastic emission picture in chapter 3. As input the experimental multiplicity and pseudo-rapidity distributions are used and the particles are randomly emitted. The M C sample is generated exactly as large as the experimental sample. For each experimental sample the generation is repeated N times and the standard deviation of the slope, $\phi_q$, is estimated as

$$\sigma(\phi) \simeq \sqrt{\langle\phi^2\rangle - \langle\phi\rangle^2} \qquad (3)$$

with an uncertainty of $(2\cdot N)^{-½}$. For N = 100 we have thus an uncertainty of about 7%. These error estimates can be considered as lower limits since the small contribution of non-statistical fluctuations tend to make them larger and this is accentuated for the higher moments.

4.2. The High-Multiplicity Tail

If the denominator of eq (2) is taken outside of the sum we see that the definition of $F_q$ contains a normalization factor times a sum. The normalization factor is the same for all events and the high multiplicity events give the main contribution to the sum. Thus the intermittency index is mainly determined by the high multiplicity tail when a broad multiplicity distribution is studied, again accentuated for the higher moments. This effect was investigated on our combined (biased towards high multiplicities) $^{16}O$+Em sample at 200 A GeV. Out of a total sample of 913 events a multiplicity cut, $n > n_{cut}$, was applied and the results where compared as a function of $n_{cut}$. For small $n_{cut}$ the slopes and the error estimates came out identical, whereas the intercepts decreased due to the changing normalization factor. A noticeable change in slopes (and errors) was only detected when less than 20% of the events remained. In heavy-ion experiments with widely varying multiplicities there is thus nothing to gain by treating full samples, it is rather preferred to divide the sample into subsamples with a limited multiplicity range or centrality cut.

4.3. Scaling Rules and Trivial Components

There are several scaling rules proposed for the intermittency indices, and I will mainly concider two of them. Seibert[10] has proposed that $\phi_q$ will be proportional to the inverse of the average particle density, i e $1/\langle\rho\rangle$. His main idea is that two-particle correlations give the main contribution to the intermittency effect. The results of a test of this scaling rule are shown in fig 4, where $\phi_2$ is plotted versus $\langle\rho\rangle$ for the 10

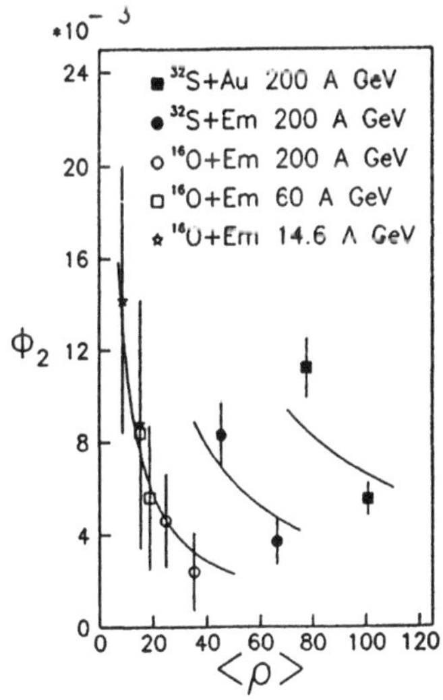


Fig 4. Intermittency index of the second order as a function of average particle density for different systems and centrality cuts.

subsamples. As can be seen in the figure, data points from the same colliding system are rather well described by a 1/<ρ>-curve independent of energy and centrality cut. Normalized to the same particle density a larger system, however, shows a larger slope. When furthermore old data from proton induced interactions[11] are reanalyzed, the obtained slopes are close to the extrapolation of the 1/<ρ>-curve for the oxygen-induced interactions.

The other scaling rule[4] which I will dwell on is

$$\phi_q = \binom{q}{2} \cdot \phi_2 \tag{4}$$

Factorial moments can be seen as integrals of multi-particle correlation functions, and higher order correlation functions contain trivial components from lower orders. This will correspondingly influence the intermittency indices. Equation (4) can thus be seen as the first expression in a series continuing with

$$\phi_q^{(3)} = \binom{q}{3} \cdot \phi_3 - (q - 3) \cdot \binom{q}{2} \cdot \phi_2 \tag{5}$$

Here the bracketed index correspond to the highest order considered for the components. If we define the normalized slope

$$\zeta_q \equiv \phi_q \,/\, \binom{q}{2} \tag{6}$$

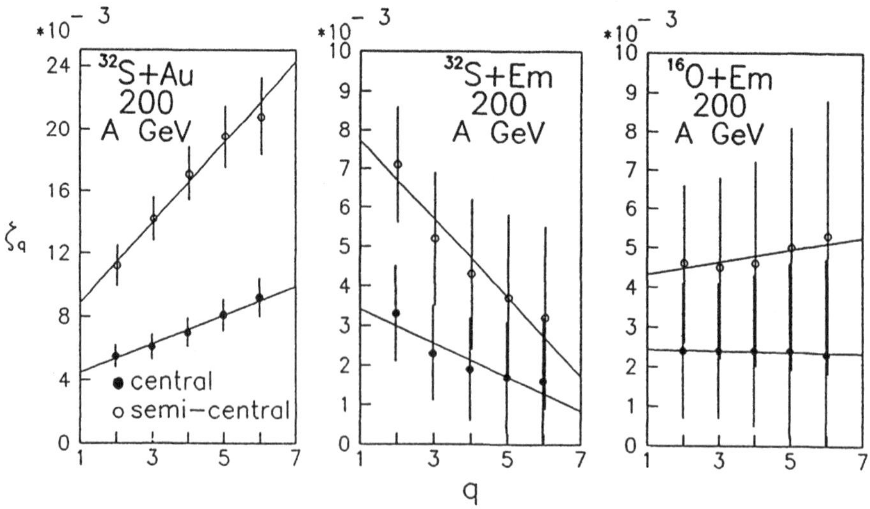


Fig 5. Normalized slopes, $\zeta_q$, as a function of q for the three different 200 A GeV samples.

a more general formula can be written as

$$\zeta_q^{(r)} = \sum_{i=2}^{r} (-1)^{r-i} \cdot \binom{q-2}{i-2} \cdot \binom{q-i-1}{r-i} \cdot \zeta_i \qquad 2 \leq r < q \qquad (7)$$

We can now study differences like $\zeta_q - \zeta_q^{(q-1)}$ both for the data samples and by the M C method discussed in section 4.1.

In fig 5 I show the normalized slopes, $\zeta_q$, for the six 200 A GeV samples as a function of the order, q. The lines are there mainly to guide the eyes, but show a rather nice linear dependence, as would be expected if only two- and three-particle correlations were present. The M C calculation shows, however, that for all samples, with the possible exception of the $^{32}S$+Au data, $\zeta_q$ is consistent with being constant. A small statistical offset at q = 3 will carry over to the higher moments thereby producing an apparent linear behaviour. Our data is thus consistent with the scaling rule described by eq (4). Actually the differences $\zeta_q - \zeta_q^{(q-1)}$ will follow a $1/\langle\rho\rangle^{q-1}$-behaviour, using the same arguments as Seibert[10].

### 4.4. Uncertainties Due to Production of Electron Pairs

A possible source of systematic errors are electron pairs produced either by a direct Dalitz decay or through $\gamma$-conversion. The influence on the slopes from both these processes has been carefully studied by Fritiof-simulations and are found to follow the same scaling rules as discussed in section 4.3. Dalitz production alone is found to be negligible and the combined effect can be parametrized as

$$\phi_{q,pair} \simeq 0.033 \cdot \binom{q}{2} \cdot \frac{p}{\langle\rho\rangle} \qquad (8)$$

where p is the percentage of converted gammas. For all data samples the percentage is small enuogh to make this systematic error smaller than the quoted statistical errors, and for the $^{32}S$+Au samples where p is of the order 3% these corrections will lower the slopes 10-20%.

## 5. CONCLUSIONS

In a limited $\delta\eta$-range (0.1 - 2) the intermittency index is constant. In a wider $\delta\eta$-range (down to 0.01) $\phi_q$ decreases with decreasing $\delta\eta$. For a given system $\phi_2$ follows a $1/\langle\rho\rangle$-dependence, but normalized to the same density a larger system shows a larger slope. The data seem to be consistent with the scaling rule given by eq (4), with the possible exception of the $^{32}S$+Au sample. $\phi_q$ versus $\langle\rho\rangle$ will look similar to $\phi_2$ versus $\langle\rho\rangle$ but with increased experimental errors. Finally, it will be hard, if not impossible, to study factorial moments of orders larger than 3, with the emulsion technique, since the contributions from the second and third order drown the signals from the higher orders.

## 6. ACKNOWLEDGEMENTS

The financal support from Kungliga Fysiografiska Sällskapet i Lund and from the Swedish Natural Science Research Concil is gratefully acknowledged.

## 7. REFERENCES

1. For recent reviews see: L Van Hove, Mod Phys Lett **A4**(1989)1867; B Buschbeck and P Lipa, Mod Phys Lett **A4**(1989)1871.
2. See, for instance, contributions to this workshop.
3. S Garpman et al, Nucl Instr Meth **A269**(1988)134; M I Adamovich et al, Phys Lett **201B**(1988)397; Phys Rev Lett **62**(1989)2801.
4. A Bialas and R Peshanski, Nucl Phys **B273**(1986)703; **B308**(1988)857.
5. M I Adamovich et al, Lund preprint, LUIP 8912, to appear in Phys Lett B.
6. K Fialkowski et al, Acta Physica Polonica **B20**(1989)639.
7. M I Adamovich et al, Seattle preprint UWSEA PUB 90-4, submitted to Phys Rev Lett.
8. B Nilsson-Almquist and E Stenlund, Conp Phys Comm **43**(1987)387.
9. See, for instance, R Holinski et al, Phys Rev Lett **62**(1989)733.
10. D Seibert, Minesota preprint Nov, 1989, to appear in Phys Rev D.
11. I Otterlund et al, Nucl Phys **B142**(1978)445; J Babecki, private communication.

# REPRESENTING DISTRIBUTION FROM $e^+e^-$ ANNIHILATION EXPERIMENTS BY SHIFTED POISSON TRANSFORMS

A.H. Chan and C.K. Chew

Department of Physics
National University of Singapore
Kent Ridge, Singapore 0511

**ABSTRACT**

*In the past, it was already known that the Negative Binomial Laguerre and Partially Coherent distributions could be obtained from a positive weight superposition of Poisson distributions. Such statistical formulae had been used to fit high energy annihilation data with varying success.*

*In this paper, we present a simple distribution which is obtainable from a shifted Poisson transform and it fits the recently available $e^+e^-$ data exceedingly well.*

## Introduction

By postulating a suitable f(x) and using an ordinary Poisson transform the distribution $P_n$ can be written as follows [1,2]

$$P_n = \int_0^\infty f(x) \frac{(x\bar{n})^n e^{-x\bar{n}}}{n!} dx \qquad (1)$$

where f(x) needs not always be positive [3], but the fundamental relations

$$\sum_{n=0}^{\infty} P_n = 1,$$

$$\sum_{n=0}^{\infty} n P_n = \bar{n}, \qquad (2)$$

must be satisfied. These imply the following normalization conditions

$$\int_0^\infty f(x)\, dx = 1,$$

$$\int_0^\infty xf(x)\, dx = 1. \qquad (3)$$

The Poisson distribution has been widely used in photo-countingstatistics. As on example, by assuming a Poisson weight f(x) = $e^{-x}$, one can obtain $P_n = \bar{n}^n/(1 + \bar{n})^{n+1}$ which is basically the Bose-Einstein statistics. This simple derivation makes no use of thermodynamics.

A more general Bose-Einstein distribution can be derived by considering the ways to distribute n particles in k Bose-Einstein cells of equal measuring occupancy $\bar{n}/k$. The result is the famous negative binomial distribution (NBD) found by Planck many years ago [3]

$$P_n^k = \sum_{\substack{n_1 n_2 \ldots n_k \\ \Sigma n_i = n}} \Pi_i \frac{(\bar{n}/k)^{n_i}}{(1 + \bar{n}/k)^{n_i + 1}}$$

$$= \frac{\Gamma(n + 1)}{\Gamma(n + 1)\,\Gamma(k)} \frac{(\bar{n}/k)^n}{(1 + \bar{n}/k)^{n+k}} \,. \tag{4}$$

This distribution which is now an important distribution in high energy particle physics can be attained by proposing a gamma distribution for f(x)

$$f(x) = \frac{k^k}{\Gamma(k)} x^{k-1} e^{-kx} \tag{5}$$

## Formalism

Lately, it was reported that NBD fails to describe new $p\bar{p}$ [4] at 900 GeV and $e^+e^-$ [5] for large central intervals. Here we propose an additional degree of freedom k′ to give a more general form of NBD and greater freedom for $\chi^2$/pfits.

Firstly we hypothesize a certain gamma function shifted by k′ as

$$f(x) = \frac{(k + k')^{k'+k}\, x^{k'+k-1}\, e^{-(k'+k)x}}{\Gamma(k + k')} \tag{6}$$

and a Poisson transform weight shifted by the same amount. We can now write the distribution namely,

$$P_n^{k,k'} = \int_0^\infty \frac{(k+k')^{k'+k}\, x^{k'+k-1}\, e^{-(k'+k)x}}{\Gamma(k+k')} \frac{[x(\bar{n}-k')]^{n-k'}\, e^{-x(\bar{n}-k')}}{(n-k')!}\, dx \tag{7}$$

The final solution obtained by the compound shifted Poisson distribution will be

$$P_n^{k,k'} = \frac{(n+k-1)!}{(n-k')!\,(k'+k-1)!} \left(\frac{\bar{n}-k'}{k+\bar{n}}\right)^{n-k'} \left(\frac{k+k'}{k+\bar{n}}\right)^{k+k'} . \tag{8}$$

Of course one could use ordinary Poisson with a different weight f(x) in Eq. (1) to obtain Eq. (8), in particular,

$$f(x) = \left[\frac{(\bar{n}-k')^{n-k'}\, n!}{\bar{n}^n (n-k')!}\right]\left[\frac{(k+k')^{k'+k}\, x^{k-1}\, e^{-kx}}{\Gamma(k'+k)}\right]. \tag{9}$$

However Eq. (9) will not exhibit the nice KNO scaling form i.e. by expressing Eq. (8) in terms of scaled variable, $Z = n/\bar{n}$ and $\bar{n}P^{k,k'} \sim \psi_{k,k'}(n/\bar{n})$ at

$$\psi_{k,k'}(z) = \frac{(k+k')^{k'+k}\, z^{k'+k-1}\, e^{-(k'+k)z}}{\Gamma(k'+k)} \tag{10}$$

which is Eq. (6) (independent of $\bar{n}$, $n \gg 1$, $\bar{n} \gg 1$).

It should be obvious now that by setting $k' = 0$, eq.(8), the generalized NBD will be reduced to NBD eq. (4) and the corresponding $f(x)$, eq.(6) reducedto eq.(5). However fixing $k = 0$, one can arrive at a Fury-Yule distribution(FYD). This distribution has also been proposed to fit Sp$\bar{p}$S and ISR data [6]. The scaling form is also a gamma function, replacing k by k' in eq.(5).

## Further Analysis Related to Shifted Poisson Transform

Suppose we consider the following probability generating function

$$G(\lambda) = \sum_{n=0}^{\infty} \lambda^n P_n \qquad \text{where} \qquad P_n = \frac{[x(\bar{n} - k')]^{n-k'} \, e^{-x(\bar{n}-k')}}{(n - k')!} \tag{11}$$

we get

$$G(\lambda) = \lambda^{k'} \, e^{(\lambda-1)x(\bar{n}-k')} \tag{12}$$

Now one can write, for the class of shifted- Poisson transform, the result in terms of a modified-Laplace transform as

$$L_m[G(\lambda)] = \int_0^{\infty} f(x) \, G(\lambda) \, dx \tag{13}$$

In particular using KNO scaling function, eq.(10) for $f(x)$ we obtain the generating function

$$L_m[G(\lambda)] = \lambda^{k'}\left[\frac{(k + k')}{(k' + k) - (\lambda - 1)(\bar{n} - k')}\right]^{k+k'} . \qquad (14)$$

By letting $\bar{n} - k' = (k + k')(e^{At} - 1)$, one could recover the generating function of GMD, namely $G(\lambda) = \lambda^{k'}[\lambda(1 - e^{At}) + e^{At}]^{-(k+k')}$ [7].

As expected $G'(1) = \bar{n}$, one could work out the following relations to facilitate the use of $C_q$ moments to quantify the curve (eq.(8))

$$\overline{n^2} = G''(1) + G'(1) \qquad (15)$$

$$\overline{n^3} = G'''(1) + 3G''(1) + G'(1) \qquad (16)$$

$$\overline{n^4} = G'^{V}(1) + 6G'''(1) + 7G''(1) + a'(1) \qquad (17)$$

$$\overline{n^5} = G^{V}(1) + 10G'^{V}(1) + 25G'''(1) + 15G''(1) + G'(1) \qquad (18)$$

The first five moments $C_q = \overline{n^q}/\bar{n}^q$ can be found in ref [8].

## Results and Discussion

Eq.(8), commonly known as the generalized multiplicity distribution (GMD)[7], has already been employed to fit both pp and $p\bar{p}$ collision data [8] and the fits are promising. Table[1] gives the possible k and k' values for $e^+e^-$collision at each $\sqrt{s}$ energies. $\bar{n}$ is obtained from experiment [5] and the bestfits are attempted, corresponding to minimum $\chi^2$/pts. The fits are very good indeed.

The mean hadron multiplicity $\bar{n}$ increases with k′ while k decreases. From the $D = \left(\overline{n^2} - \bar{n}^2\right)^{1/2}$ moment, one observes an increase in the spread of the distribution with $\sqrt{s}$. In earlier works [7,8] when fitted to pp and $p\bar{p}$ datas similar features were exhibited except that k values were much smaller. This is consistent because $k = m\tilde{A}/A$ where m is the initial number of quarks, $\tilde{A}$ the probability of $q \to q + g$ process and A, the probability $g \to g + g$ branching. We expect quark branching to be dominant in $e^+e^-$, hence the high k value. k′ the initial number of gluons is about the same for all $e^+e^-$, pp and $p\bar{p}$ annihilation [8,9].

It is more meaningful to interpret k and k′ stochastically using the Markov branching [7,8], however, the above formalism is used to highlight the statistics properties of GMD. Using a shifted Poisson by k′ values also helps to build in GMD, a condition that requires $n \geq k'$. For $k = 0$, FYD necessitated $n \geq k'$ from GMD.

We note from Fig.[5], KNO scaling becomes less violated with increasing energies. This is not so in pp and $p\bar{p}$ collisions [8]. In fact KNO scaling becomes increasingly violated as $C_q$ moments increases with energies rather than being energy independent as expected from such a scaling. In view of the present trend of $C_q$ moments being more independent at higher energy, it would mean future $e^+e^-$ multiplicity curves would fall on the same distribution using KNO scaling variables and would have completed one's study of $e^+e^-$ annihilation. However, it is still too early to tell and entails more data at higher energies. Figs.[1 to 4] give $e^+e^-$, GMD for 14, 22, 34.8 and 43.6 GeV energies.

It is interesting to point out that when $C_q$ moments are plotted against pseudorapidity $\eta_c$ for $p\bar{p}$ data at 200, 546 and 900 GeV [8], trends similar to Fig.[5] show up. Perhaps when studying $p\bar{p}$ data in limited pseudorapidity ranges, a more fundamental aspect will be revealed as $e^+e^-$ collisions are considered as basic annihilation.

## Conclusion and Comments

It is clear that GMD also provides a good theoretical distribution for $e^+e^-$ data. For a meaningful description we still maintain a stochastic approach through Markov branching processes in an attempt to comprehend the multiplicity data of High Energy Collisions. From this perspective, the detailed discussion given in [7,8,9] provides an expression for the mean hadron multiplicity for GMD.

$$\bar{n}(t) = (k' + k)e^{At} - k \tag{19}$$

where t is the evolution parameter. Ref.[7] gives the following equation for $C_2$ moments (GMD)

$$\begin{aligned} C_2 = {} & k(k + 1)\left(\frac{1 - k'/\bar{n}}{k' + k}\right)^2 \\ & + 2k'(k + 1)\frac{(1 + k/\bar{n})(1 - k'/\bar{n})}{(k' + k)^2} \\ & + k'(k' - 1)\left(\frac{1 + k/\bar{n}}{k' + k}\right)^2 + \frac{1}{\bar{n}} \end{aligned} \tag{20}$$

However, for t = 0, $\bar{n} = k'$ would imply $C_2 = 1$. Noting that $D^2/\bar{n}^2 = C_2 - 1$, the dispersion D = 0 simply means that the GMD is all squeezed up at n = k'.

This suggests that after $e^+e^-$ annihilation and before t evolutes at t = 0, the GMD is actually a Dirac delta function, peaking at $n = k' = \bar{n}$

$$P_n^{k,k'} = \begin{cases} 1 & n = k' \\ 0 & n \neq k' \end{cases} .$$

As t evolutes t > 0, the GMD will be realized and D the dispersion will increase.

**References**


[1] G. Giovannini, Nuovo Cimento 15A (1973) 543.

[2] M. LeBellac et al., Nucl. Phys. B62 (1973) 350.

[3] P. Carruthers and C.C. Shih, Int. J. Mod. Phys. 2A (1987) 1447.

[4] R.E. Ansorge et al., preprint CERN-EP/88-172.

[5] W. Braunschweig et al., preprint Desy 89-038 ISSN 0418-9833.

[6] R.C. Hwa and C.S. Lam, Phys Lett B173 (1986) 346.

[7] C.K. Chew et al., Phys Lett. B186 (1987) 411.

[8] A.H. Chan and C.K. Chew, Nuovo Cimento 101A (1989) 409.
A.H. Chan and C.K. Chew, Phy. Rev. D41 (1990) 851.

[9] C.K. Chew and Y.K. Lim, Phys. Lett. 163B (1985) 257.

**Table** [1] k′, k, $C_q$ and $\chi^2$/pts are computed using GMD. Values given in brackets ar experimental data.

| $\sqrt{s}$ GeV | $\bar{n}$ | D | | k′ | k | | $C_2$ | $C_3$ | $C_4$ | $C_5$ | | $\chi^2$/pts |
|---|---|---|---|---|---|---|---|---|---|---|---|---|
| 14 | 9.30 | 3.08 | | 1.65 | 38.00 | | 1.11 | 1.33 | 1.73 | 2.41 | | 1.10(13) |
| | (± 0.16) | | | | | | (1.11 ± 0.04) | (1.35 ± 0.06) | (1.78 ± 0.10) | (2.55 ± 0.17) | | |
| 22 | 11.30 | 3.57 | | 1.58 | 32.00 | | 1.10 | 1.31 | 1.67 | 2.28 | | 0.27(14) |
| | (± 0.20) | | | | | | (1.10 ± 0.04) | (1.31 ± 0.06) | (1.68 ± 0.09) | (2.31 ± 0.15) | | |
| 34.8 | 13.59 | 4.08 | | 1.80 | 25.00 | | 1.09 | 1.29 | 1.63 | 2.20 | | 1.03(18) |
| | (± 0.10) | | | | | | (1.09 ± 0.01) | (1.29 ± 0.02) | (1.64 ± 0.04) | (2.23 ± 0.06) | | |
| 43.6 | 15.08 | 4.52 | | 2.00 | 21.00 | | 1.09 | 1.28 | 1.62 | 2.17 | | 0.80(19) |
| | (± 0.21) | | | | | | (1.09 ± 0.03) | (1.29 ± 0.04) | (1.63 ± 0.06) | (2.18 ± 0.10) | | |

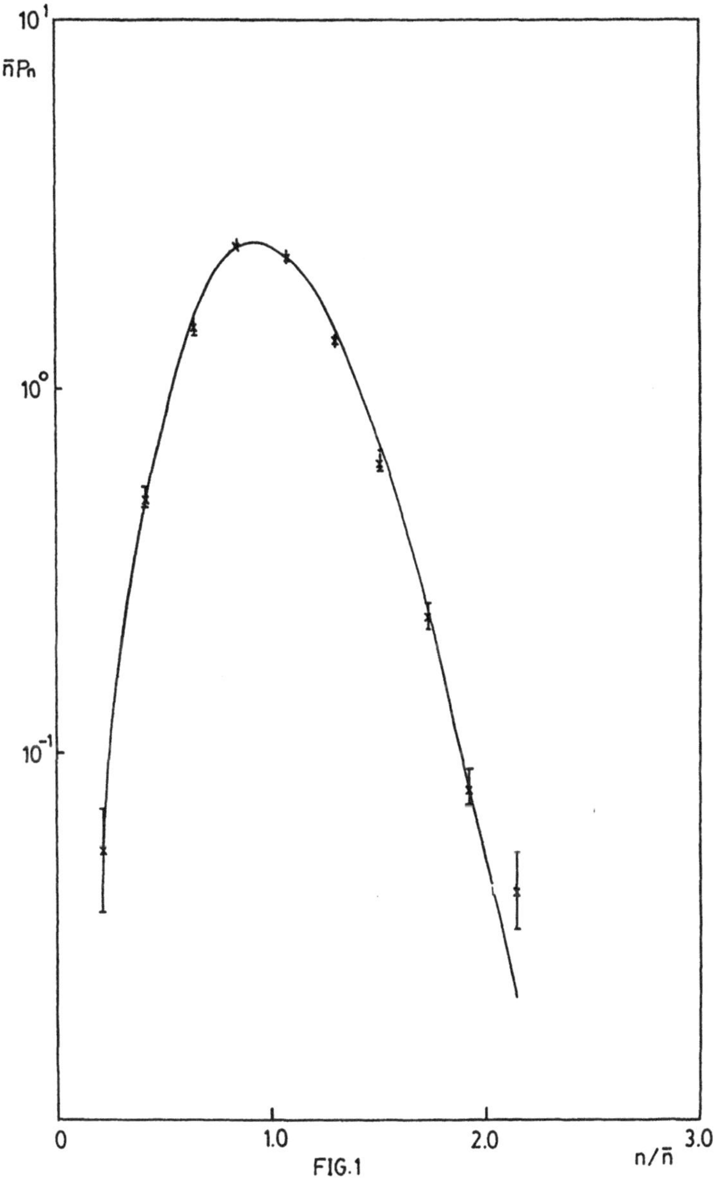


FIG.1

Comparison of GMD with $e^+e^-$ annihilation data at 14 GeV.

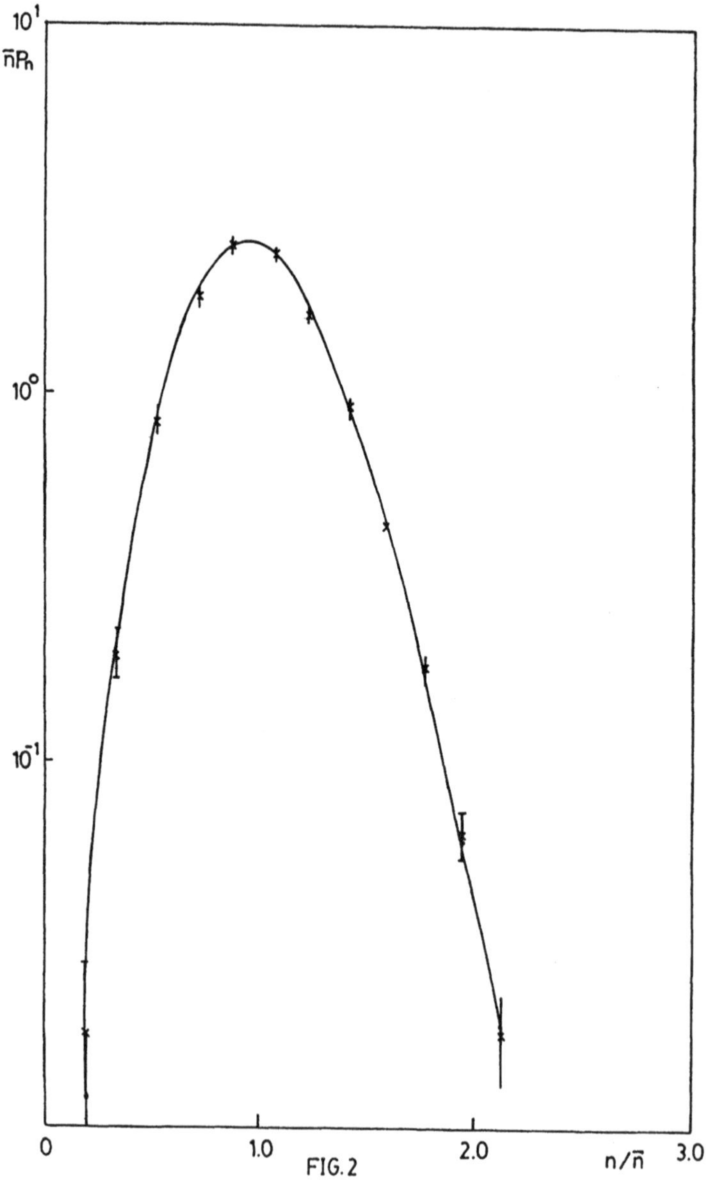


FIG. 2

Comparison of GMD with $e^+e^-$ annihilation data at 22 GeV.

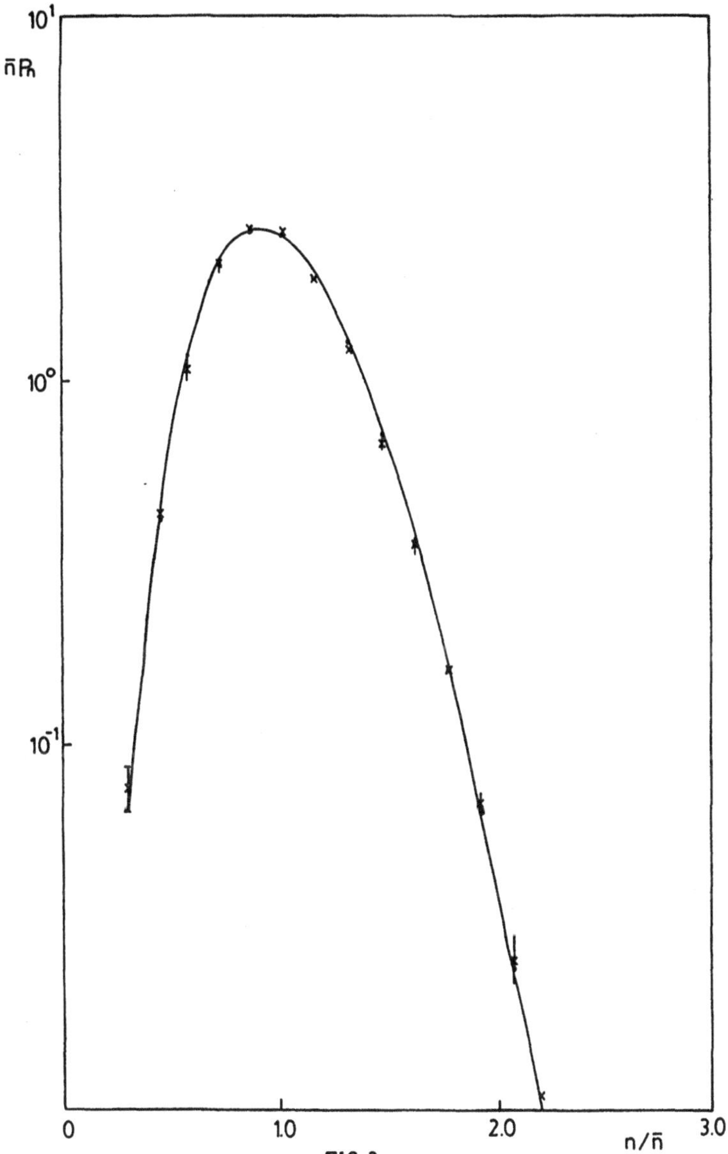


FIG. 3

Comparison of GMD with $e^+e^-$ annihilation data at 34.8 GeV.

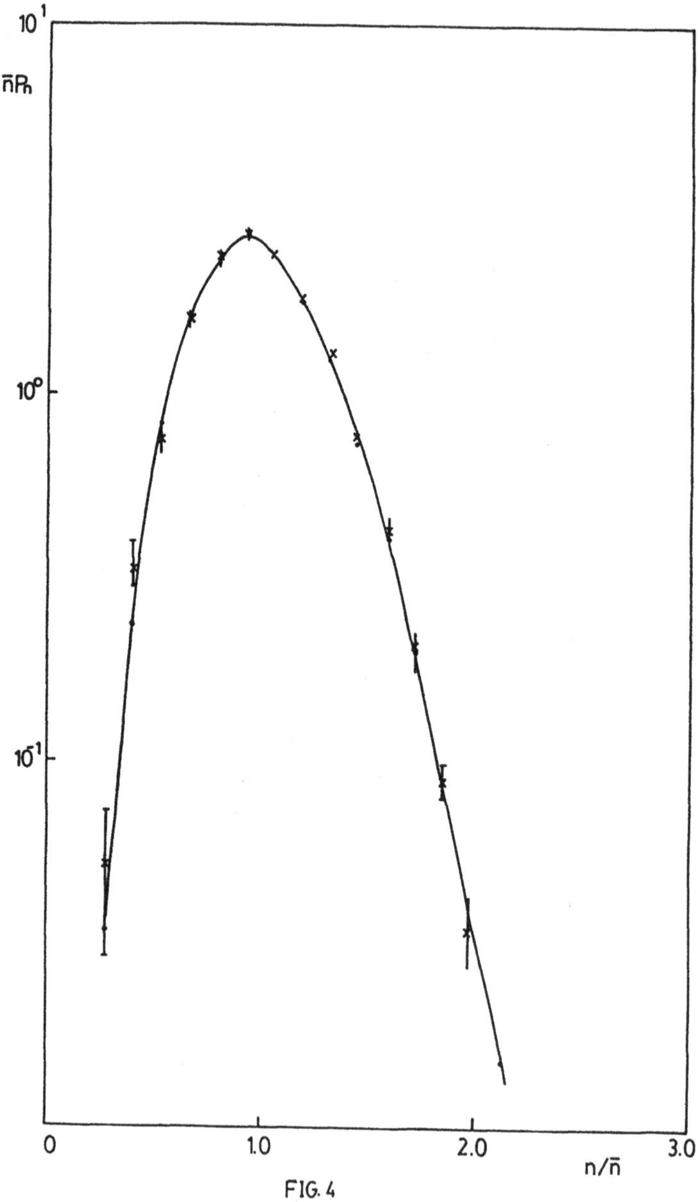


FIG. 4

Comparison of GMD with $e^+e^-$ annihilation data at 43.6 GeV.

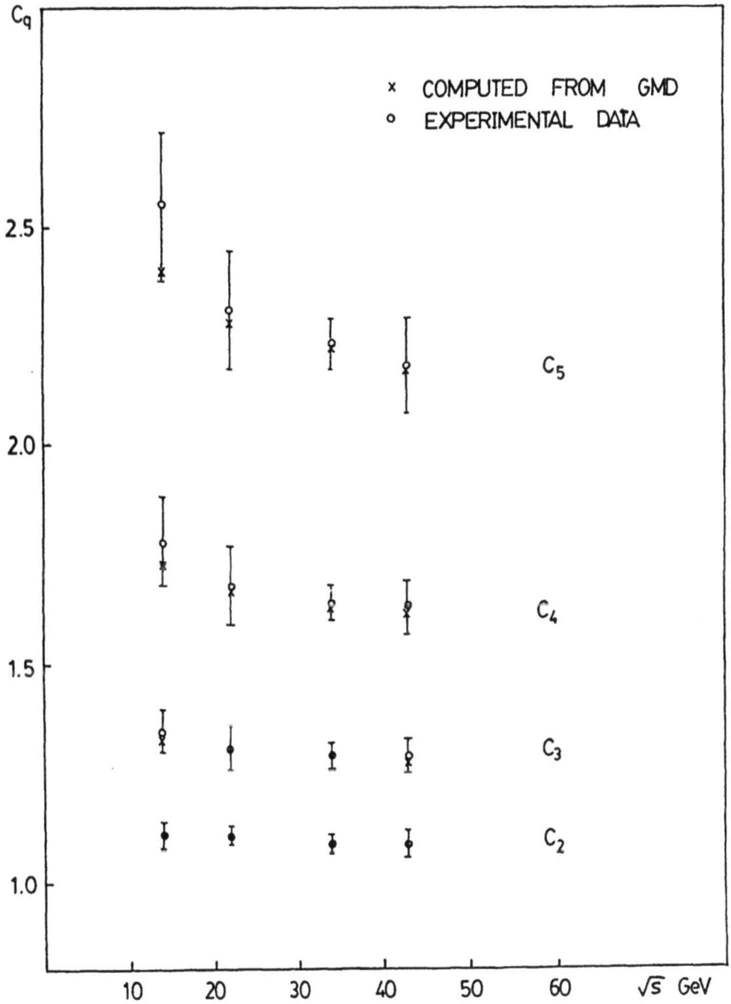


FIG 5

$C_q$ plotted against c.m.s. $\sqrt{s}$ GeV, annihilation energy.

# III. Round Table (Fluctuations)

*Chairman: P. Carruthers*

# Three Topics in Intermittency

A.Bialas[†]
CERN, Geneva

**Abstract**

We discuss three recent developments in studies of intermittency: (a) evidence for large fluctuations in 3-dimensional momentum distributions, (b) evidence for collective effects in nuclear collisions, (c) speculations on the possible existence of different phases in intermittent patterns observed in high-energy collisions.

## 1. Multidimensional analysis of intermittency

Observation of "spikes" in some high-energy experiments [1-3] prompted the hypothesis of "intermittency" [4] which was shown to be effectively investigated by testing the power law behaviour of the scaled factorial moments

$$F_q(l) = F_q(L)(L/l)^{f_q} \qquad 1$$

where $F_q(l)$ is the scaled factorial moment[1] of rank q at the resolution $l$ [4,5].

The existing data confirm the validity of (1) for one-dimensional spectra in rapidity and azimuthal angle, as well as in these two variables combined. [2] Moreover, the data show that the effect is much stronger in two-dimensional than in one-dimensional spectra. This observation led Ochs [7] to study models of intermittency in more than one dimension. He found that by projection of a multidimensional distribution on a sub-space of lower dimension the intermittency effect is strongly reduced and may even completely disappear at sufficiently high resolution.

---

[*] Contribution to the panel discussion on intermittency at the Marburg CAMP workshop (May 1990).

[†] On leave from Institute of Physics, Jagellonian University, Cracow

[1] Our notation follows closely that of [6] where also an extensive list of references on intermittency can be found.

[2] For a recent review of the data, see [6].

Following this observation, the effects of projections were studied in [8] by considering a 3-dimensional two-particle inclusive distribution described by a singular inclusive density

$$\rho_2(p_1^\mu, p_2^\mu) = Q^{-f_2} exp[6(m_{1t} + m_{2t})] \qquad 2$$

where $Q^2 = -(p_1^\mu - p_2^\mu)^2$ and the exponential factor provides the transverse momentum limitation. Since the factorial moments are integrals of the inclusive distribution functions, singularity in (2) is equivalent to the scaling law (1). Thus (2) describes an intermittent system in three dimensions.

Using (2) it is possible to calculate (by a straightforward multidimensional integration) the factorial moment $F_2$ for the distribution (2) projected on any subspace in 3-dimensional momentum space. The results are shown in Fig.1. One sees that the effects of projection are indeed dramatic: the calculated intermittency slopes are reduced by more than an order of magnitude. One sees also that (a) the calculated two-dimensional intermittency is much stronger than the one-dimensional, in agreement with experimental observations, and (b) a cut in transverse momentum also boosts intermittency, in qualitative agreement with the data of Ref.[9].

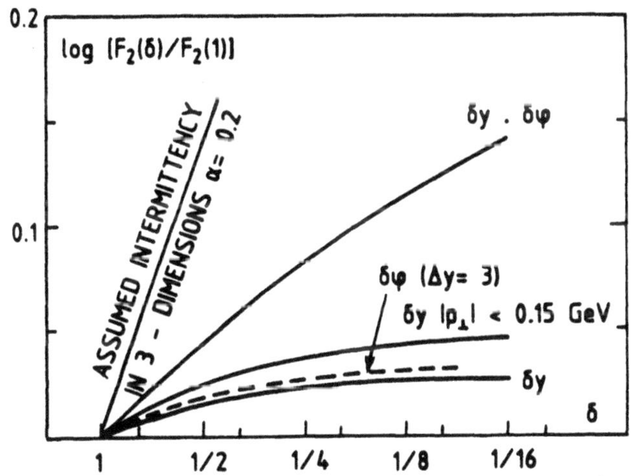


Fig.1. Projections of 3-dimensional intermittency [8].

These results may be of importance for further studies of intermittency, because they indicate that (a) the effect is likely to be much stronger than hitherto suspected, (b) it is essential to undertake serious studies of multidimensional spectra, and (c) the observed flattening of factorial moments at high resolution need not be a consequence of some limiting scale but simply the result of projection. If these preliminary conclusions are confirmed, the emerging picture of intermittency may turn out to be quite different from the present one.

## 2. Collective effects in nuclear collisions

The second topic I should like to discuss is the possibility of using the intermittency measurements as a tool for studying collective phenomena in nuclear collisions [10-12]. The argument can be summarized as follows.

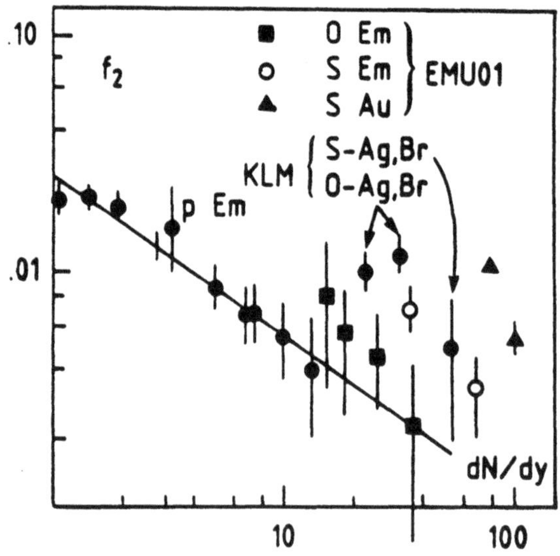


Fig.2. $f_2$ for pp and for nuclear collisions [6,12-14].

A convolution of several "sources", each one emitting an intermittent spectrum of particles, reduces the observed intermittency effect because the fluctuations are smeared out. Quantitatively, the effect on intermittency exponents can be approximated by the formula

$$f_q(N) = f_q(1)/N \qquad 3$$

where $N$ is the number of sources. If sources are identical, $N$ must be proportional to the observed rapidity density and we obtain

$$f_q = (dn/dy)^{-1} \tag{4}$$

This formula was found to be consistent with pp data [12] and it is thus very interesting to check if the nuclear data also follow the same pattern: deviations from the simple rule (4) indicate a collective mechanism at work.

A comparison of pp [12] and nuclear [13,14] data is shown in Fig.2 [6,12]. One sees that the data for nuclear collisions deviate substantially from the pp data, and that the deviation increases with increasing nuclear number. This is a rather strong hint that some collective phenomenon takes place. Although its nature is still unclear, it is interesting to speculate that it may be related to production of quark-gluon plasma [15].

## 3. Phase-space structure of self-similar cascades

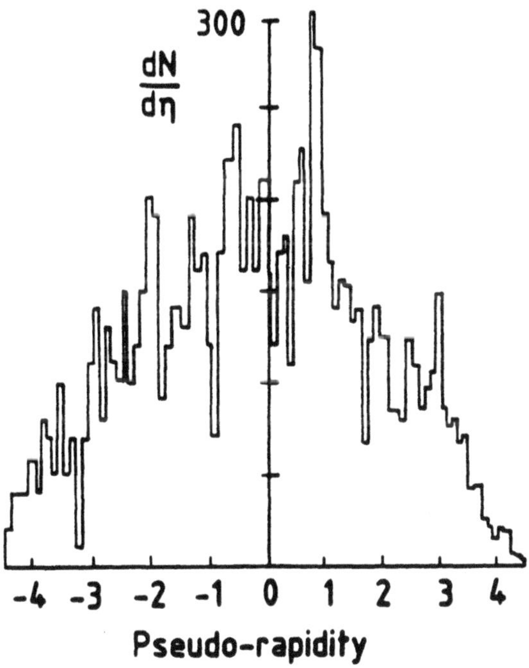


Fig.3. The JACEE event [1].

My last point concerns the possibility raised recently by Peschanski [16] that intermittent spectra may occur in very different patterns, resembling different phases of the spin-glass systems. The relevance of this problem to particle physics is illustrated by the two well-known "intermittent" events shown in Figs 3 and 4. One sees that they are rather different: the JACEE [1] event contains many "spikes" and "holes", whereas the NA22 event [2] consists of just one spike. The question then arises if this is only a statistical accident, or if there are indeed different kinds of "intermittent" events [16-20].

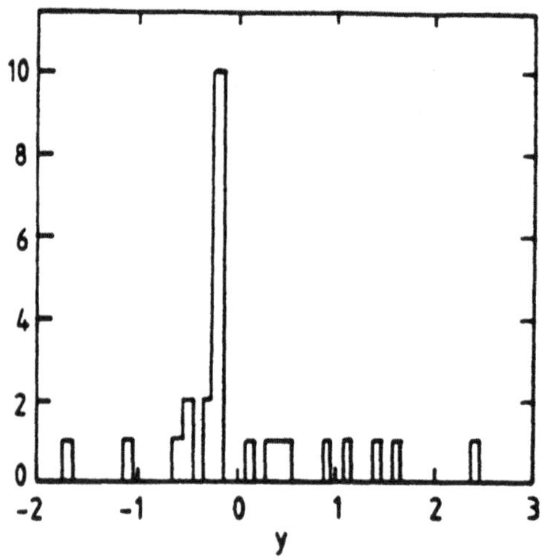


Fig.4. The NA22 event [2].

The theoretical condition for the existence of such different phases of a self-similar cascade is that the function

$$\lambda_q = (f_q + 1)/q \qquad 5$$

has a minimum at some value of q (not necessarily an integer) [16,17].

In Fig.5 $\lambda_q$, as measured in several high-energy experiments [9,13,21,22], is plotted versus $q$. For $\lambda_q$ derived from rapidity spectra integrated over azimuthal angle and transverse momentum [13,21,22] one sees a suggestive flattening of the plot but there is no evidence (yet?) for a minimum. On the other hand, recent data of the NA22 coll. [9], where the distribution is cut off at $|p_t| < 0.15 Gev$, do show a marked minimum in $\lambda_q$. Thus we may expect that at low transverse momentum the intermittent events should indeed split into separate classes. To learn more, one should now study (a) fluctuations of the moments of the distribution [17] and (b) the correlated factorial moments [5,20] which are the order parameters characterizing different phases of self-similar cascades. It would be interesting to perform these measurements not only for low transverse momentum tracks, but also for 2- and 3-dimensional spectra, where, as we have already explained, intermittency is expected to be much better seen [7,8].

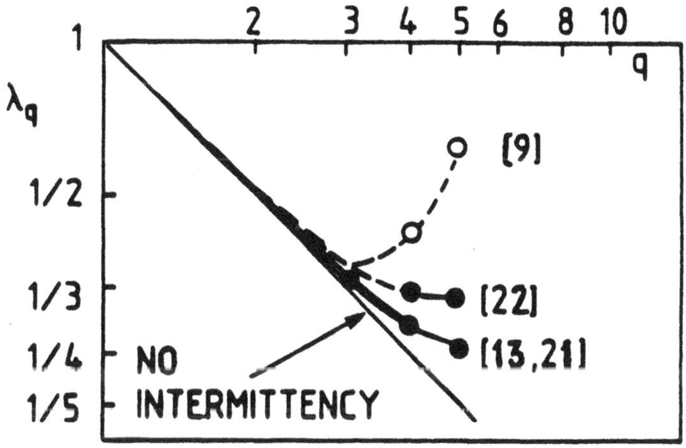


Fig.5. $\lambda_q$ vs $q$. Data from [9,13,21,22].

## Acknowledgements

I would like to thank R.Weiner and S.Raha for kind hospitality at the Marburg CAMP workshop.

## References

1. JACEE collab., T.H.Burnett et al., Phys.Rev. Letters 50 (1983) 2062.
2. NA22 collab., M.Adamus et al., Phys.Letters B185 (1987) 200.
3. UA5 collab., G.J.Alner et al., Phys.Rep. 154 (1987) 247.
4. A.Bialas and R.Peschanski, Nucl.Phys. B273 (1986) 703.
5. A.Bialas and R.Peschanski, Nucl.Phys. B308 (1988) 857.
6. A.Bialas, report at "Quark Matter '90", Menton (May 1990), CERN-TH.5791/90.
7. W.Ochs, contibution to the Santa Fe workschop (March 1990), Munich preprint MPI-PAE/PTh 26/90.
8. A.Bialas and J.Seixas, CERN-TH.5757/90.
9. NA22 collab., W.Kittel, contribution to Santa Fe workshop (March 1990), Nijmegen preprint.
10. A.Bialas, in Festschrift L.Van Hove, eds. A.Giovannini and W.Kittel (World Scientific 1989).
11. A.Capella, K.Fialkowski and A.Krzywicki, Phys.Letters B230 (1989) 271; Festschrift L.Van Hove, eds. A.Giovannini and W.Kittel (World Scientific 1989), p.413.
12. UA1 collab., B.Buschbeck and P.Lipa, contibution to Santa Fe workshop (March 1990), Vienna preprint HEPHY-PUB 535/90.
13. KLM coll., R.Holynski et al., Phys.Rev. C40 R2449 (1989); B.Wosiek, INP Cracow preprint 1496/PH and private communication.
14. EMU01 collab., E.Stenlund, report at "Quark Matter '90", Menton (May 1990), to be published.
15. A.Bialas and R.Hwa, CERN-TH.5754/90.
16. R.Peschanski, Nucl.Phys. B327 (1989) 144.
17. A.Bialas, A.Szczerba and K.Zalewski, Z. Phys. C46 (1990) 163.
18. A.Bialas and K.Zalewski, Phys.Letters B238 (1990) 413.
19. P.Brax and R.Peschanski, Saclay preprint, SPhT/89-203 (1989).
20. P.Brax and R.Peschanski, Saclay preprint, SPhT/90-044 (1990).
21. NA22 collab., I.V.Ajinenko et al., Phys.Letters B222 (1989) 306.
22. I.Derado, G.Jancso, N.Schmitz and P.Stopa, Munich preprint MPI-PAE/Exp El. 221.

# HOW TO BUILD HADRONIC MULTIPLICITY MOMENTS FROM TWO-PARTICLE CORRELATIONS

P. Carruthers

Department of Physics, University of Arizona
Tucson, Arizona 85721

## ABSTRACT

The usual factorial moments $F_p$ (averaged over rapidity bins of width $\delta y$) typically increase for increasing $p$ (fixed $\delta y$) and decreasing $\delta y$ (fixed $p$). Whatever the functional dependence of the moments $F_p(\delta y)$, they must satisfy the identities expressing their values in terms of the rapidity density correlation functions $\rho_p$. These functions, and hence $F_p$, contain significant non-dynamical background contributions. When the $F_p$ are expressed in terms of the physically relevant cumulant moments we find dominant combinatoric contributions from $F_2$ itself, and a simple recursive structure relating the $F_p$ moments. These relations are valid independent of their assumption of translation invariance. In each order $p$, the residual cumulant seems to be well approximated by the linked pair approximation (LPA) in constituent 2-particle correlations. This structure seems to coincide with negative binomial moments for the hadronic collisions studied to date, after the unknown constants are determined. In conclusion, the quantitative behavior of the moments $F_p$ seems explicable by a simple scheme in which combinatoric contributions supplemented by the LPA account for all observed effects. Current data are not decisive concerning the behavior of $F_p$ for small $\delta y$, specifically whether they obey power-law behavior or "saturate" (tend to a constant).

## 1. INTRODUCTION

The bin-averaged hadronic factorial moments $F_p\ (\delta y)\ \ (p = 2, 3, 4, \ldots)$ have recently been measured for a wide variety of projectiles and targets.[1] Apart from the magnitude of the fluctuations a very interesting question concerns the functional dependence on the rapidity bin width $\delta y$ for high resolution ($\delta y \to 0$). In particular, a power-law behavior $F_p \propto (\delta y)^{-\alpha_p}$ should indicate an irregular fractal character of the rapidity distribution.[2] Such fractal behavior[3,4] has become known as "intermittency" in nuclear and high energy physics.

A convenient measure of this effect was proposed by Białas and Peschanski:[5] the bin-averaged factorial moment $F_p$ (averaged over $M$ adjacent bins of width $\delta y$)

$$F_p \equiv \frac{1}{M} \sum_{i=1}^{M} \frac{\langle n_i(n_i - 1) \cdots (n_i - p + 1)\rangle}{\bar{n}^p} \qquad (1-A)$$

with $\bar{n}$ the average number per bin.

Most experiments present data for charged multiplicities, although more detailed positive and negative charge correlations are becoming available. For one species of particle (charged in this case) we can express Eq. (1) as an average of the $p$–th order density correlation function $\rho_p(y_1, y_2 \cdots y_p) = (d^p\sigma/dy_1, dy_2 \ldots dy_p)/\sigma_I$ as follows[6–8]

$$F_p = \frac{1}{M(\delta y)^p} \int_\Omega dy_1, \ldots dy_p\ \frac{\rho_p(y_1, y_2, \cdots y_p)}{(\bar{\rho}_1)^p}\ , \qquad (2-A)$$

where $\bar{\rho}$ is the average number density, $\bar{\rho}_1 = \bar{n}/\delta y$. The integration domain $\Omega$ is indicated in Fig. (1) for $p = 2$. Hence, knowledge of the correlation function completely determines[6–10] $F_p$ (as well as other more general bin correlations[10]).

Definition (1-A) is known as the "horizontal" average. Instead of normalizing to the total particle density $\bar{n}$, one can consider normalizing in each rapidity bin to $\overline{n_i}^p$, obtaining the "verticle average:"

$$F_p^V \equiv \frac{1}{M(\delta y)^p} \sum_m \frac{\langle n_i(n_i - 1) \ldots (n_i - p + 1)\rangle}{\overline{n_i}^p} \qquad (1-B)$$

In terms of the p-density correlation the verticle moment is given by

$$F_p^V = \frac{1}{M(\delta y)^p} \sum_m \int_{\Omega_m} dy_1 \ldots dy_p \frac{\rho_p(y_1 \cdots y_p)}{(\rho_m)^p} \qquad (2-B)$$

The particular advantage of this definition is that the $\rho_1$ factors in (5) exactly cancel, leading to Eqs. (9) as exact equations largely determined by $F_2$, independent of the assumptions of translation invariance.

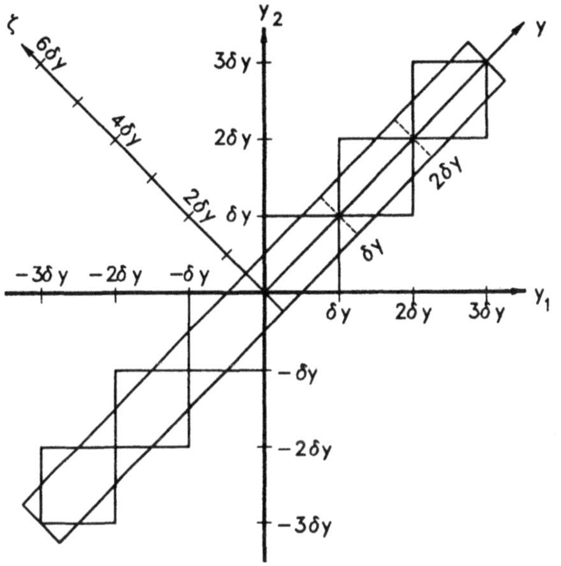


Fig. 1 The integration domain is shown for the bin-averaged factorial moment $F_2$. The diagonal strip approximation to the squares provides an easy evaluation of translation-invariant integrands.

A particularly interesting case occurs when $\rho_p$ scales,

$$\rho_p(\lambda y_1, \lambda y_2, \ldots) = \lambda^{-\alpha_p} \rho(y_1, y_2, \ldots y_p) \, . \tag{3}$$

As is easily seen from Eq. (2), (3) implies the moment scaling relation

$$F_p(\lambda \delta y) = \lambda^{-\alpha_p} F_p(\delta y) \, , \tag{4}$$

which is solved by $F_p(\delta y) \propto (\delta y)^{-\alpha_p}$.

Current experiments are inconclusive with regard to power-law behavior. In many cases a "saturation" occurs following an initial increase of $F_p$ as $\delta y$ is decreased. In addition, it is possible that true scaling in higher dimensions (e.g. $\vec{p}_\perp$ and $y$, or $\eta = -\ln \tan \frac{\theta}{2}$ and $\phi$) seems to saturate when projected to lower dimensions.[11] Whatever the behavior, however, the $F_p$ must conform to Eq. (2), which is actually an identity. Therefore we must analyze the structure of the density correlation $\rho_p$.

## 2. ANALYSIS IN TERMS OF CUMULANT CORRELATIONS

As is well known, typical density correlations contain background contributions of physically trivial significance. The physically interesting correlations are expressed by the cumulant moments $C_p(y_1, \ldots, y_p)$

$$\begin{aligned} C_2(y_1,y_2) &= \rho_2(y_1,y_2) - \rho(y_1)\rho_1(y_2) \,, \\ C_3(y_1,y_2,y_3) &= \rho_3(y_1,y_2,y_3) - \rho_2(y_1,y_2)\rho_1(y_3) - \rho_2(y_2,y_3)\rho_1(y_1) \\ &\quad - \rho_2(y_3,y_1)\rho_1(y_2) + 2\rho_1(y_1)\rho_1(y_2)\rho_1(y_3) \,, \end{aligned} \tag{5}$$

etc. The integrals of (5) over $\Omega$ (Eq. (2)) are just the factorial cumulants $\delta_p$:

$$\begin{aligned} &\int_\Omega dy_1 \ldots dy_p \; C_p(y_1 \ldots y_p) = \sum_i f_p^{(i)} \\ &f_2 = \langle n(n-1)\rangle - \langle n\rangle^2 \\ &f_3 = \langle n(n-1)(n-2)\rangle - 3\langle n(n-1)\rangle\langle n\rangle + \langle n\rangle^3, \ldots \end{aligned} \tag{6}$$

For Poisson statistics all $f_p (p \geq 2)$ vanish. The $C_p$ cannot vanish if the counting statistics deviate from a Poissonian definition of statistical independence. For example, the conventional definition of "no correlation" i.e., $\rho_2(y,y') = \rho_1(y)\rho_1(y')$ implies a Poissonian character for $f_2$ in the relevant phase space interval. In general, if a variable $y_j$ becomes statistically independent in $\rho_p$, the associated cumulant vanishes.

Since the interesting physics is expressed by cumulants it is essential to rewrite the $\rho_p$ in terms of the $C_p$, thus inverting the expansions (6). In order to simplify notations we use "reduced" moments

$$\begin{aligned} r_p &\equiv \rho_p/\rho_1(y_1)\rho_1(y_2)\ldots\rho_1(y_p) \\ k_p &\equiv C_p/\rho_1(y_1)\rho_1(y_2)\ldots\rho_1(y_p) \,. \end{aligned} \tag{7}$$

We have, for example[6]

$$\begin{aligned} r_2 &= 1 + k_2 \,, \\ r_3 &= 1 + \sum_{(3)} k_2(i,j) + k_3 \,, \\ r_4 &= 1 + \sum_{(6)} k_2(i,j) + \sum_{(3)} k_2(i,j)k_2(l,m) + \sum_{(4)} k_3(i,j,k) + k_4 \,, \end{aligned} \tag{8}$$

etc.[12] Here the numbers standing beneath the summation signs give the number of contributing permutations.

Eqs. (8) express the density correlations systematically in terms of lower order ("uncorrelated", "background", "unlinked") $k_n (n < p)$ plus the true correlation

$k_p$. We note that even if all cumulants with $p > 2$ vanish, the number of terms involving $k_2$ which contribute to $r_p$ rapidly increase. Indeed for low orders these terms are numerically predominant when applied to hadronic data, with the highest cumulant $k_p$ typically comprising 20% or less of the total.

## 3. EXPRESSION OF FACTORIAL MOMENTS IN TERMS OF FACTORIAL CUMULANTS

Next we integrate (2-B) using (8) and define the bin-averaged factorial cumulant $K_p$ by Eq. (2) with $C_p$ replacing $\rho_p$. The result is (no approximations involved)

$$
\begin{aligned}
F_2 &= 1 + K_2 \\
F_3 &= 1 + 3K_2 + K_3 = 1 + 3(F_2 - 1) + K_3 \\
F_4 &= 1 + 6K_2 + 3\overline{K_2^2} + 4K_3 + K_4 \\
F_5 &= 1 + 10K_2 + 15\overline{K_2^2} + 10\overline{K_3 K_2} + 10K_3 + 5K_4 + K_5 \ .
\end{aligned}
\tag{9}
$$

Here the bar superscript indicates a vertical bin average $\overline{AB} = \sum_m A_m B_m / M$. Often this quantity is close to the horizontal average. Note that given $F_p$ we can determine all $K_p$ recursively. An important feature of these equations is that they depend only on the integrated correlation functions, and are less sensitive to errors that may be present in the latter. For example we can eliminate $K_3$ in favor of $F_2$, $F_3$ and $F_4$.

Before discussing the structure of the higher cumulant moments consider what happens if we completely ignore the correlations $K_3$, $K_4$ in (9): In this case $F'_p$ are

$$
\begin{aligned}
F'_3 &= 1 + 3(F_2 - 1) \\
F'_4 &= 1 + 6(F_2 - 1) + 3(F_2 - 1)^2 \\
F'_5 &= 1 + 10(F_2 - 1) + 15(F_2 - 1)^2 \\
F'_6 &= 1 + 15(F_2 - 1) + 45(F_2 - 1)^2 + 15(F_2 - 1)^3 \ .
\end{aligned}
\tag{10}
$$

In the case of two-particle (all-charge) correlations $F_2$ is typically larger than 1, less than 2. Hence the increasing powers $(F_2 - 1)^n$ decrease in (10) but are nonetheless significant.

In Table 1 we compare (10) with data from the NA22 experiment. Although this approximation falls short of the measured moments, the "predictions" are not bad, and certainly better than typical event generators often used as theoretical reference points.

Table 1. The ratio $R_n = F'_n/F_n$ of the two-particle combinatoric concentration of Eq. (10) to the experimental $F_n$ is given for the NA22 experiment (vertical averaging, as in Ref. 1).

| $-\ell n$(delta y) | $R_3$ | $R_4$ | $R_5$ |
|---|---|---|---|
| -1.386 | 0.981 | 0.950 | 0.914 |
| -0.693 | 0.980 | 0.957 | 0.939 |
| 0.00 | 0.955 | 0.866 | 0.729 |
| 0.693 | 0.942 | 0.804 | 0.567 |
| 1.012 | 0.946 | 0.846 | 0.720 |
| 1.609 | 0.941 | 0.798 | 0.566 |

## 4. LINKED PAIR APPROXIMATION FOR THE CUMULANT MOMENTS

Having established the kinematic/combinatoric content of the $F_p$ moments, we try to improve the description of the data by a study of the cumulant ($G_p$) moments. Unlike the classical theory of liquids, there does not yet exist a systematic formulation of correlations in terms of dynamics. However we have found[6] that a suggestive structure – the linked pair approximation (LPA) – seems to give a reasonable quantitative account of the data. (The same structure successfully[13] accounts for galaxy correlations to 5-th order; in this case the 2-particle correlation scales.)

The pairs refer to the labels of the two-particle cumulant $k_2$, and the linkage implies that each rapidity label $y_i$ occurs twice. Specifically,

$$\begin{aligned} k_3(1,2,3) &= \frac{A_3}{3}\left[k_2(1,2)k_2(2,3) + k_2(2,3)k_2(3,1) + k_2(3,1)k_2(1,2)\right] \\ k_4(1,2,3,4) &= \frac{A_4}{12}\sum_{\text{perm.}} k_2(1,2)k_2(2,3)k_2(3,4) \\ k_5(1,2,3,4,5) &= \frac{A_5}{60}\sum_{\text{perm.}} k_2(1,2)k_2(2,3)k_2(3,4)k_2(4,5)\,. \end{aligned} \tag{11}$$

Each term in the sums in (11) gives identical contribution to $K_p$. These can be done numerically, here we give the result in the limit of translation invariance:

$$K_p = A_p(K_2)^{p-1} = A_p(F_2 - 1)^{p-1}\,. \tag{12}$$

This result expresses the entire factorial moment $F_p$ in terms of powers of $K_2 = F_2 - 1$:

$$\begin{aligned} F_3 &= 1 + 3(F_2 - 1) + A_3(F_2 - 1)^2 \\ F_4 &= 1 + 6(F_2 - 1) + (3 + 4A_3)(F_2 - 1)^2 + A_4(F_2 - 1)^3 \\ F_5 &= 1 + 10(F_2 - 1) + (15 + 10A_3)(F_2 - 1)^2 + 5A_4(F_2 - 1)^3 + A_5(F_2 - 1)^4 \,. \end{aligned} \tag{13}$$

In each order we fit $A_3$ (checking the $\delta y$ dependence) and insert into the next higher moment. The result (Table 1) is quite satisfactory. We are presently testing (12) for a variety of different reactions. An interesting result of our preliminary fits is that the coefficients $A_n$, at least for hadronic collisions, approximate the values $A_3 = 2$, $A_4 = 6$, $A_5 = 24$, which convert the $F_p(\delta y)$ to negative binomial forms[14] with $K_2 = F_2 - 1$. Apparently the negative binomial is closely connected with the linked pair approximation.

Finally consider the consequences of our approach when the 2-particle cumulant $k_2$ scales. (Note that as $\delta y \to 0$, $r_2$ also "scales" but in practice the background terms in $r_n$ tend to obscure cumulant scaling, which is more natural on physical grounds.) For $k_2 \propto (1/\delta y)^{\nu-1}$ we find $F_2 - 1 \propto (\delta y)^{-\nu}$; in this case the cumulant $K_p$ behaves as

$$K_p \propto \left(\frac{1}{\delta y}\right)^{(p-1)\nu} \tag{14}$$

and $F_p$ is of the form $\Sigma_{n=1}^{p-1}(\delta y)^{-n}C_n$. We can note that (14) corresponds to $\alpha_p = (p-1)\nu$, corresponding[15] to a "homogeneous fractal" of dimension $\nu$.

## 5. CONCLUSIONS

In conclusion we collect our main points:

1. The principal contribution to the $F_n$ moments is due to physically trivial background correlations described by the empirical value of $F_2$.
2. The cumulant moments are well described by the linked pair approximation.
3. The fitted coefficients $A_n$ describing the strength of the cumulant moments seem to agree with the negative binomial distribution.
4. Since $|F_2 - 1| < 1$ the power structure of $K_p \sim (F_2 - 1)^{p-1}$ provides a natural framework for ordering the numerical values of $K_p$. Moreover, since $F_2 - 1$ increases with decreasing $\delta y$, the principal effect of the higher order correlations occurs for $\delta y \to 0$.
5. In the case of scaling, the dominant contributions come from the highest order moment $K_p \sim (\delta y)^{-(p-1)\nu}$. Due to the large background terms, it is important to subtract the lower order terms to assess the validity of scaling.
6. The assumption of translation invariance used in our first work[6] is not necessary for the validity of Eqs. (9), which becomes exact for the "verticle-average" moments.

## 6. ACKNOWLEDGEMENTS

The author acknowledges support from the Divisions of High Energy and Nuclear Physics of the U. S. Department of Energy, as well as the Lounsberg and Alexander von Humboldt Foundations. Among many colleagues who have helped clarify these issues I wish to especially thank A. Bokan, H. C. Eggers, G. Jancos, I. Sarcevic and C. C. Shih.

## REFERENCES


1. W. Kittel, Santa Fe Workshop on Intermittency in High Energy Collisions, March 18-21, 1990 (World Scientific, to be published).

2. P. Carruthers, "Fractal Structures and Correlations in Hadronic Multiparticle Distributions," in Proceedings of the Perugia Workshop on Multiparticle Dynamics, June 1988, Perugia, Italy, ed. R. Hwa, G. Pancheri and Y. Srivastava (World Scientific, Singapore, 1989).

3. B. Mandelbrot, "The Fractal Geometry of Nature" (W. H. Frieman and Co., N. Y., 1982).

4. T. Vičsek, "Fractal Growth Phenomena" (World Scientific, Singapore, 1989).

5. A. Białas, and R. Peschanski, Nucl. Phys. B273, 703 (1986); ibid B308, 857 (1988).

6. P. Carruthers and I. Sarcevic, Phys. Rev. Lett. 63, 1562 (1989).

7. P. Carruthers, Int. J. Mod. Phys. A4, 5587 (1989).

8. P. Carruthers, E. M. Friedlander, C. C. Shih and R. M. Weiner, Phys. Lett. 222B, 487 (1989).

9. A. Capella, K. Fiałkowski and A. Krzywicki, Phys. Lett. 230B, 149 (1989).

10. P. Carruthers, H. C. Eggers, Q. Gao and I. Sarčević, Int. J. Mod. Phys. A (to be published).

11. W. Ochs, preprint MPI-PAE/PTH 35/90, Santa Fe Workshop on Intermittency in High Energy Collisions (World Scientific, to be published).

12. The expansion of moments in terms of cumulants to 10th order (and vice versa) can be found in Vol. 1 of Kendall and Stuart, "Advanced Theory of Statistics," (Oxford University Press, 1974).

13. P. J. E. Peebles "The Large-Scale Structure of the Universe" (Princeton University Press, 1980).

14. E. DeWolf, Acta Physica Polonica (to be published).

15. H. G. E. Hentschel and I. Procaccia, Physica D8, 435 (1983).

# Entropy, dimensions and other multifractal characteristics of multiplicity distributions

V.Šimák
Institute of Physics, Czechoslovak Academy of Sciences,
Na Slovance 2, CS 180 40 Prague, Czechoslovakia

M.Pachr, M.Šumbera and I.Zborovský
Nuclear Physics Institute, Czechoslovak Academy of Sciences,
CS 250 68 Řež near Prague, Czechoslovakia

## 1. Introduction

Recently a large number of experimental data accumulated in hadron-hadron and electron-positron interactions stimulated a new development in multiparticle dynamics. A new methods introduced originally for description of chaotic dynamical systems, diffusion-limited growth of structures and other nonlinear phenomena [1] have been also applied to the multiplicity distributions [2-4]. Several authors have studied intermittent behaviour of mutiplicity distributions in a small rapidity windows . The exponential increase of a scaled factorial moments with decreasing window width was interpreted a signal of multifractal property of (a chaotic) dynamics of multiparticle production [2].

In [5] we have suggested to study the multiplicity distributions using their information entropy. The linear increase of the entropy in the full phase space with a maximum rapidity of produced hadrons and its scaling properties in windows were found. In the present paper we will show that the entropy per maximum rapidity when interpreted as a special case of a whole set of (fractal) generalized dimensions characterizing the multiplicity distribution allows one to establish a new type of multifractality in multiparticle dynamics.

The underlying mechanism leading to the multifractal behavior of the multiplicity distribution is studied using the QCD based Monte Carlo calculations.

## 2. Entropy and generalized dimensions.

In our previous paper [5] we have shown that for multiplicity distributions $P_n$ from pp and $p\bar{p}$ interactions above c.m.s. energy $\sqrt{s} > 20$ GeV [6] their information entropy

$$S = - \Sigma P_n \ln P_n \tag{1}$$

increases linearly with $\ln\sqrt{s}$ (fig 1a).Its intercept being approximately equal to $\ln(m_\pi)$. Hence

$$S/Y_{max} = \text{const.} \tag{2}$$

where $Y_{max} = \ln(\sqrt{s}/m_\pi)$ is just a maximum rapidity of produced hadrons. For $e^+e^-$ annihilations multiplicity data [8] the same behaviour is observed (fig. 1b).

The Rennyi order-q information entropy

$$I_q = \frac{1}{1-q} \ln \Sigma (P_n)^q \tag{3}$$

serves as a generalization of the standard information entropy S as follows from the relation $I_1 = S$. The experimental values of the quantity $D_q = I_q / Y_{max}$ for various positive values of q (fig.3 a,b) seem to behave in a similar way as for q=1 (i.e. they are approximately energy independent). The energy averaged values of $D_q$'s are summarized in table.

| $D_q$ | q=1 | q=2 | q=3 | q=4 | q=5 | q= ∞ |
|---|---|---|---|---|---|---|
| $pp/p\bar{p}$ | .413<br>±.006 | .392<br>±.007 | .381<br>±.008 | .374<br>±.008 | .370<br>±.009 | .339<br>±.012 |
| $e^+e^-$ | .397<br>±.010 | .367<br>±.010 | .352<br>±.010 | .343<br>±.010 | .337<br>±.010 | .300<br>±.009 |

Thus we have shown that the Rennyi entropies $I_q$ are the linear functions of $\ln(\sqrt{s}/m_\pi)$ with proportionality factor given by $D_q$:

$$I_q = D_q \ln(\sqrt{s}/m_\pi) \tag{4}$$

This result can be directly translated into the language of fractal geometry as used when studying the growth structures [1]. The fractal with overall extent $\sqrt{s}$ is composed of $N = \sqrt{s}/m_\pi$ discrete sites each of size $m_\pi$. To the n-th site we assign a probability $P_n$. Hence $\delta = 1/N$ can be considered as relative resolution of the multiplicity distribution structure. The spectrum of fractal dimensions can be now obtained by studying the dependence of $\Sigma(P_n)^q$ on improving relative resolution $\delta$:

$$Z(q,\delta) = \sum (P_n)^q \sim \delta^{-\tau(q)} \tag{5}$$

where $\tau(q)$ is defined as:

$$\tau(q) = -\lim_{\delta \to 0} \frac{\ln Z(q,\delta)}{\ln \delta} \tag{6}$$

Multifractality is a property of $\delta$-independence of $\tau(q)$ for $\delta$ small enough [1]. This is observed in data in fig. 2a,b (note that the region $q<0$ is very sensitive to the statistical errors). It is also in agreement with previously found $\delta$-independence of quantities $D_q$ (generalized dimensions) which are related to the function $\tau(q)$ by:

$$D_q = \frac{\tau(q)}{1-q} \tag{7}$$

Another test of multifractality may be performed by studying q-dependence of the ratios

$$R(\sqrt{s}_1,\sqrt{s}_2) = \frac{\ln[\,Z(q,\delta_1)]}{\ln[\,Z(q,\delta_2)]} = \frac{\ln\delta_1}{\ln\delta_2} \tag{8}$$

The data in fig.4 a,b are indeed compatible with expected independence on q.

The meaning of the function $\tau(q)$ is more transparent after applying Legendre transformation from independent variables $\tau$ and q to the variables $\alpha$ and f:

$$\alpha(q) = -\frac{d\tau(q)}{dq}$$

$$f(\alpha(q)) = q\alpha(q) + \tau(q) \tag{9}$$

Then $f(\alpha)$ is a fractal dimension of subset composed from bins whose occupancy probability is around $P \approx \delta^{\alpha}$ [1]. It is connected with the frequency $N(\alpha,\delta)$ of occurrence of various probabilities at resolution $\delta$ by relation:

$$N(\alpha,\delta) \sim \delta^{-f(\alpha)} \tag{10}$$

As $\tau(q=1)$ is defined to be zero for normalized probability distributions, from (9) follows that $f(\alpha)$ has a slope of unity where $\alpha$ and $f(\alpha)$ are equal ( $\alpha(q=1) = f[\alpha(q=1)]$ ). From explicit dependence of $\alpha(q)$ on $P_n$ and $\delta$:

$$\alpha(q) = \lim_{\delta\to 0}\frac{\sum P_n^{\,q}\ln P_n}{\left(\sum P_n^{\,q}\right)\ln\delta} \tag{11}$$

we have $\alpha(q=1) = D_1 = -S/\ln\delta$. For all other points $f(\alpha)$ is less then $\alpha$. Hence the total probability $N(\alpha,\delta)\delta^{\alpha}$ accumulated in bins with $P \approx \delta^{\alpha}$ is maximal for the subset with scaling exponent $\alpha = D_1$. In other words with decreasing $\delta$ the total probability gradually concentrates on subsets with fractal dimension close to the entropy dimension $-S/\ln\delta$.

$f(\alpha)$ curve for $pp/p\bar{p}$ data is presented in fig.5, for $e^+e^-$ fig.6. For both reactions the scaling is best around the region $\alpha(q=1)$. This may indicate the relevance of some iterative and/or selfsimilar particle production mechanism. To understand it better we have used Monte-Carlo QCD-based generator HERWIG 3.0 to calculate $f(\alpha)$ curve and $D_1, D_3$ and $D_\infty$ for corresponding set of energies (fig.7 a,b). We see that within the errors the model reproduces multifractal behaviour.

## 3. KNO scaling violation and multifractality.

The observed energy dependence of the KNO function $\psi(z=n/\langle n\rangle) = \langle n\rangle P_n$ at $Sp\bar{p}S$ Collider energies [7] contrasts with the entropy scaling which seems to be valid in a wider energy interval. Using the multifractality we are able to derive a bound on the energy evolution of the maximum of the function $\psi(z)$: $\psi_{max} = \langle n\rangle P_{max}$, where $P_{max} = \max_n\{ P_n\}$.

From (11) we have:

$$P_{max} \sim \delta^{-\alpha(q=\infty)} \qquad (12)$$

Using the entropy bound [5] $\ln\langle n\rangle + 1 \le S = -\alpha(q=1)\ln\delta$ we have:

$$\psi_{max} \sim \left(\frac{\sqrt{s}}{m_\pi}\right)^{\alpha(q=1)-\alpha(q=\infty)} \qquad (13)$$

From unequality $\alpha(q=\infty) \le \alpha(q) \le \alpha(q=-\infty)$ (see eq.12) one concludes that together with observed increase of the width of $\psi$ function its maximum will increase too. It is interesting whether this nontrivial behaviour can be reproduced by other phenomenological approaches.

References:

[1] T.C.Halsey et al., Phys. Rev.A33(1986)1141.
J.Feder: Fractals. Plenum Press, New York and London, 1988.
G.Paladin and A.Vulpiani, Phys. Rep. 156(1987)147.

H.G.Schuster: Deterministic chaos, VCH, Weinheim, 1988.
[2] A.Bialas: Intermittency and multiparticle dynamics. preprint TPJU 15/90, Inst. Phys. Cracow, 1990.
A.Bialas and R.Peschanski, Nucl.Phys. b273(1986)703.
[3] P.Carruthers: Multihadron rapidity distributions, fractals and entropy. In Proc. of the Peruggia Workshop on Multiparticle Dynamics, June 1988 (World Scientific)
[4] I.Dreamin, Mod.Phys.Let. A5(1990)281.
R.Hwa, Phys.Rev D41(1990)1456.
H.Satz, preprint CERN-TH 5589/89.
[5] V.Šimák et al., Phys.Let. B206(1988)159.
[6] NA22 Collab., M.Adamus et al., Phys.Let. B177(1986)239.
K.Alpgard et al., Phys.Let. B115(1982)71.
[7] UA5 Collab., G.J.Alner et al., Phys.Let. B138(1984)304.
UA5 Collab., R.G.Ansorge et al., Z.Phys. C43(1989)357.
[8] PLUTO Collab. (9.4,12.0,13.0,17.0,22.0,27.5,30.6 GeV) private communication.
TASSO Collab., W.Braunschweig et al.,preprint DESY 89--038(1989) (14.0,22.0,34.8,43.6 GeV).

Figure captions:

Fig. 1. Energy dependence of the entropy S for a) pp and $p\bar{p}$, b) $e^+e^-$ interactions.

Fig. 2. Generalized fractal dimensions $D_q$ as a function of $Y_{max}$ for $pp/p\bar{p}$ (a) and $e^+e^-$ (b) data.

Fig. 3. Function $\tau(q)$ for $pp/p\bar{p}$ (a) and $e^+e^-$ (b) interactions.

Fig. 4. Ratios (8) normalized to highest available energy a) $pp/p\bar{p}$ (900GeV) b) $e^+e^-$ (43.6 GeV).

Fig. 5 $f(\alpha)$ curve for $pp/p\bar{p}$ interactions. Statistical errors are indicated at the endpoints of curves.

Fig. 6. $f(\alpha)$ for $e^+e^-$ interactions.

Fig. 7. HERWIG 3.0 results for $e^+e^-$interactions: a) $f(\alpha)$ curve, b) generalized dimensions $D_1, D_3$ and $D_\infty$ .

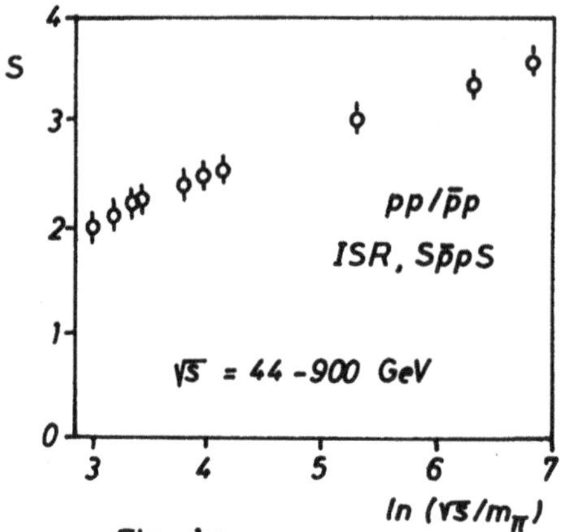


Fig. 1a

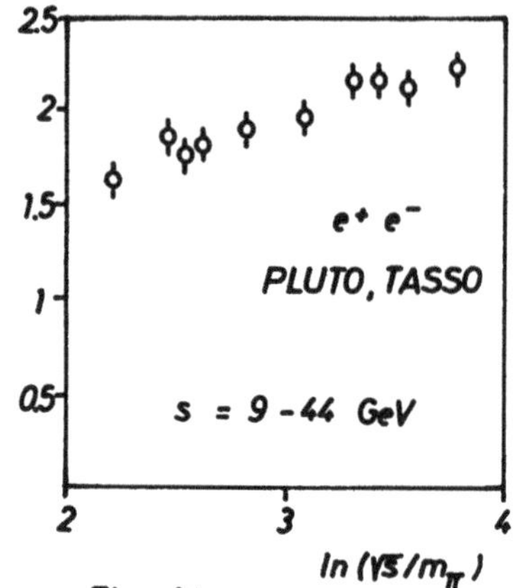


Fig. 1b

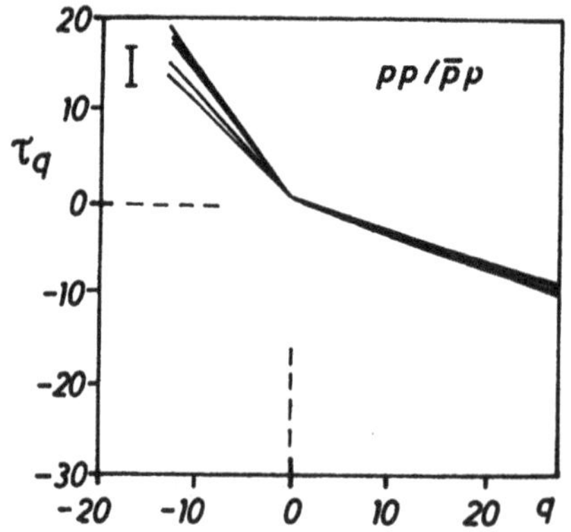


Fig. 3a

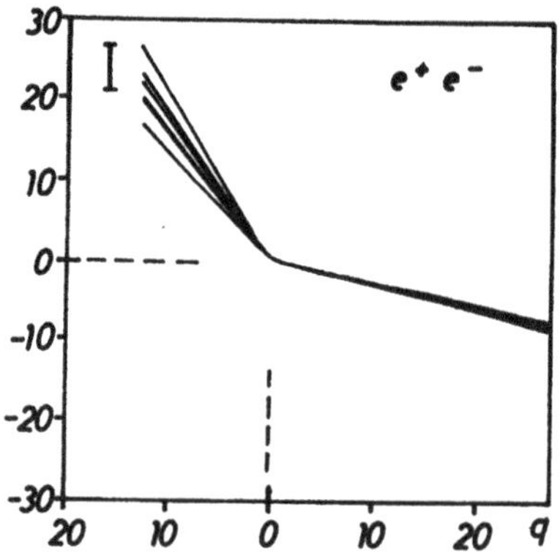


Fig. 3b

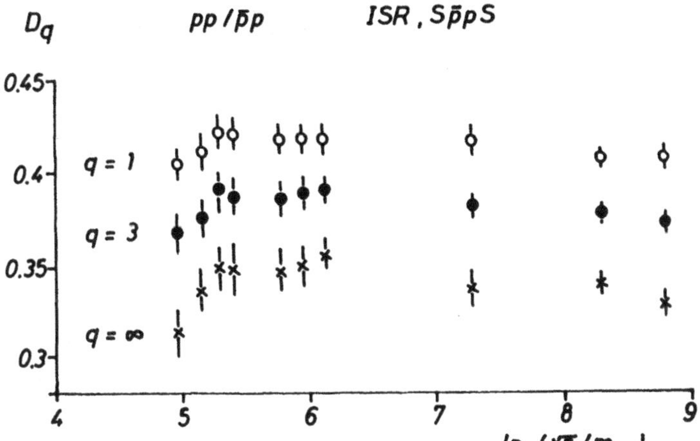
$D_q$
pp / p̄p
ISR , Sp̄pS
0.45
0.4
0.35
0.3
q = 1
q = 3
q = ∞
4
5
6
7
8
9
ln (√s / $m_\pi$ )

Fig. 2a

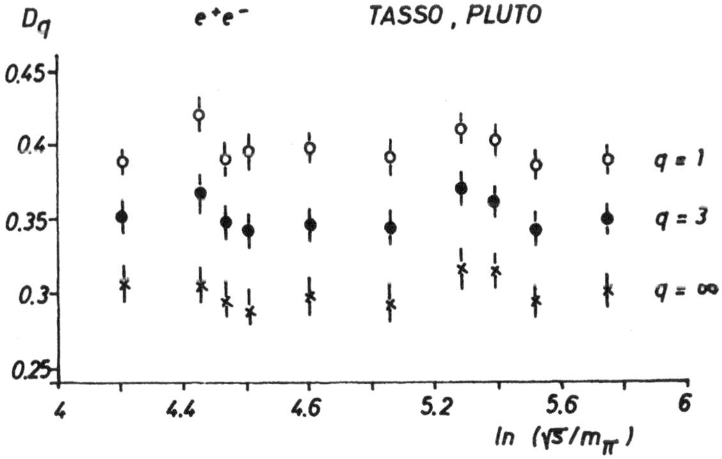
$D_q$
e⁺e⁻
TASSO , PLUTO
0.45
0.4
0.35
0.3
0.25
q = 1
q = 3
q = ∞
4
4.4
4.6
5.2
5.6
6
ln (√s / $m_\pi$ )

Fig. 2b

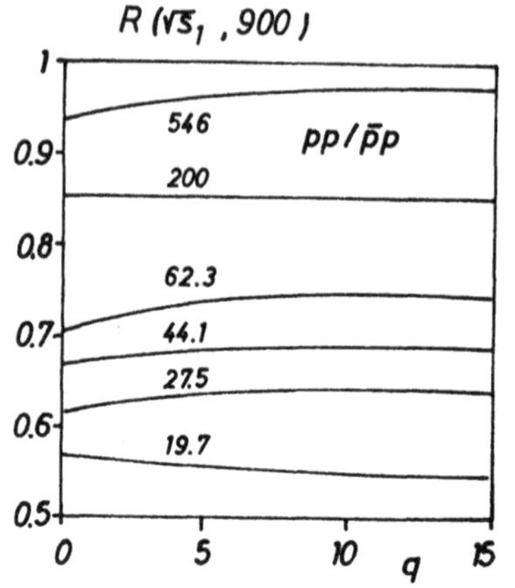


Fig. 4 a

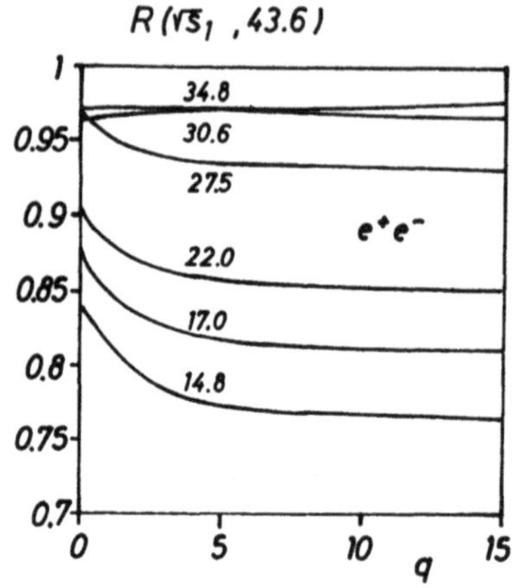


Fig. 4 b

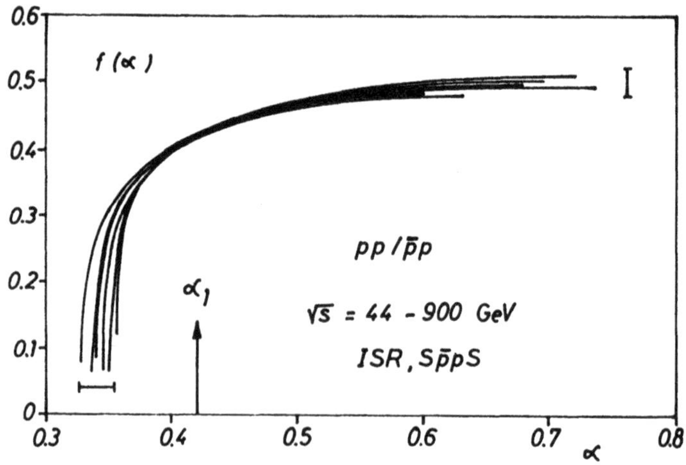


Fig. 5

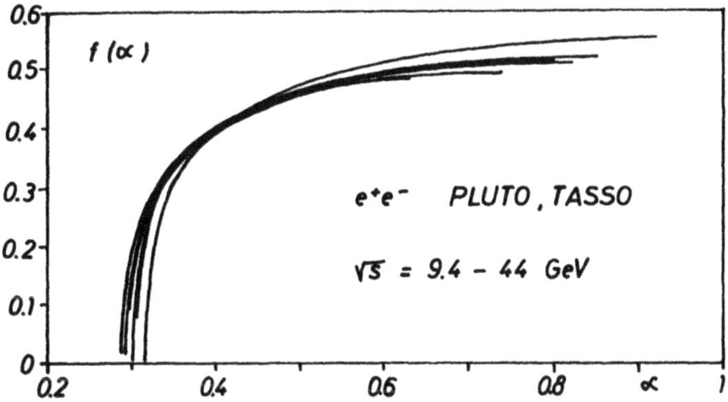


Fig. 6

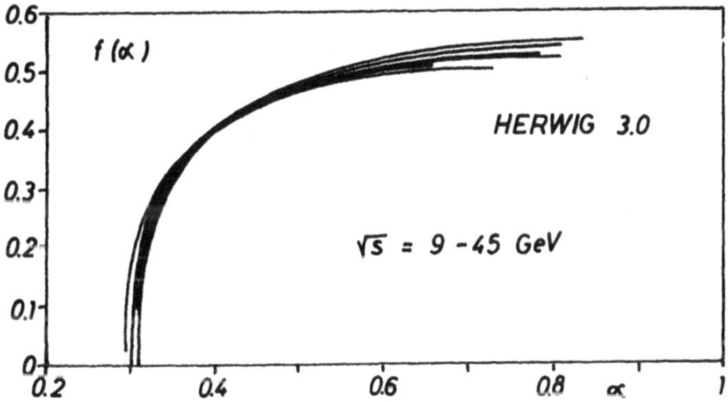


Fig. 7a

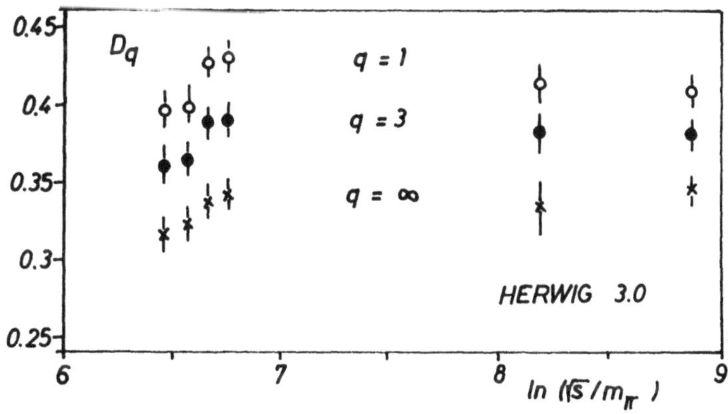


Fig. 7b

# THE NUMBER OF SOURCES IN MULTIPARTICLE PRODUCTION PROCESSES*)

G.Wilk[1], Z.Włodarczyk [2] and R.M.Weiner[3]

[1]Soltan Institute for Nuclear Studies, Warsaw, Poland
[2]Inst. of Physics, Pedagogical University, Kielce, Poland
[3]Dept. of Physics, University of Marburg, Marburg, FRG

The number of sources $k$ in multiparticle production has been estimated as a function of inelasticity and energy in the interacting gluon model. We find that $k$ levels off at the value of $7 \div 8$ if the inelasticity continues to decrease as indicated by the present trend.

## 1. Introduction

Multiparticle production processes are a dominant part of high energy collisions. In searches for "interesting physics" they are often regarded as a troublesome background. However, multiparticle production is a vital source of information about soft strong interactions and this is "interesting physics" as well, as this workshop has proved.

For the time being, multiple hadronic production can only be treated in a phenomenological way. Usually it is visualised as proceeding in two steps. The first step consists of the formation of a number of more or less defined intermediate objects, followed by their hadronization. Here we would like to concentrate on this first step only. In what follows we shall call these objects *sources*. Their number, $k$, has important applications and implications in multiparticle phenomenology. This applies, among other things to the statistical [1-4] and quantum statistical [5] models including as a particular case the negative binomial (NB) [6] parametrization, branching models [7,8] and also string models [9]. In statistical models considered so far various

*) presented by G.Wilk

values of $k$ appear. These vary from $k=1$ [1], $k=2$ [2] or $k=3$ [3] to an undefined number of "fireballs-inside-fireballs" in the statistical bootstrap model [4]. An analysis of early UA5 data in a cluster model lead to $k$ values which saturated at a value $k \simeq 8$ with growing energy [6]. This result was confirmed in a more refined analysis in ref. [7]. In branching models $k$ enters essentially through the assumed initial conditions for a given branching process whereas in string models $k$ corresponds to the number of strings formed in the first stage of collision and therefore reflects the dynamics of the interaction process [9]. These are only some examples of the explicit occurences of $k$ in phenomenological models.

Before proceeding to its concrete evaluation let us briefly illustrate the influence $k$ exerts on the multiplicity distributions $P(n)$ (which is the quantity most sensitive to $k$). In order to ensure the maximal possible model independence and clearly exhibit all assumptions, we shall use for this purpose the maximization of the information entropy method [10]. In this case, knowledge of only the mean multiplicity $\langle n \rangle$ implies that the most probable distribution is the geometrical one:

$$P_G(n) = \frac{1}{1 + \langle n \rangle} \cdot \left( \frac{\langle n \rangle}{1 + \langle n \rangle} \right)^n \tag{1}$$

whereas <u>additional</u> <u>information</u> that all particles in question are <u>indistinguishable</u> changes it immediately to a Poisson distribution

$$P_P(n) = \frac{\langle n \rangle^n}{n!} \cdot \exp( -\langle n \rangle ). \tag{2}$$

On the other hand, if we know that there are on the average $\langle n \rangle$ <u>indistinguishable</u> particles emerging from $k$ <u>independent</u> <u>sources</u>, we are immediately led [11] to the famous NB distribution

$$P_{NB}(n;k) = \binom{n+k-1}{n} \cdot \left( \frac{\langle n \rangle}{k} \right)^n \cdot \left( 1 + \frac{\langle n \rangle}{k} \right)^{-(n+k)} \tag{3}$$

which obviously interpolates between $P_G$ and $P_P$ (if $1<k<\infty$).

## 2. The number of sources in the Interacting Gluon Model

In all but one examples presented above the number $k$ was either determined from experiment [2,6,7] or it was simply postulated [1,3,8]. Only in some string models (cf. ref.[9]) one can, in principle, express $k$ via the quark structure functions of the incoming hadrons and the cross sections of the reaction.

Here we shall present a method for the determination of $k$ based on the model for the inelasticity distribution $\chi(K)$ which we have proposed recently (the inelasticity $K$ is defined as the fraction of the total c.m.s. energy of the reaction, $\sqrt{s}$, which is available for production of secondaries). In this model, to be briefly summarized below, high energy collisions are assumed to be dominated by the gluonic interactions (hence IGM for "Interacting Gluon Model")[12-13] and there exists a direct one-to-one correspondence between the distribution of the number of sources $f(k)$ and $\chi(K)$. Therefore, knowing $\chi(K)$, one can compute $f(k)$ (and vice versa). The number of sources $k$ fluctuates from event to event around a mean $\langle k\rangle=\Sigma f(k)\cdot k$. It should be remembered that it was always $\langle k\rangle$ which entered all our considerations above.

The IGM can be summarized in the following way. When two highly energetic hadrons collide, they interact mainly through their gluon components ( which include also all $q\bar{q}$ sea pairs). The valence quarks together with the gluons which escape from the interaction region form the two leading particles (LP). The interacting glue produces (through gluon-gluon fusion) a number $k$ of interaction centers ( which we call mini-fireballs, MF's, and which are our sources). The number $k$ varies from event to event.It was tacitly assumed that the MF's eventually form a lump of ( gluonic ) matter, which we called central fireball (CF), and which decays into secondaries observed in the central region of collision. Until now we were mainly interested in the energy $M=\sqrt{xys}$ deposited in the central region by the incoming hadrons ( $x$, $y$ being fractions of their energy momenta ) or in the inelasticity $K=M/\sqrt{s}$ of the reaction ( or, equivalently in the LP spectra: $x_L=1-x$ ) [12,13]. Therefore the actual form and dynamical properties of the CF were not important. In particular in the basic formula for the probability to form a CF by depositing fractions $x$ and $y$ of the energy momenta of the incoming hadrons, the MF's were summed over:

$$\chi(x,y) = \sum_{\{k_i\}} \delta\Big[ x - \sum_i k_i x_i \Big]\cdot\delta\Big[ y - \sum_i k_i y_i \Big]\cdot\prod_{\{k_i\}} P(k_i) \qquad (4)$$

( all masses and transverse momenta are neglected in eq.(4); notice that $\chi(K)=\int_0^1 dx\int_0^1 dy\chi(x,y)\delta(\sqrt{K}-xy)$ ). Assuming now that the number distribution of MF's, $P(k_i)$, is Poissonian and expressing delta functions via Fourier integrals one can perform all summations and arrive at the general formula [12]:

$$\chi(x,y) = \frac{1}{(2\pi)^2}\int_{-\infty}^{+\infty} dt \int_{-\infty}^{+\infty} du\cdot\exp\Big\{ i(xt+yu)+$$

$$+ \int_0^1 dx'\int_0^1 dy'\frac{d\langle k(x',y')\rangle}{dx'dy'}\cdot\Big[ e^{-i(x't+y'u)} - 1 \Big]\Big\}. \qquad (5)$$

Its main ingredient is the spectral function of produced MF's, i.e. the mean number of MF's at given $x$ and $y$ which is proportional to the number of $gg$ interactions:

$$\frac{d\langle k\rangle}{dxdy} = \frac{\sigma_{gg}(x,y)}{\sigma^{in}_{hN}(s)}\cdot G_h(x)\cdot G_N(y)\cdot\Theta( xy - K^2_{min} ). \qquad (6)$$

From it, on the average, we have

$$\sigma^{in}_{hN}(s)\langle k \rangle = \int_0^1 dx \int_0^1 dy\cdot\Theta(xy-K^2_{min})\cdot\sigma_{gg}(x,y)G_h(x)G_N(y) \qquad (7)$$

where $\sigma_{hN}^{in}(s)$ is the ( experimental ) hadronic inelastic cross section for c.m. energy of reaction $\sqrt{s}$, $\Theta$ is the Heavyside function, $K_{min} = \frac{2\mu}{\sqrt{s}}$ is the minimal inelasticity corresponding to the lightest MF produced, $G(x)$ is the gluonic structure function and $\sigma_{gg}(x,y)$ is the gluon-gluon cross section. This last quantity is parametrized as follows:

$$\sigma_{gg}(x,y) = \frac{4\mu^2\alpha}{M^2} + \delta \ln\left[\frac{M^2}{4\mu^2}\right]. \tag{8}$$

There are only 3 parameters in the model: the mass of the lightest MF produced $2\mu$ which defines the phase space, $\alpha$ which gives the $\sigma_{gg}$ at the boundary of the phase space and $\delta$ which fixes the asymptotic growth of $\sigma_{gg}$ with $M^2=xys$. With such a parametrization we can easily account for any energy dependence of $\langle k \rangle$.

Different terms in $\sigma_{gg}$ determine different asymptotic energy dependences of $\langle k \rangle$:

$$\begin{aligned} \sigma_{gg} &\simeq \ln[M^2/(4\mu^2)] &&\longrightarrow \sigma_{hN}^{in}(s)\langle k \rangle \simeq \ln^3 s \\ \sigma_{gg} &\simeq const. &&\longrightarrow \sigma_{hN}^{in}(s)\langle k \rangle \simeq \ln^2 s. \\ \sigma_{gg} &\simeq M^{-2} &&\longrightarrow \sigma_{hN}^{in}(s)\langle k \rangle \simeq \ln s \end{aligned} \tag{9}$$

To obtain eq.(9) we have used the standard small $x$ behaviour of the gluonic structure function, i.e. $G(x)\simeq\frac{1}{x}$ ( in the full numericalcalculations the simplest form $G(x) = \frac{p(n+1)}{x}(1-x)^n$ was used with $n=5$ and $p=0.5\div0.6$, cf. [12] ).

It can be shown that effectively we have a Gaussian-like distribution for $\chi(x,y)$ [12]:

$$\chi(x,y) \simeq \exp\left[ - \frac{(x - \langle x\rangle)^2}{2\langle x^2 \rangle} - \frac{(y - \langle y\rangle)^2}{2\langle y^2 \rangle} \right] \tag{10}$$

where

$$\langle x^n y^m \rangle = \int_0^1 dx\ x^n \int_0^1 dy\ y^m\ \frac{d\langle k\rangle}{dxdy}. \tag{11}$$

In our case $\langle x\rangle = \langle y\rangle$ ( due to symmetry of NN collision). Therefore, because the inelasticity $K=\frac{M}{\sqrt{s}} = \sqrt{xy}$, this means that $\langle K\rangle \simeq \langle x\rangle$. From eqs.(6) and (11) one can now show that under the same conditions as in eq.(9)

$$\langle\ x\ \rangle \simeq \frac{\langle\ k\ \rangle}{\ln s}, \tag{12}$$

i.e. we expect that approximately

$$\langle k(s)\rangle \simeq \langle\ K(s)\ \rangle\cdot \ln s. \tag{13}$$

This is our main result; it is also presented in exact form in Figs.1,2 below.

## 3. Numerical results

The hitherto comparison of IGM with data concentrated on the distributions and the energy dependence of inelasticity, on spectra of leading particles ( in both nucleon -nucleon and nucleon-nucleus collisions ) and on the distribution of the energy density attainable in the scatterings of heavy ions [12-13]. It turns out that the decrease of $\langle\ K\ \rangle$ with energy ( from the value of $\langle\ K\ \rangle\approx 0.5$ at ISR to $\langle\ K\ \rangle\approx 0.3$ at SPS energy range, cf. refs. [14-17] ) fixes the parameter $\delta$ at a value $\delta\approx 0\div 0.01$. The parameters $\mu$ and $\alpha$ are, on the other hand, correlated to some extent. Therefore, in all applications we have found it useful to keep $2\mu$=0.35 GeV and to fix $\alpha$ accordingly. Its value depends on the type of reaction and on the particular implementation of the IGM used, i.e. either the approximate, analytical version as presented above (cf. ref.[12]) or a Monte Carlo event generator code which we employed for present calculations. (The first simple version of the Monte Carlo code was presented in ref.[13]). For completeness we sketch the algorithm used:

(i) For both projectile (P) and target (T), choose the energy-momentum fractions $\{v_i^P, v_j^T\}$ of the valence quarks from some suitable joint valence quark distribution [18]. (In this way the amount of the energy-momentum allocated to gluons, i.e. the parameters $p_P=1-\Sigma\nu_i^P$ and $p_T=1-\Sigma\nu_j^T$ fluctuate from event to event, which substantially improves the LP spectra).

(ii) Sample (according to the gluonic structure functions introduced above) the energy-momentum fractions $\{x^P, y^T\}$ for gluons in (P) and (T) until $x_1^P + \cdots + x_{NP}^P \le p_P$ and $y_1^T + \cdots + y_{NT}^T \le p_T$ ( notice that $x,y \in (K^2_{min},1)$ for all $x$'s and $y$'s).

(iii) Check which of the $N_P$ gluons from (P) and $N_T$ gluons from (T) interact and form MF's (i.e. our sources). This is done by checking for consecutive $i \in [1,N_P]$ and $j \in [1,N_T]$ if

$$[\textit{random number } r \in (0,1)] < p_{ij}, \qquad p_{ij} = \mathrm{Min}[1, \sigma_{gg}(x,y)/\sigma_{hN}^{in}] \tag{14}$$

where $\sigma_{gg}(x,y)$ and $\sigma_{hN}^{in}$ are gluonic and hadronic cross sections respectively as defined before. If (14) is not satisfied, the check continues with the next values of $i$ and $j$, if (14) is satisfied, a new MF with mass $M_{ij}=(x_i y_j s)^{1/2}$ and rapidity (in the c.m. frame of the colliding particles) $\Delta_{ij}=\frac{1}{2}\ln(x_i/y_j)$ is formed; the check continues but now without corresponding $x$ and $y$. As a result we get a set of $k$ MF's (together with their masses and rapidities) which can be now used for some more detailed calculations [19].

After repeating this procedure a number of times we finally get $f(k)$. The inelasticity distribution $\chi(K)$ is obtained by collecting the values of inelasticity in a given event, i.e. by collecting $K = [(\Sigma x_i^P)(\Sigma y_j^T)]^{1/2}$.

To exhibit the interrelation between the inelasticity $\langle K \rangle$ and the number of emitting sources $\langle k \rangle$, as given by eq.(13), we plot in Fig. 1 the mean inelasticity $\langle K \rangle$ versus the mean number of MF's $\langle k \rangle$ for different energies. The parameter $2\mu$ was kept constant ( $2\mu=0.35$ GeV ), while the parameters $\alpha$ and $\delta$ were varied in order to get different values of $\langle K \rangle$ for different energies. Notice that because the percentage of the energy-momenta of colliding hadrons allocated to gluons is limited to

$0.5 \div 0.6$, $\langle K \rangle$ is accordingly limited from above. Fig. 2 illustrates in detail four particular scenarios for the $s$-dependence of $\langle K \rangle$:

(*i*) $\langle K \rangle$ decreases continously ( as demanded by some recent analyses of data [14-17], $\alpha$=60 mb, $\delta$=0. ). In this case $\langle k \rangle$ at first increases and then levels off at a value $\langle k \rangle = 7 \div 8$ at the SPS energy range where eqs.(9,13) become fully justified.( We would like to stress that this value coincides with that found in refs. [6,7] for completely different reasons ).

(*ii*) $\langle K \rangle$ is constant = 0.5 ( $\alpha$=71 mb and $\delta = 2 \cdot 10^{-4}$ mb ).

(*iii*) $\langle K \rangle$ is constant = 0.3 ( $\alpha$=20 mb and $\delta = 7 \cdot 10^{-5}$ mb ).

In cases (*ii*) and (*iii*), $\langle k \rangle$ increases continously.

(*iv*) $\langle K \rangle$ first decreases between $\sqrt{s}$=20 and $\sqrt{s}$=500 GeV from $\simeq$0.5 to $\simeq$0.3 and then remains constant. In this case $\langle k \rangle$ changes its slope at $\sqrt{s} \simeq$200-600 GeV but continues to grow indefinitely.

A corollary of these results is that if the saturation of $\langle k \rangle$ at the value $7 \div 8$ derived in ref. [3] is to be taken seriously, the decrease of inelasticity with energy is a necessary consequence. On the other hand one can also argue that the IGM could provide a necessary dynamical background for all statistical models in which the concept of number of sources is used [1-5]. Actually one can make an even stronger statement, namely that IGM should in fact be part of any serious statistical model in which the initial energy available, i.e. $K\sqrt{s}$ and the number of sources $k$ (with their masses and rapidity rapidity distributions) is important.

This work was supported in part by the Heraeus-Stiftung, Hanau, the Federal Minister for Research and Technology (BMFT) under the contract number 06 MR 713 and the Gesellschaft fuer Schwerionenforschung, Darmstadt.

## References

[1] E.Fermi, Prog.Theor.Phys. 5, 570 (1950);
I.Pomeranchuk, Dokl.Akad.Nauk SSSR, 78, 889 (1951).

[2] P.Ciok, T.Coghen, J.Gierula, R.Hołyński, A.Jurek, M.Mięsowicz, T.Saniewska and J.Pernegr, Nuovo Cim. 10, 741 (1958).

[3] Cai Xu, T.C.Meng and C.S.Huang, Phys.Rev. D33, 1287 (1986).

[4] R.Hagedorn, Nuovo Cim. Suppl., 3, 147 (1965).

[5] G.N.Fowler et al., Phys.Rev. Lett. 57, 2119 (1986) and Phys.Rev. D37, 3127 (1988); cf. also: P.Carruthers and C.C.Shih, Int.J.Mod. Phys. A2, 1544 (1987).

[6] G.Espong, Proc. 16$^{th}$ Int.Symp. on Multiparticle Dynamics, Kiryat Anavim, 1985, ed.J.Grunhaus, Editions Frontieres, Gif- sur-Yvette, 1985, p.309.

[7] A.Giovannini and L.Van Hove, Z.Phys. C30, 391 (1986) and Acta Phys.Polon. B19, 917 and 931 (1988).

[8] C.Iso and K.Mori, Z.Phys. C46, 59 (1990).

[9] K.Werner, Phys.Rev.Lett. 61, 1050 (1988).

[10] Cf. for example: H.Haken, *Information and self-organization*, Springer-Verlag, 1988 and references therein.

[11] B.Carazza and A.Gandolfi, Lett. Nuovo Cim. 15, 553 (1976).

[12] G.N.Fowler et al., Phys.Rev.Lett. 55, 173 (1985), Phys.Rev. D35, 870 (1987) and Phys.Rev.C40, 1219 (1989).

[13] G.Wilk and Z.Włodarczyk, V$^{th}$ISVHECRI, Łódź, Poland, 29 August -7 September 1988, Univ. Łódź Publishers., Łódź 1988, p.171.

[14] J.Wdowczyk and A.W.Wolfendale, Nature, 306, 347 (1983); Nuovo Cim. 45A, 433 (1979) and J.Phys. G10, 257 (1983).

[15] E.M.Friedlander and R.M.Weiner, Phys.Rev. D28, 2903 (1983).

[16] UA5 Collab., G.J.Alner et al., Z.Phys. C33, 1 (1986).

[17] Y.T.Chou and C.N..Yang, Phys.Rev. D32, 1692 (1985); K.Watanabe and S.Tone, Phys.Rev. D39, 195 (1989).

[18] Cf. for example: J.Kuti and V.F.Weisskopf, Phys.Rev. D4, 3418 (1971), E.Takasugi and X.Tata, Phys.Rev. D26, 120 (1982).

[19] Cf. talk by Z.Włodarczyk, these proceedings.

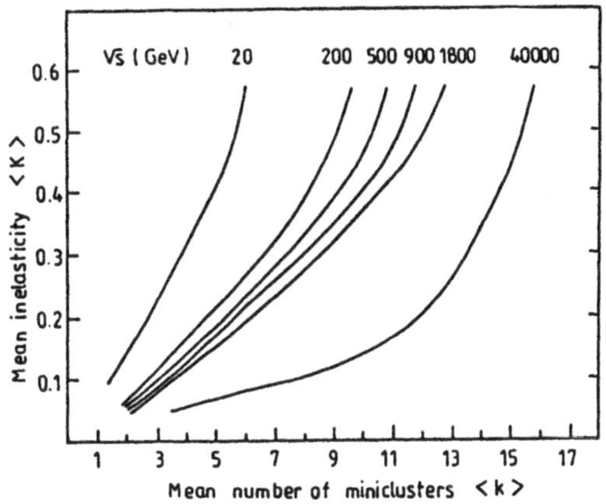


Fig. 1 The mean inelasticity ⟨ $K$ ⟩ versus the mean number of MF's for different energies.

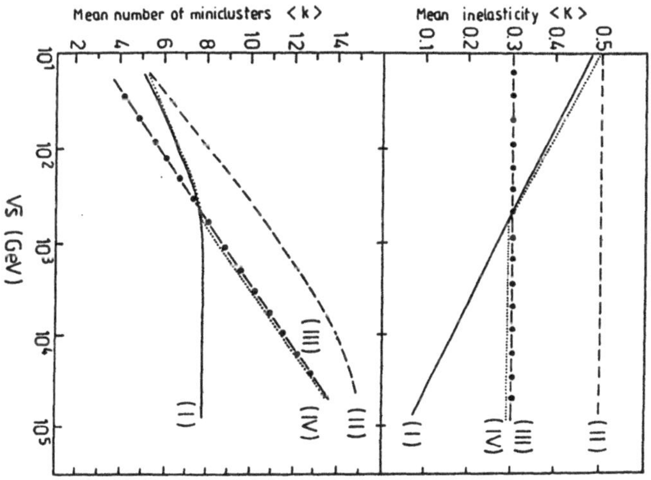


Fig. 2 The energy dependence of the mean number of MF's ⟨$k$⟩ for different scenarios of the energy dependence of mean inelasticity ⟨ $K$ ⟩. The open circles represent the mean number of sources from UA5 data (interpreted in terms of NB distribution)[6].

# IV. Round Table (Correlations)

Chairman: *R. M. Weiner*

# STATUS AND PRESENT DEVELOPMENTS OF FERMION CORRELATION STUDIES

D. ARDOUIN*
Laboratoire de Physique Nucléaire,
IN2P3 and Université de Nantes
2, rue de la Houssinière, 44072 Nantes Cedex 03, FRANCE

Abstract: A brief summary of two-fermion correlation studies is presented together with some present developments undertaken in heavy-ion collisions in order to assess the importance of quantum and source-lifetime effects.

## I - General remarks

A general remark which can be put up about two fermion correlations, as compared to meson correlation studies, is the fact that only a few and recent experiments have been performed. The first indication of a two-proton correlation effect was found by Siemarczuk et al[1] in 1967 at Dubna followed by a similar observation at Berkeley by the group of J. Carroll[2] using Ar + KCl interactions at 1.8 AGeV. More recently, this technique has been also used in heavy-ion collisions at Michigan State, Oak-Ridge, Berkeley, GANIL, SATURNE and Stony-Brook[3-7].

* Paper prepared in collaboration with J.P. Alard and J. Pluta

At variance with meson studies, the correlation-functions have been shown to be mostly sensitive to final state interactions. The sensitivity to the space-time extent of the source is now observed in the height of the correlation correlated to the nuclear interaction and the Coulomb repulsion, the latter being mainly responsible for the observed depletion at low momenta. A typical example taken from ref.5 is shown in Fig. 1 for proton-proton correlations measured in the Ar + Au interaction at E/A = 60 MeV.

Some theoretical works by Koonin, Lednicky and Boal are available which treat the scattering state of two protons[8-10]. These calculations and their comparison with data usually ignore the time-dependence thus providing upper-limits for the space extent of the emitting sources. Consequently, all the dynamical aspects of the source formation and the collision process itself have not been included in such approaches.

With respect to the latter question, it is interesting to quote here some general trends of the available data :

. for fast protons (p > 0.35 GeV/C), the source-size thus determined is comparable with the target size itself (Fig.2 from ref.11)

. the low energy protons systematically correspond to larger sizes reflecting a possible lifetime effect

. larger effective sizes have been observed when triggering on large particle multiplicities : see Fig.3 from ref.4 (Plastic Ball data) and Fig.4 from ref.12 (GANIL data). If one can make a link between multiplicity and impact parameter, such geometrical variations are not predicted by the available theoretical descriptions of ion collisions. Thus, one can again propose a collision-time effect with the impact parameter variation.

II - Search for lifetime effects

Fig.5 shows some recent proton-proton data obtained[13] at GANIL for the $^{16}O$ + Au system at E = 94 A. MeV. Clearly, we then observe a disappearance of the peak at q = 20 MeV/c corresponding to the singlet S-wave interaction. This phenomenon is observed for the backward angle data and is due to the sequential nature of proton emission observed when looking at an equilibrated source. In other words, the observation of a time-scale effect, lowering nuclear final-state interaction, has been possible. From those data and similar data on the Ar + Ag system, it has been possible to extract a lifetime $\tau = (1.1 \pm 0.2) \times 10^{-21}$s assuming either pure quantum statistics or three-body coulomb interactions[13-14].

In such conditions, it is also possible[8-9,15] to look for a directional dependence of the correlation function. Emission of a long-lived sources gives rise to a an elongated shape in the longitudinal direction with a subsequent reduction of the Pauli suppression at low relative momenta. Thus an enhancement may be observed as compared to a transverse observation of the correlation function. A weak effect seems to have been observed for the Ar + Ag system[16] for backward angle measurements (Fig.6), in agreement with evaporative calculations, indicating a lifetime effect. Detailed comparisons clearly require high accuracy data, especially at low relative momenta in order to establish the exact influence of quantum statistical effects. The main general conclusion of this type of approach is the necessity of exclusive measurements or chosen kinematics (use of backward hemisphere for an asymmetric system in the present case).In such cases, the observation of a known equilibrated source of proton emission allows to assess the respective influence of space and time extents.

III - Outlook over some present prospects

On the theoretical point of view, I first want to mention current dynamical calculations by Schuck, Remaud and collaborators[17] using a semi-classical approach of heavy-ion collisions. They seem to be able to predict a Pauli suppression at low relative momenta when looking at the dynamical evolution at the dynamical evolution of a fermionic system.

Experiments are currently being prepared in order to measure specific correlation-functions at very low momenta ; I want to briefly list some of them in which some colleagues or myself are involved :

. measurement of (p,p), (p,d), (d,d) data at q = 1 - 50 MeV/c using a detector designed for very small relative angle measurements. The two last systems can help to quantify the importance of Coulomb effects.

. measurement of (n,n) and (p,n) correlation data. Such measurements made in a LUND/SARA collaboration are under analysis while a project is undertaken with the DEMON detector[18]. The first attempt by Bayukov et al[19] to perform such measurements clearly indicates a very interesting sensitivity to the source lifetime.

. measurement of photon-photon correlations. An experiment has been recently performed in the GANIL/TAPS collaboration looking at $\gamma - \gamma$ correlations for the Xe + Au system. Preliminary results and simulations indicate the feasibility of such an experiment providing us with a unique measurement of boson correlations.

IV - Conclusion

I have shown that we are witnessing an increasing interest for light-particle correlation measurements in nucleus or hadron collisions. Under specific conditions, these measurements and their theoretical description offer a unique way of approaching the dynamics of such collisions. By comparison with meson correlations, they might be able, in a near future, to assess the importance of lifetime, quantum and coherence effects in observed particle correlations.

I thank E. Gerbaud for a careful and quick typing of this manuscript.

REFERENCES

1) Siemarczuk et al, Phys. Lett. **24 B** (1967) 675
2) Z. Zarbakhsh et al, Phys. Rev. Lett. **46** (1981) 1268
3) W. Lynch et al, Phys. Rev. Lett. **51** (1983) 1850
4) H. Gustafson et al, Phys. Rev. Lett. **53** (1984) 544
5) J. Pochodzalla et al, Phys. Rev. **C 35** (1987) 1695
6) P. De Young et al, Phys. Rev. **C 39** (1989) 128
7) P. Dupieux et al, Phys. Lett. **B 200** (1988) 17
8) S. Koonin, Phys. Lett. **70 B** (1977) 43
9) R. Lednicky et al, Sov. J. Nucl. Phys. **35 (5)** (1982) 770
10) D. Boal et al, Phys. Rev. Lett. **57** (1985) 2901
11) J. Pluta, DUBNA JINR 88 - 754
12) A. Kyanowski et al, Phys. Lett. 181 (1986) 43
13) D. Ardouin et al, Nucl. Phys. A 495 (1989) 57 c.
J. Québert, Proceedings of the XXVII Bormio Meeting (1989)
14) B. Erazmus and N. Carjan (to be published)
15) S. Pratt et al, Phys. Rev. **C 36** (1987) 2390
16) D. Goujdami et al (to be published)
17) P. Schuck, B. Remaud (private communication)
18) B. Heusch, Internal Report Strasbourg/PN 90-10
19) Y. Bayukov et al, Phys. Lett. **189 B** (1987) 291

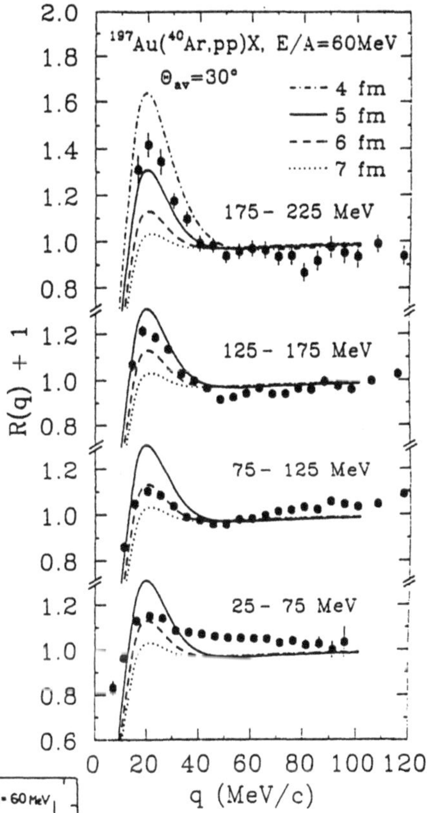


Fig.1
Two-proton correlation functions. data are from the Ar + Au → p + p + X reaction. Calculations are from ref. 10 using different source-sizes and zero lifetime assumptions.

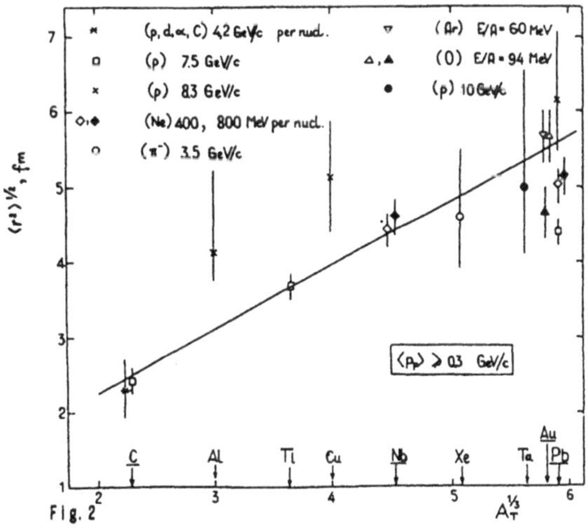


Fig. 2
Dependence of the source size on the target mass number for fast protons and various colliding systems (from ref. 11)

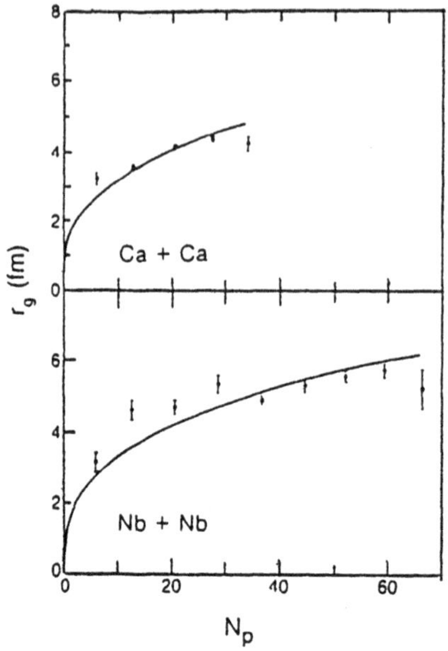


Fig. 3
Dependence of the proton source size determinations on the Plastic-Ball multiplicity for the Ca + Ca and Nb + Nb collisions (from ref. 4) under zero lifetime assumption.

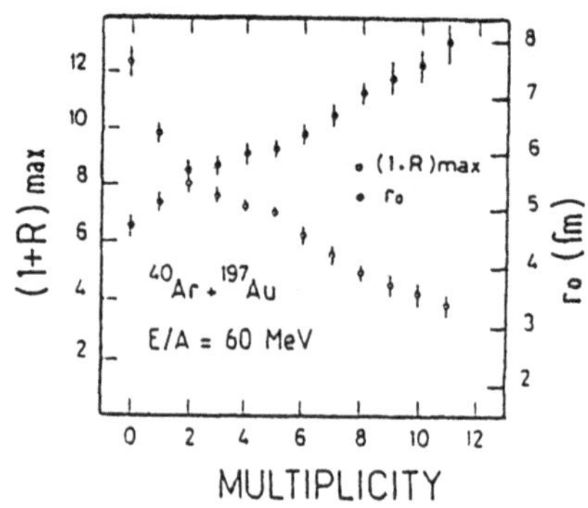


Fig. 4
Evolution of the d-$\alpha$ correlation-function (open circles) and size parameter $r_0$ (full circles) as a function of the forward multiplicity (from ref. 12)

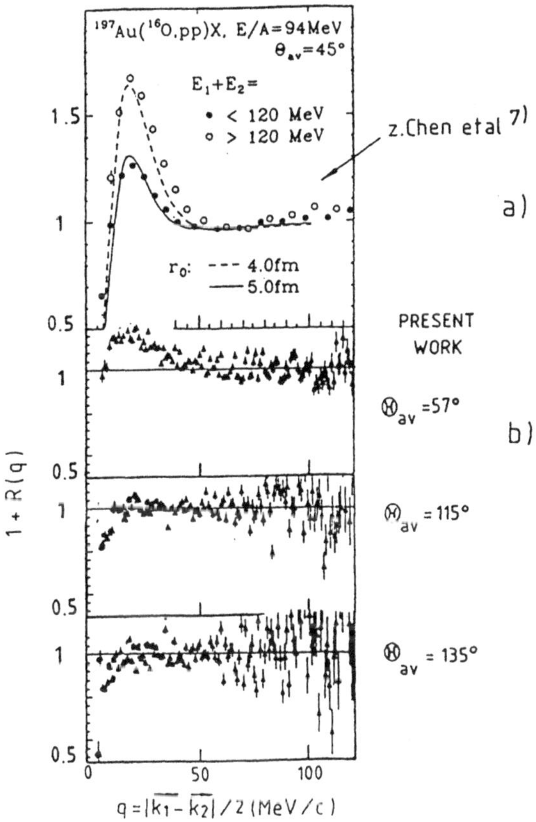


Fig. 5
Comparison of two-proton correlation-functions measured at various laboratory angles for the O + Au reaction at E = 94 A MeV.

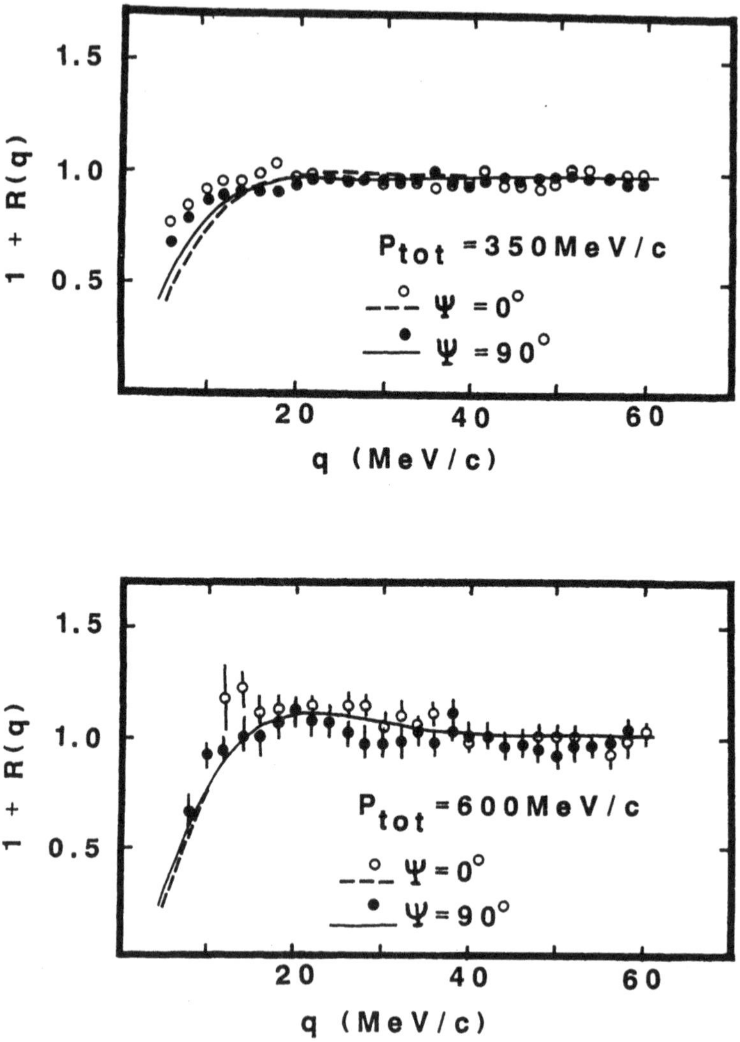


Fig. 6
Two-proton correlation functions for the Ar + Ag reaction (ref.16). Data are shown for relative momenta parallel ($\psi$ = 0°) or transverse ($\psi$ = 90°) direction with respect to the total momentum $P_t$. Two selections on $P_t$ values are shown.

# The GGLP effect alias Bose-Einstein effect alias H-BT effect

Gerson Goldhaber

Lawrence Berkeley Laboratory and Department of Physics,

University of California

Berkeley, California

This work was supported by the Director, Office of Energy Research, Office of High Energy and Nuclear Physics, Division of High Energy Physics, of the U.S. Department of Energy under Contract No. DE-AC03-76SF00098.

What is clear is that we have been working on this effect for thirty years. What is not as clear is that we have come that much closer to a **precise** understanding of **the** effect.

In response to the convener of this roundtable discussion, I will present the historical development from the point of view of particle physics, as well as some results from $e^+e^-$ interactions.

This subject has already been discussed in part in my talks at LESIP I [1] and LESIP II [2]. Other excellent reviews of the entire field have recently been given by Bill Zajc, Bengt Loerstad, Werner Hofmann, and a series of incisive papers by Mike Bowler[3].

**The historical development from the particle physics point of view.**

**1959** The empirical observation. G. Goldhaber, et al.[4]

This was a $\bar{p}p$ propane bubble chamber experiment at the Bevatron, designed to look for the $\rho^0 \rightarrow \pi^+\pi^-$ decay. In this connection, we compared $\pi^+\pi^-$ mass distributions with the $\pi^\pm\pi^\pm$ distributions. While the statistical accuracy was not adequate to establish the existence of the $\rho^0$, we did observe a marked and very significant difference in angular opening angle for LIKE and UNLIKE pion pairs.

The lessons from this experiment are:

- Be sure to get enough statistics to achieve the desired goal!
- Be alert for unexpected (unexplained) effects!

**1960** The interpretation of the empirical observation in terms of the symmetrization of the identical particle wave functions.

Gerson Goldhaber, Sulamith Goldhaber, Wonyong Lee, and Abraham Pais[5] were able to reproduce the empirical angular distribution by a detailed explicit multipion phase space calculation in which LIKE particle wave functions were symmetrized.

It must be remembered that this work was carried out **before** the detailed Monte Carlo phase space simulations were invented. We did not have LUND 6.3 available to carry out these integrations! All multiparticle integrals were numerically evaluated on a comparatively primitive computer (IBM 650). In fact, we had to invent our own Monte Carlo methods for these integrations.[5]

The result was clear. **The** effect could be interpreted as the consequence of the Bose-Einstein nature of LIKE pions. See Figure 1.

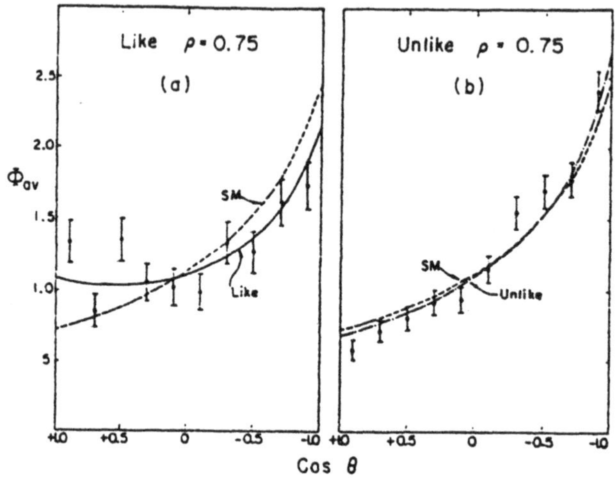


The functions $\Phi_{av}(\cos\theta)$ computed at $\rho = 0.75$ are compared with the experimental distribution of angles between pion pairs. Figures 6(a) and 6(b) give the distributions for like and unlike pions respectively. Also shown in each is the curve for $\Phi_{av}^{SM}(\cos\theta)$, the statistical distribution, without the effect of correlation functions. Here $\Phi_{av}$ represents an average of $\Phi_4$, $\Phi_5$, and $\Phi_6$, weighted according to the individual charge channels. The experimental data comes from reference 1 (see also Table I, footnote a). Here $\rho$ is in units of $\hbar/\mu_\pi c$.

Figure 1

The consequence of this analysis was the introduction of the expression:

$$R = 1 + e^{-r^2Q^2}$$

$$\text{where } Q^2 = -(p_1 - p_2)^2$$

$$= M_{12}^2 - (m_1 + m_2)^2$$

is the invariant four momentum difference of the two like pions, R is a "correlation function" and r represents the radius of some volume in four space. The numerical value we found to fit the data was $r \simeq 1.0$ fm. While one can argue that we should have used 3-momenta, the main justification for 4-momenta was that transformations and integrals were simpler with 4-vectors!

**1960-1975** Many experimental confirmations of **the** effect in various particle physics experiments.(6)

- Biswas, et al. and M. Deutschmann, V. Cocconi, D. Morrison, and others of CERN who carried out a major study on a variety of interactions.(6) An important result was the realization that the effect did not appear to come up to its maximal value, and that a factor $\lambda$ was needed to fit the experimental data. This changed the expression for the correlation function to

$$R = 1 + \lambda\, e^{-r^2Q^2}$$

In the meantime, the new resonances (beyond the $\Delta(1236)$) were discovered. The $Y^*(1385) \equiv \Sigma(1385)$, the $K^*(890)$, the $\rho(770)$, $\omega(780)$, $\eta(550)$, $f(1280)$, and many others.(7) These extremely important particle physics discoveries led to confusion about the interpretation of the effect." The effect was not really due to Bose-Einstein statistics; why not just a reflection of resonances?" Does this mean we discovered resonances - and did not know it? I am afraid not!

**1970-1975** Can the effect become a precise tool in pion interferometry? Does the answer lie in the stars?

The cast of characters changes. Enter particle physicists who have heard of Astronomy[8-11]. Flashback to 1956. Meanwhile, what have Hanbury-Brown and Twiss been doing? R. Hanbury-Brown has been thinking about how to measure stellar radii, or at least the angle $\theta$ subtended by nearby stars. He enlisted R. Q. Twiss to develop the mathematical theory of Intensity interference, or second-order interference[12]. What is more, they actually built the equipment to carry out the experiment. There is an interesting similarity to our experience. Skepticism that such an effect can exist! I quote from the very readable book by R. Hanbury-Brown[13],

> "Another stream of objections about photons were both instructive and entertaining. Our whole argument was based on the idea that the fluctuations in the outputs of two photoelectric detectors must be correlated when they are exposed to a plane wave of light. We had shown that this must be so by a semi-classical analysis in which light is treated as a classical wave and in this picture there is no need the worry about photons - the quantization is introduced by the discrete energy levels in the detector. However, if one must think of light in terms of photons then, if the two pictures are to give the same result, one must accept that the times of arrival of these photons at the two separated detectors are correlated - they

Figure 2

Figure 3

tend to arrive in pairs. Now, to a surprising number of people, this idea seemed not only heretical but patently absurd and they told us so in person, by letter, in publications, and by actually doing experiments which claimed to show that we were wrong. At the most basic level they asked how, if photons are emitted at random in a thermal source, can they appear in pairs at two detectors? At a more sophisticated level, the enraged physicist would brandish some sacred text, usually by Heitler, and point out that the number $n$ of quanta in a beam of radiation and its phase $\phi$ are represented by non-commuting operators and that our analysis was invalidated by the uncertainty relation. "

Hanbury-Brown and Twiss developed a theory for intensity interferometry treating the stars in a first approximation as luminous disks. This led to the expression of

$$\Gamma^2(d) = [2\, J_1(x)/x]^2 \text{ where } x = \pi\theta d/\lambda.$$

Here $\Gamma^2$ is the "normalized correlation," $J_1(x)$ a Bessel function, $\lambda$ the mean wave length of the light observed, d the distance between the two mirrors, and $\theta$ the resulting angle subtended by the star as approximated by a disk.

The essence of their equipment consisted of two roughly paraboloid mirrors, each of which focused the light from a star onto a photomultiplier tube. Figure 2 shows such a mirror. An essential feature of the device is the "correlator. " An electronic circuit that receives the signals from both mirrors and "multiplies them." A technique that allows to distinguish between noise and correlated signals. An interesting feature is that the mirrors need

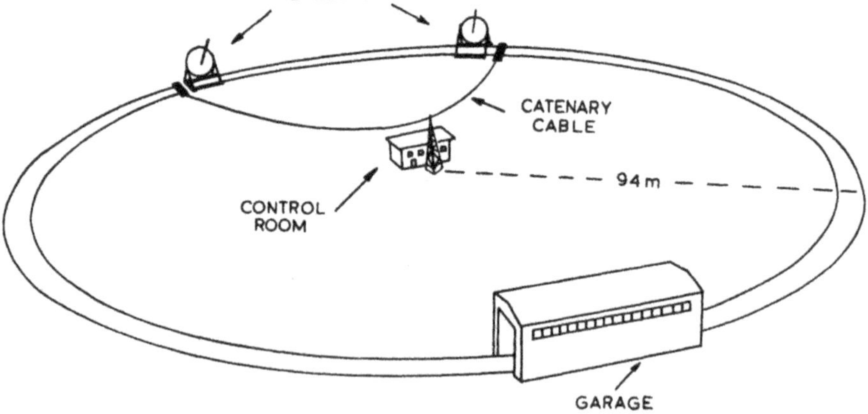


Figure 4

not be of the usual high optical quality associated with telescope mirrors. To quote:

> "It took us six months to realize that although we should certainly need two very large telescopes, they could be extremely crude by astronomical standards. Their function would simply be to collect the light from the star like rain in a bucket and pour it on to the detector there was no need to form a conventional image. In practice this meant that the whole problem was transformed into one of reasonable cost since our telescopes need only be like the paraboloids used for radio-astronomy, but with light-reflecting surfaces. The necessary precision of these surfaces would be governed by the maximum permissible field of view and not, as at radio wavelengths, by the beamwidth; for bright stars a field of view of several minutes of arc would be tolerable and this could be achieved with the sort of structures which were used by radio-astronomers for microwaves."

Figure 3 gives a photograph and a sketch of the complete interferometer at Narrabri in Australia. Figures 4 and 5 show examples of the Interference distributions and the measured angles θ subtended by the various stars.
The awareness by particle physicists (8-11) of the work by Hanbury-Brown and Twiss on stellar interferometry led to attempts and suggestions to place "pion interferometry" on a more respectable footing.

**Summary of Further Developments**

**~ 1975 - 1982** M. Deutschmann et al. CERN

- λ as an empirical parameter.
- Event Mixing for reference samples.

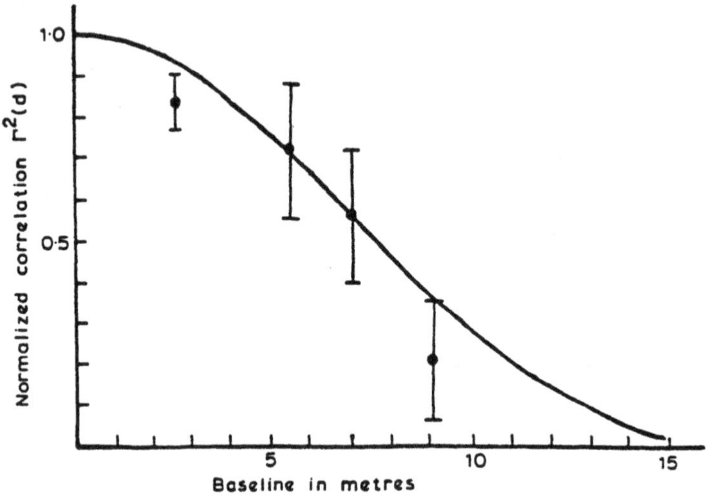


Correlation versus baseline for Sirius. The full line shows the theoretical variation of $\Gamma^2(d)$ the normalized correlation for an angular diameter of $6{\cdot}9 \times 10^{-3}$ seconds of arc. The points show the experimental results together with their probable errors. From Hanbury Brown and Twiss (1958 b).

Figure 5

Measurement at Jordell Bank.

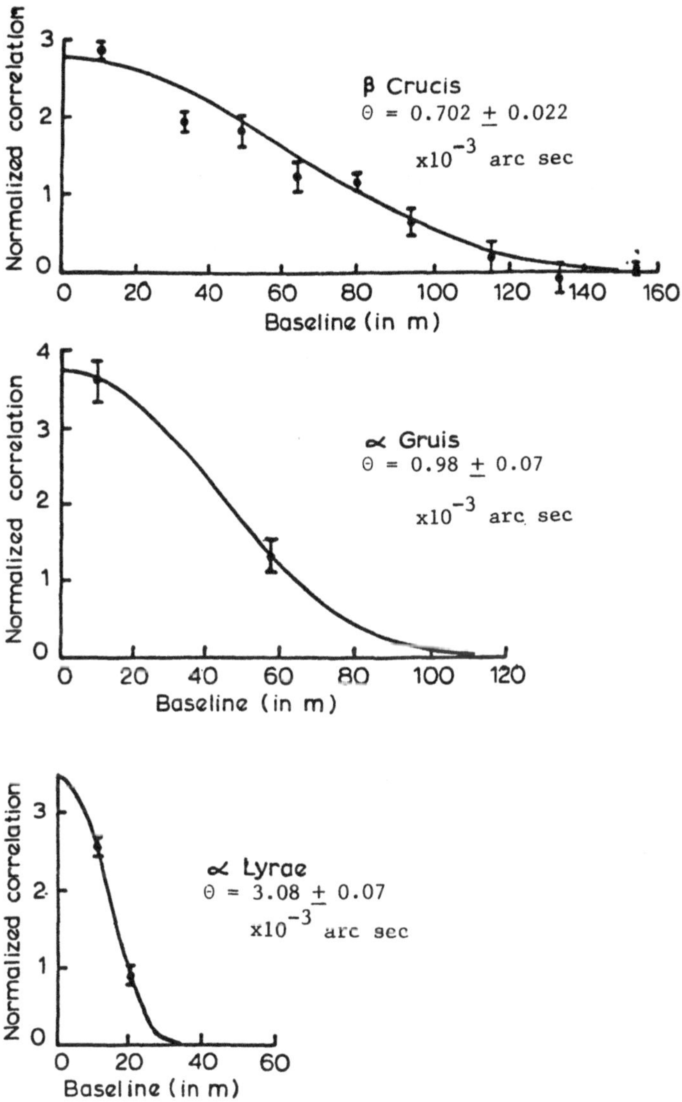


Correlation versus baseline measured for three stars of different angular size. The full lines were fitted to the points as described in the text. From Hanbury Brown, Davis, Allen and Rome (1967).

Figure 6

**1977** G. N. Fowler and R. M. Weiner

also A. Giovannini and G. Veneziano

$\lambda$ is Chaos/Coherence Parameter

$\lambda = 1$ Chaotic Source

$\lambda = 0$ Pion Laser

**1977** P. Grassberger

Effect of resonances.

Pion carried by "long lived" resonance. eg. $\omega^o$

$$L = \beta\gamma c\tau$$

$$= \frac{\langle p \rangle}{M} c\tau \qquad \tau = \frac{\hbar}{\Gamma}$$

$$= \frac{\langle p \rangle}{M\Gamma} \hbar$$

$$= \frac{\langle p \rangle}{M\Gamma} \; 197 \text{ MeV fm}$$

Typical values

| Res: | $\rho$ | $K^*$ | $\omega^o$ | $\eta$ |
|---|---|---|---|---|
| L fm | 2 | 4 | 20 | 400 |

Typical Radius observed : 0.7 fm

For Particles:

| | $D^o$ | $D^+$ | B |
|---|---|---|---|
| $L \simeq c\tau$ | $10^{11}$ fm | $3 \times 10^{11}$ fm | $3 \times 10^{11}$ fm |

Thus resonances, and certainly particles.
should reduce $\lambda$

**1986** B. Andersson and W. Hofmann

also M. Bowler

Introduction into Lund model of 2 pion symmetrization.
Interchange on color string

1987 M. Suzuki $\pi\pi$ final state interaction are expected to reduce $\lambda$ by 20%.

## Conclusions on the Effect From $e^+e^-$ Experiments

We have carried out four distinct experiments with $e^+e^-$ interactions, in the Mark II detector. Juricic, et al.[(14)]

We can describe these 4 Data Sets, which differ in energy and production mechanisms as:

- SPEAR $J/\psi$ – 3 Gluon annihilation, E = 3.1 GeV.
  Below charm threshold.
- SPEAR $q\bar{q}$ - Beginning of JET production, E = 4 - 6.5 GeV.
  Above charm production threshold.
- PEP $q\bar{q}$ - 2 and 3 JET production, E = 29 GeV.
  Above charm and bottom production threshold.
- PEP $\gamma\gamma$ – VDM and hard scattering, $\langle E \rangle \simeq 5$ GeV.
  Mostly below charm threshold.

From these we observed the following:

- $r \simeq 0.7$ fm is nearly the same in all four cases.

From this we conclude that here r is a measure of the size of the **local region** rather than the **entire** source size.

- $\lambda$ is nearly maximal $\approx 1$ in the two cases below charm threshold.
  **Thus** resonance effects are not observed. Would increase $\lambda$ by ~50% for the $\rho$, $K^*$ $\omega^0$, $\eta$ etc. **if we correct for it**

- Also the final state interaction effect predicted by M. Suzuki is not observed.
  Would increase $\lambda$ by ~20% **if corrected for.**
- The reduced value of $\lambda$ ($\approx$ 0.6 - 0.5) observed for the SPEAR $q\bar{q}$ and PEP $q\bar{q}$ data sets can be explained by the onset of charm and charm + bottom, respectively.
- Thus there is apparently no room for the reduction of $\lambda$ due to coherence effects.

**Acknowledgements**

This work was supported by the Director, Office of Energy Research, Office of High Energy and Nuclear Physics, Division of High Energy Physics, of the U.S. Department of Energy under Contract No. DE-AC03-76SF00098.

**References**


1. G. Goldhaber, LESIP I, Bad Honnef, Germany, Editors D. K. Scott and R. M. Weiner, World Scientific Publishing Co., p. 115 (1984). See references quoted therein.

2. G. Goldhaber and I. Juricic, Hadronic Matter in Collision, Proceedings of the Second International Workshop on Local Equilibrium in Strong Interaction Physics, Santa Fe, New Mexico (1986).

3. See papers by these authors at this conference for references.

4. G. Goldhaber, W. B. Fowler, S. Goldhaber, T. F. Hoang, T. E. Kalogeropolous, and W. M. Powell, Phys. Rev. Lett. **3**, 181 (1959).

5. G. Goldhaber, S. Goldhaber, W. Lee, and A. Pais, Phys. Rev. **120**, 300 (1960).

6. See reference quoted in (1) above.

7. For a discussion of the discovery of resonances see, for example, R. Cahn and G. Goldhaber, *The Experimental Foundations of Particle Physics* Cambridge University Press, 1989.

8. E. V. Shuryak, Phys. Lett. **44B**, 387 (1973). See also paper by Grishin.

9. G. I. Kopylov and M. I. Podgoretsky, Sov. J. Nucl. Phys. **19**, 215 (1974); G. I. Kopylov, Phys. Lett. **50N**, 472 (1974).

10. G. Cocconi, Phys. Lett. **49B**, 459 (1974).

11. M. Gyulassy, S. K. Kaufmann, and L. W. Wilson, Phys. Rev. **C20**, 2267 (1979).

12. R. Hanbury-Brown and R. Q. Twiss, Philos. Mag. **45**, 663 (1954), Nature **178**, 1046 (1956); E. M. Purcell, Nature **178**, 1449 (1956).

13. R. Hanbury-Brown, *The Intensity Interferometer, Its Application to Astronomy*. London; Taylor and Francis, Ltd. (1974).

14. I. Juricic, et al. Phys. Rev. **D39**, 1 (1989).

# TO THE QUESTION ON THE INFLUENCE OF SCATTERINGS ON TWO-PARTICLE CORRELATIONS OF IDENTICAL PIONS

M.I.Podgoretsky

Laboratory of High Energies, Joint Institute for Nuclear Research, Dubna

ABSTRACT

Moving after their generation inside nuclear matter, pions can undergo one or several scatterings (inelastic in the general case) prior to the stage of free propagation. It is commonly assumed that pair correlations of identical pions are determined by a relative position of points of the last scatterings which thereby play a role of sources. This argument seems to be directly clear and fully natural.

On the other hand, the structure of the one-particle amplitude of pion generation with subsequent scattering differs markedly from the structure of the generation amplitude not accompanied by scattering. Therefore, the question can arise to compare these two statements which seem at first glance to be not wholly consistent. Such a comparison is made in this paper. It is shown that despite a particular structure of one-particle scattering amplitude, the points of the last scatterings really play a role of full and equal pion sources in two-particle correlations.

As known, pair correlations of identical pions are determined by the space-time size of the region in which generation occurs. More precisely, the question concerns distances between the points of "the last interactions", i.e. between those points after which pions propagate freely. These can be the pion sources themselves if there are no scatterings later on or the points at which the last scatterings take place on condition that such scatterings occur actually. Although the aforesaid about the role of scatterings is conventional and very natural, it is worth-while to discuss this question more carefully confining ourselves, for definiteness, to the simplest model of independent one-particle sources.

If a one-particle source is at point $\vec{r}$ and emits a pion with momentum $\vec{p}$ and energy $\varepsilon$ at time $t$, then in the absence of subsequent scatterings the amplitude

$$A(\vec{p}) \sim e^{-i(\vec{r}\vec{p}-\varepsilon t)} \tag{1}$$

corresponds to this process. Let two of such sources placed at points $\vec{r}_1$ and $\vec{r}_2$ generate at $t_1$ and $t_2$ identical pions with momenta and energies ( $\vec{p}'$, $\varepsilon'$) and ( $\vec{p}''$, $\varepsilon''$). Then the two-particle amplitude can be written as

$$A(\vec{p}'\vec{p}'') \sim \left\{ e^{-(\vec{r}_1\vec{p}'-t_1\varepsilon')} \cdot e^{-(\vec{r}_2\vec{p}''-t_2\varepsilon'')} + e^{-i(\vec{r}_1\vec{p}''-t_1\varepsilon'')} \cdot e^{-i(\vec{r}_2\vec{p}'-t_2\varepsilon')} \right\} \tag{2}$$

Let us look what changes when scattering exists. Let a pion with momentum and energy ( $\vec{P}_a$, $\varepsilon_a$) be emitted at point $\vec{r}_a$ at the time $t_a$. Now let a scatterer be placed at point $\vec{r}_b$ . For simplicity it is assumed to be massive and rigidly fixed. Then in the act of scattering, energy should be conserved while momentum can change. In particular, equalities $|\vec{P}_a| = |\vec{P}_b|$ and $\varepsilon_b = \varepsilon_a$ should be valid in elastic transition $\vec{P}_a \to \vec{P}_b$ . If the distance between the points $\vec{r}_a$ and $\vec{r}_b$ is great in comparison with the wave length ($|\vec{r}_a - \vec{r}_b| \cdot |\vec{P}_a| \gg 1$), the amplitude of all the process including generation and subsequent scattering can be written in the form

$$A(\vec{p}_a,\vec{p}_b) \sim e^{-i(\vec{r}_a\vec{p}_a-t_a\varepsilon_a)} \cdot e^{-i\vec{r}_b\vec{q}} \tag{3}$$

with $\vec{q} = (\vec{p}_b - \vec{p}_a)$ the momentum change in the scattering. It is natural that the coordinate $\vec{r}_b$ enters into formula (3) without the corresponding moment $t_b$ as a monochromatic wave, figuratively speaking, is scattered "all the time". Formula (3) can be written in another form:

$$A(\vec{p}_a,\vec{p}_b) \sim e^{+i(r_{ab}p_a+t_a\varepsilon_a)} \cdot e^{-\vec{r}_a\vec{p}_b}, \quad r_{ab}=|\vec{r}_a-\vec{r}_b|, \quad p_a=|\vec{p}_a| \qquad (4)$$

As seen, the structure of expressions (3) and (4) differs markedly from that of formula (1). So, at this stage it is not quite clear whether the scattering point can identify with a "new source"; as a minimum, some further discussion is required in this case.

Let us consider a somewhat more complicated case of inelastic scattering when some other particles taking some part of initial energy $\varepsilon_a$ originate at the point $\vec{r}_b$ in addition to the scattered pion with momentum $\vec{p}_b$ and energy $\varepsilon_b$. The amplitude $A(\vec{p}_a, \vec{p}_b)$ is again given by formula (3) or (4), but now $\varepsilon_a = \varepsilon_b + E_b$, where $E_b$ is the energy carried away by additional particles. As the energy $\varepsilon_a$ does not correspond to the momentum of the scattered particle $\vec{p}_b$, there arises an additional difference of the scattering point from a "full and equal" new source.

However, the above differences have no effect on pair correlations of identical particles which occur just as in the situation in which the points of elastic and even inelastic scattering are the most real sources of particles. For clarity the proof is accompanied by the figure which shows the discussed events schematically (fig. 1). The one-particle sources emitting pions at $t_1$ and $t_2$ are placed at points $\vec{r}_1$ and $\vec{r}_2$ and the massive scatterers at points $\vec{r}_3$ and $\vec{r}_4$. Additional particles which carry away energies $E_3$ and $E_4$, respectively, are produced in the general case besides identical pions with momenta and energies $(\vec{p}', \varepsilon')$ and $(\vec{p}'', \varepsilon'')$. The mode in which the first of the pions is emitted from the point $\vec{r}_3$ and the second from $\vec{r}_4$ is presented on the left hand of the figure, the crossing mode on the right side.

The momenta and energies of primary pions which, generally speaking, do not coincide with ( $\vec{p}'$, $\varepsilon'$) and ( $\vec{p}''$, $\varepsilon''$) are also given in the figure. In particular, from the energy conservation law we have the following equalities:

$$\varepsilon_1' = \varepsilon' + E_3, \quad \varepsilon_2'' = \varepsilon'' + E_4, \quad \varepsilon_1'' = \varepsilon'' + E_3, \quad \varepsilon_1'' = \varepsilon' + E_4 \tag{5}$$

Using formula (4), the two-particle amplitude of the process can be expressed as

$$A(\vec{p}', \vec{p}'') \sim \left\{ e^{i(\tau_{13} p_1' + t_1 \varepsilon_1' - \vec{\tau}_3 \vec{p}')} \cdot e^{i(\tau_{24} p_2'' + t_2 \varepsilon'' - \vec{\tau}_4 \vec{p}'')} + \right. \\ \left. + \; e^{i(\tau_{13} p_1'' + t_1 \varepsilon'' - \vec{\tau}_3 \vec{p}'')} \cdot e^{i(\tau_{24} p_2' + t_2 \varepsilon_2' - \vec{\tau}_4 \vec{p}')} \right\} \tag{6}$$

The probability of the process is

$$W \sim 1 + \cos\delta \tag{7}$$

with the phase

$$\delta = \tau_{13}(p_1' - p_1'') + t_1(\varepsilon_1' - \varepsilon_1'') + \tau_{24}(p_2'' - p_2') + t_2(\varepsilon_2'' - \varepsilon_2') - (\vec{\tau}_3 - \vec{\tau}_4)(\vec{p}' - \vec{p}'') \tag{8}$$

The momenta $\vec{p}_1'$ and $\vec{p}_1''$ are close in value and parallel, the same refers to $\vec{p}_2''$ and $\vec{p}_2'$. As known, under these conditions the following equalities are valid:

$$p_1' - p_1'' = \frac{\varepsilon_1' - \varepsilon_1''}{V_1}, \qquad p_2'' - p_2' = \frac{\varepsilon_2'' - \varepsilon_2'}{V_2} \tag{9}$$

where $V_1$ and $V_2$ are the velocities of primary pions emitted from the points $\vec{r}_1$ and $\vec{r}_2$. Using (9), we obtain

$$\tau_{13}(p_1' - p_1'') + t_1(\varepsilon_1' - \varepsilon_1'') = \left(t_1 + \frac{\tau_{13}}{V_1}\right)(\varepsilon_1' - \varepsilon_1'') = t_3(\varepsilon_1' - \varepsilon_1''), \tag{10}$$

$$\tau_{24}(p_2'' - p_2') + t_2(\varepsilon_2'' - \varepsilon_2') = \left(t_2 + \frac{\tau_{24}}{V_2}\right)(\varepsilon_2'' - \varepsilon_2') = t_4(\varepsilon'' - \varepsilon'),$$

where $t_3 = t_1 + \frac{\tau_{13}}{V_1}$ and $t_4 = t_2 + \frac{\tau_{24}}{V_2}$ are the "moments of arrival" of the corresponding pions at the scattering points $\vec{r}_3$ and $\vec{r}_4$. Further, from equalities (5) we conclude that

$$\varepsilon_1' - \varepsilon_1'' = \varepsilon' - \varepsilon'' , \quad \varepsilon_2'' - \varepsilon_2' = \varepsilon'' - \varepsilon' \qquad (11)$$

Finally, for the probability of the process we get

$$W(\vec{p}', \vec{p}'') \sim 1 + \cos\{(\vec{r}_3 - \vec{r}_4)(\vec{p}' - \vec{p}'') - (t_3 - t_4)(\varepsilon' - \varepsilon'')\} \qquad (12)$$

If independent one-particle sources generating identical pions with momenta and energies ( $\vec{p}'$, $\varepsilon'$) and ( $\vec{p}''$, $\varepsilon''$) were placed at points $\vec{r}_3$ and $\vec{r}_4$ instead of scatterers, the two-particle amplitude would be written in the form similar to (2) and the final result would be exactly the same as (12). As would be expected, the conventional viewpoint is true in the above-considered simplest situation.

However, in more complex cases the presence of scattering can lead to the change of the picture of pair correlations of identical pions. As assumed previously, the wave emitted at point $\vec{r}_1$ is scattered only at point $\vec{r}_3$ and the wave from point $\vec{r}_2$ only at point $\vec{r}_4$. However, the conditions are possible in which each of these waves is scattered at both points $\vec{r}_3$ and $\vec{r}_4$ . Then the height of the correlation maximum can be shown to decrease, i.e. the factor $\lambda < 1$ appears in front of the interference term in formula (12). This question is specially considered in paper/1/. Some analogy with interference correlations for multiparticle sources of identical pions arises here (see/1-4/).

It should be also noted that interference correlations in the pairs of unidentical pions may arise from scatterings. Let some two-particle source of a ( $\pi^+\pi^-$ ) pair, e.g. $\rho^0 \to \pi^+\pi^-$, be placed at point $\vec{R}$ . Two modes leading to the appearance of $\pi^-$-meson with momentum $\vec{p}'$, and $\pi^+$-meson with momentum $\vec{p}''$ after their scatterings at points $\vec{r}_1$ and $\vec{r}_2$ are schematically shown in fig. 2. The interference of the corresponding two amplitudes results in the correlations between $\pi^+$ and $\pi^-$ which are similar in their nature to those characteristic of the pairs of identical pions.

The author expresses his gratitude to V.L.Lyuboshitz for significant remarks.

REFERENCES

1. Podgoretsky M.I., JINR, P2-81-325, Dubna, 1981.
2. Lednicky R. et al., JINR, E2-82-725, Dubna, 1982.
3. Lednicky R. et al., YaF 38, 251 (1983).
4. Paul H. Rev. Mod. Phys. 58, 209 (1986).

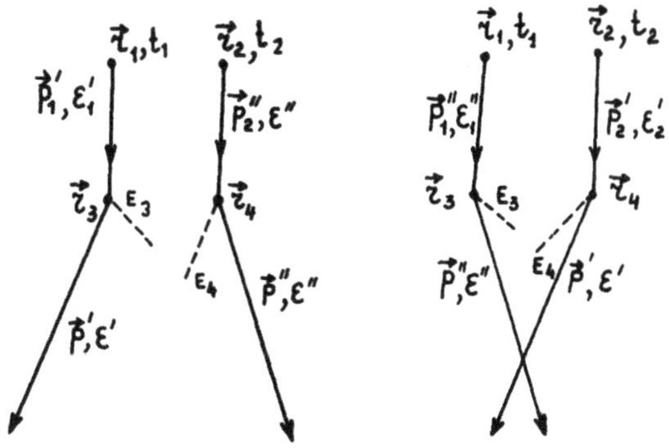
$\vec{r}_1, t_1$
$\vec{r}_2, t_2$
$\vec{p}_1', \varepsilon_1'$
$\vec{p}_2'', \varepsilon''$
$\vec{r}_3$
$\varepsilon_3$
$\vec{r}_4$
$\varepsilon_4$
$\vec{p}'', \varepsilon''$
$\vec{p}', \varepsilon'$
$\vec{r}_1, t_1$
$\vec{r}_2, t_2$
$\vec{p}_1'', \varepsilon_1''$
$\vec{p}_2', \varepsilon_2'$
$\vec{r}_3$
$\varepsilon_3$
$\vec{r}_4$
$\vec{p}'', \varepsilon''$
$\varepsilon_4$
$\vec{p}', \varepsilon'$

Fig. 1

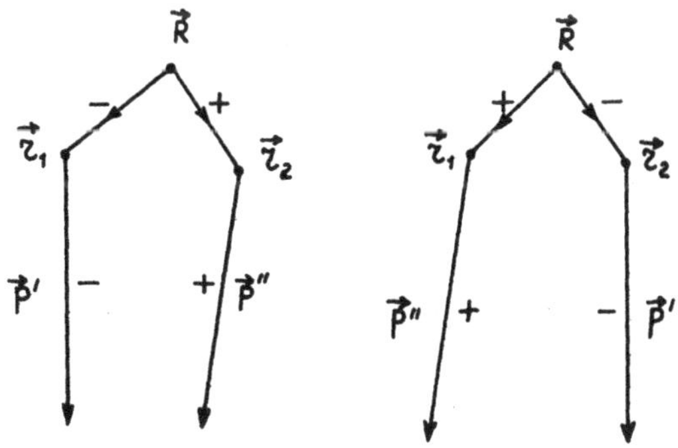
$\vec{R}$
$-$
$+$
$\vec{r}_1$
$\vec{r}_2$
$\vec{p}'$
$-$
$+$
$\vec{p}''$
$\vec{R}$
$+$
$-$
$\vec{r}_1$
$\vec{r}_2$
$\vec{p}''$
$+$
$-$
$\vec{p}'$

Fig. 2

# Bose-Einstein Correlations and Quantum Statistics

R.M. Weiner

Physics Department, University of Marburg

Wave function approach and quantum statistics; two types of correlations.
The discovery of Bose-Einstein correlations (BEC) in particle physics due to G.G. Goldhaber, S. Goldhaber, W. Lee and A. Pais came about in connection with the observation of differences in the distribution of the opening angles of identical and unidentical pion pairs [1]. It was therefore natural at that moment to interpret this effect by constructing the wave function $\phi$ of the pair of pions and considering the difference which occurs by the symmetrization of $\phi$, due to Bose statistics:

$$\phi^{unlike}(1,2) = \phi_1(1)\phi_2(2) \tag{1}$$

$$\phi^{like}(1,2) = \frac{1}{\sqrt{2}}[\phi_1(1)\phi_2(2) + \phi_1(2)\phi_2(1)] \tag{2}$$

The two-particle correlation function for identical bosons is then

$$C_{(2)}(k_1, k_2) \sim \int \rho(x_1)\rho(x_2)|\phi^{like}|^2 dx_1 dx_2 \tag{3}$$

where $\rho(x)$ is the space-time distribution of the source and $k_i$ are the momenta of the observed particles. In this way $C_2(k_1, k_2)$ depends only on the length (time) scales which enter $\rho$. If we parametrize in particular $\rho$ in the form

$$\rho(x) = e^{-x^2/R^2} \tag{4}$$

where $R$ is the radius/lifetime of the source, then the BEC depends on $R$ only. This is the "traditional" approach to BEC. The "modern" and more general approach to BEC is based on quantum statistics (QS). As is well known, QS contains automatically the appropriate statistics for identical particles, which enters through the commutation or anticommutation rules of the creation and annihilation operators. Thus for bosons we have

$$[a_k, a_{k'}] = [a_k^+, a_{k'}^+] = 0 \tag{5}$$

$$[a_k, a_{k'}^+] = \delta_{kk'} \tag{6}$$

where [ ] denote the commutation. $a_k$ are the annihilation operators of a boson of momentum $k$. In terms of these $a_k$ one defines the coherent states $|\alpha_k>$

$$a_k|\alpha_k>=\alpha_k|\alpha_k> \tag{7}$$

and in terms of these coherent states the entire QS formalism is constructed. (Any state can be expanded in terms of the set of coherent states.)

In quantum statistics, on top of the geometrical length scale $R$ a new length scale appears, namely the correlation (coherence) length $L$. It is defined through the correlator for the (chaotic) fields $\pi_{ch}$

$$C(x_1,x_2)=<\pi_{ch}(x_1)\pi_{ch}(x_2)>exp(-|x_1-x_2|^2/L^2) \tag{8}$$

where $x$ is again the space-time coordinate and the brackets <> denote the ensemble average. These chaotic fields describe the fluctuations of the system and it is through them (and the associated correlation length $L$) that the link between two-body and many-body correlations is achieved. (In standard QS all higher-order correlations can be expressed through two-body correlations.)

The appearance of the two length scales $R$ and $L$ in BEC is exemplified by eq. (26) in ref. [2], from which follows that for purely chaotic fields ($p(0)=1$) the second order BEC $C_2(k_1,k_2)$ reads:

$$C_2(k_1,k_2)=1+exp[-\frac{1}{4}(R^2-R_L^2)(k_1-k_2)^2] \tag{9}$$

where

$$R_L^2=\frac{R^2L^2}{R^2+L^2} \tag{10}$$

The "traditional" wave function approach is just a particular case of QS when $L=0$.

Coherence and soft pions.

Another new element introduced by QS is the concept of coherence associated with coherent states. One of the properties of coherent fields $\pi_c$ is the fact that the two-particle correlation factorizes

$$<\pi_c(x_1)\pi_c(x_2)>=<\pi(x_1)><\pi(x_2)> \tag{11}$$

When one considers the general case of a superposition of coherent and chaotic fields

$$\pi=\pi_c+\pi_{ch} \tag{12}$$

the BEC contains now two exponentials rather than one (cf. eq. (26) of ref. [2]

$$\begin{aligned} C_2(k_1,k_2) &= 1+2p(0)[1-p(0)exp[-\frac{1}{4}(R^2/R_L^2)(k_1^2+k_2^2)]\cdot \\ &\cdot\ exp[-\frac{1}{8}(R^2/R_L^2)(k_1-k_2)^2/A(k_1)A(k_2)+ \\ &+\ p^2(0)exp[-\frac{1}{4}(R^2-R_L^2)(k_1-k_2)^2]/A(k_1)A(k_2) \end{aligned} \tag{13}$$

where

$$A(k_i) = p(0) + [1 - p(0)]exp[-\frac{1}{2}(R^2 - R_L^2)k_i^2]$$

and the chaoticity $p(k)$ is given by

$$p(k) = < |\pi_{ch}(k)|^2 > /[< |\pi_{ch}(k)|^2 >< |\pi_c(k)|^2 >]. \quad (14)$$

Thus (partial) coherence modifies not only the intercept of $C_2(k_1, k_2)$ at $k_1 - k_2 = 0$, but also the form of $C_2(k_1, k_2)$ for $k_1 - k_2 \neq 0$.

Eq.(13) has the remarkable property that the interference term between the chaotic and coherent fields (the second term on the r.h.s.) is strongly damped except for soft particles ($k^2 \sim m^2$). This feature may facilitate the detection of the two exponential feature of BEC. It has also the consequence that large momentum ($p_T$) particles are associated with a bigger effective radius (the second exponential in (13)) than the small $p_T$ particles. Hints for such a behaviour have been reported by the Na35 collaboration for O+Au reactions at 200 GeV/nucleon [3].

Some prejudices against coherent states.

Coherent states constitute an extremely powerful and useful formalism for the description of many-particle production processes. They are the alpha and omega of quantum statistics and found applications in such various fields as quantum mechanics and field theory, condensed matter physics, thermodynamics, atomic physics, nuclear physics etc. [4].

Unfortunately this subject has not always got among particle physicists the attention which it deserves. This explains some misunderstandings or prejudices like the following:

1. If coherence may be seen at particle level, it can hardly occur for a much bigger system like the nucleus. If this argument would hold, we could never have superconductors, superfluids, lasers, where the coherence extends over macroscopic systems!

2. Conservation laws like e.g. charge conservation prevent the "literal" use of coherent states for identical charged particle interferometry, because for coherent states the number of particles is undefined [5]. Statements of this kind oversee the fact that conservation laws apply only to the global system, and not to parts of it where coherent states are used.

3. "Correlations due to the symmetrization of wave functions are absent for the pions in the same coherent state."[5],[6]. Such statements are slightly misleading as follows from the considerations made above, where it was shown that coherent states lead automatically to the symmetrization of the wave functions. The point is more subtle in the sense that in a coherent state the

phase of the wave function (field) is well defined and this makes coherent fields satisfy eq. (11) rather than eq. (8).

Attempts are sometimes made to rederive results for BEC, obtained in the QS framework by inventing particular configurations of sources [5], [6]. In a certain sense these attempts, although quite ingenious, remind one of attempts to explain the Balmer formula, without the use of quantum mechanics, or the motion of planets without the Keppler equations. Not only is the QS formalism the correct and general formalism [1] to be used but it is also the simpler one. It relates, among other things, in a straightforward way correlations with multiplicity distributions $P(n)$ predicting certain effects, which could have hardly be expected or predicted without this formalism. As an example we consider the expression for the factorial cumulants of $P(n)$:

$$\mu_{(r)} = < n >^r \, [(r-1)! p^r B_r(\frac{Y}{\xi}) + r! p^{r-1}(1-p)\bar{B}_r(\frac{Y}{\xi})] \tag{15}$$

Here $Y$ is the width of the interval in $k(y)$ space over which the moment of order $r$ is measured, $\xi$ the correlation length in the corresponding space ($\xi = \tilde{R}^{-1}$ where $\tilde{R}$ is the associated "radius" (cf. eq. (3) of ref.[2]) and $B_r, \bar{B}_r$ are universal functions. The form of these functions, which are determined by QS, as such that $\mu_r$ increases with the decrease of $Y$. In particular, if $Y$ refers to rapidity, then eq. (15) predicts that the multiplicity distribution broadens when narrower and narrower symmetric (with respect to the origin) rapidity windows are considered [9]. This prediction has been confirmed first by the UA-5 collaboration [10] with charged particle data and then by the Na-22 collaboration with identical particle data [11]. (As a matter of fact, quantum statistics and therefore this prediction applies really to identical particles; it can be expected to hold also for charged particles, provided charge correlations are not too strong.)

Eq. (15) exemplifies quite clearly the relation between Bose-Einstein correlations and multiplicity distributions, because the same parameters $p$ and $\xi$ which characterize BEC, determine also $\mu_r$, and thus $P(n)$. This consequence of QS distinguishes it from all other approaches to BEC and $P(n)$ which in general treat separately these two physical quantities. While in quantum optics the relationship between mutiplicity distributions and Bose- Einstein correlations has been tested experimentally, in particle and nuclear physics this is not yet the case, because of lack of adequate data [12]. The importance of such tests can hardly be overestimated. Among other things, the link between BEC and $P(n)$ could provide a simple explanation for the observation by the Na-35 collaboration of different transverse radii in different rapidity intervals [13].

---

[1]The QS formalism contains both standard coherent states, as well as squeezed coherent states (cf. e.g. [7] whee the implications of squeezed states are discussed). Furthermore it applies both to BEC with an undefined number of particles $n$ as well as to BEC with fixed $n$ [8].

# References


[1] G. Goldhaber, The GGLP Effect from 1959 to 1989, in: Proceedings of LESIP I, Local Equilibrium in Strong Interaction Physics, eds. D.K. Scott and R.M. Weiner, World Scientific 1985, p. 115.

[2] I. Andreev and R.M. Weiner, these Proceedings.

[3] J. Bartke, in: Proceedings of the Workshop on Particle Correlations and Interferometry in Nuclear Reactins (CORINNE), Nantes 1990, to be published by World Scientific.

[4] Cf. e.g. "Coherent States", J.R. Klauder and B.S. Skagerstam, World Scientific 1985. For more recent applications in particle physics cf. also R.M. Weiner in: "Multiparticle Production" (Proceedings of the Perugia Workshop), p. 231, eds. R. Hwa, G. Pancheri and Y. Srivastava, World Scientific 1989.

[5] V.L. Lyuboshitz, these Proceedings.

[6] R. Lednicky, V.L. Lyuboshitz and M.I. Podgoretzky, Sov.J.Nucl. Phys. 38, 153 (1983).

[7] A. Vourdas and R.M. Weiner, Phys. Rev. 38D, 2209 (1988).

[8] G.N. Fowler, E.M. Friedlander, X.C. He, C.C. Shih and R.M. Weiner, to be published.

[9] G.N. Fowler and R.M. Weiner, Phys. Rev. D17, 3118 (1978).

[10] G.J. Alner et al., Phys. Rep. 154, 247 (1987).

[11] M. Adamus et al., Z. Phys. C37, 215 (1988).

[12] Cf. also E.M. Friedlander, X.C. He, C.C. Shih and R.M. Weiner, University of Marburg preprint 1990.

[13] R.M. Weiner, Physics Letters 232, 278 (1989).

# EXTRACTING PHYSICS FROM TWO-PARTICLE CORRELATIONS

William A. Zajc
Physics Department
Columbia University
New York, NY 10027

**Abstract**

The extraction of physics information from two-pion correlation functions is presented. The choice of fitting methods and fit parameterizations are discussed in the context of a simple set of "test" data. The implications of exponential parameterizations for the correlation function are presented in an appendix.

## 1 Introduction

The Conference on Correlations and Multi-Particle Production (CAMP) provided a welcome opportunity to reflect on progress in the field of hadron interferometry. Motivated by the distinguished assemblage of names appearing with me on the final summary panel, I created the below rather fanciful capsule "history" of Bose-Einstein correlations: In this table, both the dates and the authors are rough

| Year | Author | Parameter(s) |
|---|---|---|
| 1960 | Goldhaber | $R$ |
| 1970 | Podgoretskii | $R, \tau$ |
| 1980 | Weiner | $R, \tau, \lambda$ |
| 1990 | (assorted) | $R_L, R_T, \tau, \lambda$ |
| 2000 | ????? | $R_L, R_T, \tau, \lambda, u_S$? |

approximations only. The purpose of this exercise is to demonstrate that progress in this field has been characterized in a rather slow but steady rate with which quantitative information can be extracted from correlation functions. Early efforts were restricted to a radius $R$, and perhaps a lifetime $\tau$. In 1978, both for theoretical and experimental reasons, a "coherence parameter" $\lambda$ was introduced. In recent years, various authors have attempted to measure the source shapes, as given by $R_L$ and $R_T$, where $L$ and $T$ refer to Longitudinal and Transverse directions, either with respect to the collision axis or the observation axis. Finally, increasing $\sqrt{s}$ (and in the case of asymmetric heavy ion collisions, increasing ambiguity) may motivate future investigators to explore the frame-dependence of their correlation functions, and hence deduce the source velocity $u_S$.

Given such a trend to more sophisticated analyses, it is only natural to assume that the method(s) for fitting a correlation function to a given functional form are both well-known and well-established. Based on anecdotal information, various conversations, and even published remarks in the literature, it is my judgement that this is not the case. Therefore, I have decided to provide a "worked example" of fitting a correlation function. This may prove beneficial to three audiences:

1.) Beginners who wish to perform such an analysis (esp. students).

2.) Theorists, who wish to propose a measurement (and who wish to make some independent judgement as to its practicality).

3.) Any "non-expert" who must determine the feasibility of a proposed experiment (e.g, Program Advisory Committee members).

Thus, this contribution is both rather technical and rather trivial. It is technical in the sense there is a detailed discussion of various fit criteria. It is trivial in the sense that these results are well-known to those who have actually performed such analyses. Of somewhat more general interest may be the Appendix, which discusses in detail the implications of an exponential form for the correlation function.

## 2 Fitting a Correlation Function– A Worked Example

In this section I will work through the fitting procedure for the simplest possible *one-dimensional* correlation function given by

$$C_2(q) = 1 + \lambda e^{-q^2R^2} \quad , \tag{1}$$

where $q \equiv |\vec{q}|$. Technically, what is meant by "one-dimensional" is "isotropic" in that this correlation function corresponds to a source density $\rho(\vec{r})$ given by

$$\rho(\vec{r}) = \frac{1}{[2\pi R^2]^{\frac{3}{2}}} e^{-r^2/2R^2} \quad . \tag{2}$$

Starting from a set of events with like-sign pion pairs, one first forms the distribution $A(q)$, the *Actual* number of events with relative momentum $q$. This is to be compared to a *Background* distribution $B(q)$, which is constructed to contain the experimental acceptance and biases, but not the Bose-Einstein correlation. Typically this is formed from fake pion pairs where each pion within the pair is taken from a different event in the data sample.

There are several choices for the precise method whereby $A$ is compared to $B$. The goal, of course, is to vary the parameters $\lambda$ and $R$ in Eq. 1 to obtain the best agreement with the experimental correlation function given by $A(q)/B(q)$. Assume that the distributions have been binned in $N$ $q$-bins with a bin index "$i$". Naively, simply minimizing the chi-squared given by

$$\chi_C^2 \equiv \sum_{i=1}^{N} \frac{(C_i - A_i/B_i)^2}{\sigma_{A_i/B_i}^2} \tag{3}$$

should suffice (here $\sigma_{A_i/B_i}$ is the error on the quantity $A_i/B_i$). However, this is seldom the case, since the errors on the ratio of two Poisson-distributed integers (the bin contents $A_i$ and $B_i$) are not themselves Poisson-distributed, but show a long tail, particularly for the ratio of small numbers. To avoid this, it is preferable to minimize the quantity

$$\chi_A^2 \equiv \sum_{i=1}^{N} \frac{(A_i - C_i \cdot B_i)^2}{\sigma_{A_i}^2 + \sigma_{C_i \cdot B_i}^2} \quad . \tag{4}$$

Even this procedure has difficulties for small values of $A_i$ and $B_i$. (It is worth noting that the trend to analyzing correlation data in increasing number of dimensions always parallels the improved statistics of modern experiments, so that the problem of small statistics never really goes away.)

It is important to keep in mind that fitting correlation data really consists of two, somewhat independent steps:

A. *Finding* a good fit to the data

and

B. *Estimating* the errors on the found parameters.

Any procedure, including guessing, can be used to achieve step A. However, in physics, the estimated (statistical) errors are by convention quoted as the equivalent Gaussian one-standard deviation values. If the fitting quantity has a true chi-squared distribution, then the solution is straightforward. However, the chi-squared distribution for $n$ degrees of freedom is the the result of convoluting $n$ independent Gaussian variates, not $n$ independent Poisson distributions. When the number of entries per bin becomes sufficiently small that the Gaussian distribution is no longer a good approximation to the Poisson, it becomes necessary to use the Principle of Maximum Likelihood (PML) method, ***which requires that the Background distribution $B_i$ is well known.*** In that case, the actual number $A_i$ in a bin should be Poisson-distributed about the expected number $\bar{A}_i \equiv C_i \cdot B_i$, so that the probability of a given set of $A_i$'s is given by

$$\mathcal{L} = \prod_{i=1}^{N} \frac{\bar{A}_i^{A_i}}{A_i!} e^{-\bar{A}_i} \quad . \tag{5}$$

One varies the parameters in $C_i$, and thus in turn the values of $\bar{A}_i$, in order to minimize (for convenience) the negative logarithm of this function, thereby maximizing the probability that the observed counts $A_i$ come from the correlation function $C_i$ with that set of parameters. Thus, the quantity that corresponds to $\chi^2$ in Eqs. 3 and 4 is given by

$$PML_0 = \sum_{i=1}^{N}[\bar{A}_i - A_i \ln \bar{A}_i] = \sum_{i=1}^{N}[(\bar{A}_i - A_i) - A_i \ln \frac{\bar{A}_i}{A_i}] \quad , \tag{6}$$

where in the last form I have added a constant term that insures that this expression reduces to (half) the correct value for chi-squared (Eq. 4 with $\sigma_{B_i} = 0$) in the limit $(\bar{A}_i - A_i)/A_i \to 0$.

Finally, the attentive reader may be somewhat uneasy at this point, in that the PML explicitly assumes that the $B_i$ are well-known, while in practice they often have statistical uncertainties comparable to the $A_i$. If this is the case, I am unaware of any corresponding extension of the PML to handle what really amounts to a hypothesis test: Are the $A_i$ consistent with $B_i$ multiplied by a function of $q(i)$? However, there are two ways to circumvent this problem:

1.) Create as many background pairs as possible, mixing each track with several other tracks. This method was used, e.g., in Ref. [1]. As first noted there, the effectiveness of this procedure depends critically on the bin-size and the size of the phase space bite. It would appear to be most worrisome for small-aperture measurements, while negligible in large-aperture experiments[2]. Secondly, it is not clear that this procedure is even necessary, in that (see below) for any reasonable data-set the $\chi^2_A$ variable is just as good a fit criteria as PML.

2.) One can attempt to augment the PML expression Eq. 6 so that in the limit of large $A_i$ it reduces to $\chi^2_A$. It is straightforward to show that such a form (denoted by $PML_A$) is given by

$$PML_A = \sum_{i=1}^{N}[(\bar{A}_i - A_i) - (A_i + C_i \cdot B_i)\ln(\frac{\bar{A}_i + C_i \cdot B_i}{A_i + C_i \cdot B_i})] \quad . \tag{7}$$

Note that the only theoretical justification I have given for this expression is that it has Eq. 4 as a limiting form, while the usual PML follows directly from the assumption of Poisson fluctuations of the $A_i$ about the known $C_i \cdot B_i$.

In spite of my reservations concerning the $PML_A$, expression, it will also be tested versus more conventional approaches in what follows.

Armed with these tools, we can now face the task of fitting correlation data. Typical "data" will always be generated assuming the true values in Eq. 1 are $\lambda = 0.6$ and $R = 4$fm. The fits will be performed over the range $0 < |\vec{q}| < 200$ MeV/c in 20 bins of size 10 MeV/c. I will provide the results from all four fitting methods ($\chi^2_C$, $\chi^2_A$, PML, and $PML_A$). As a (completely artificial) calibration point, I begin with a correlation function that has "statistical" errors, but *no* fluctuations (Fig. 1a). The errors are distributed as if resulting from a full-phase space acceptance (that is, the number of counts in a given $q$-bin is proportional to $q^2\Delta q$) with $\sim 23K$ pairs in the region $|\vec{q}| < 200$MeV/c. As can be seen from Table 2, all methods accurately

**Table 2**

Gaussian fit results for 23K pairs (no fluctuations)

| Parameter | $\chi^2_C$ | $\chi^2_A$ | PML | $PML_A$ |
|---|---|---|---|---|
| $R$ | 4.00±0.66 | 4.00±0.66 | 4.01±0.44 | 4.00±0.64 |
| $\lambda$ | 0.60±0.21 | 0.60±0.21 | 0.60±0.13 | 0.60±0.20 |

**Table 3**

Gaussian fit results for 23K pairs (with fluctuations)

| Parameter | $\chi^2_C$ | $\chi^2_A$ | PML | $PML_A$ |
|---|---|---|---|---|
| $R$ | 3.91±0.64 | 3.77±0.42 | 3.77±0.24 | 3.73±0.37 |
| $\lambda$ | 0.55±0.18 | 0.57±0.16 | 0.56±0.04 | 0.57±0.15 |
| $\chi^2$ | 17.0 | 17.2 | 35.9 | 17.2 |

converge to the true parameters, but have different error estimates. The results of allowing fluctuations may be seen in Fig. 1b and Table 3. Here the value of "chi-squared", which is actually the final values of the expressions in Eqs. 3, 4, 6 and 7, is also listed. Since we are fitting 20 bins with a two-parameter fit, the expected value of $\chi^2$ is 18. The large values of $\chi^2$ associated with the PML fits provides the clue to smaller errors returned by this method: Since the PML does not account for the errors in the background, it "sees" fluctuations twice as large it "expects". Thus, when an error is calculated (by allowing $\chi^2$ to vary by one), this happens twice as fast for the $PML_A$ method, leading to errors that are typically $\sqrt{2}$ smaller than those returned by the $\chi^2_A$ or the $PML_A$ method. (Note also that now, with realistic fluctuations in the points, the $\chi^2_C$ method *over*-estimates the errors.)

The 23K pairs in this data set are seen to provide about a 10% accuracy in $R$ and 25% in $\lambda$. The results of increasing the statistics by a factor of 10 are shown in Fig. 1c and in Table 4. It is readily seen that the trends observed previously continue (under-estimation of errors by the PML), and that the estimated errors for all methods do decrease by roughly $\sqrt{10}$, as they should. Note also that now (and only now!) are the errors from the $\chi^2_C$ method consistent with the $\chi^2_A$ and $PML_A$ results.

Emboldened with the convergence of our fitting methods, we might hope that the 256K pairs in our sample would be sufficient to precisely determine the shape of the source (after all, that was the original goal of this effort). To this end, we can try

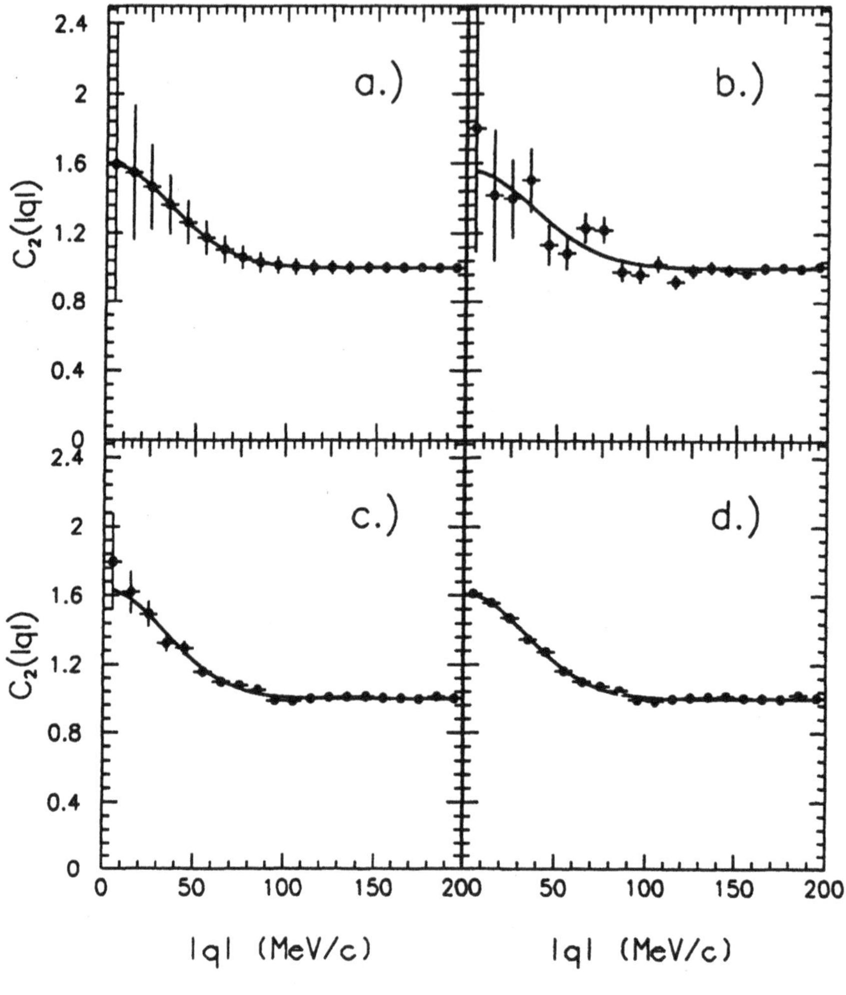


Figure 1: Fits to various quality correlation data. a.) 23K pairs, with errors, but no fluctuations. b.) 23K pairs, with both errors and corresponding fluctuations distributed as for an untriggered experiment. c.) As in b., but for 256K pairs. d.) 256K pairs with a uniform distribution of events in the interval $0 < q < 200$ MeV/c. See text for further details

**Table 4**

**Gaussian fit results for 256K pairs (with fluctuations)**

| Parameter | $\chi^2_C$ | $\chi^2_A$ | PML | $PML_A$ |
|---|---|---|---|---|
| $R$ | 4.08±0.12 | 4.07±0.13 | 4.08±0.08 | 4.07±0.13 |
| $\lambda$ | 0.63±0.04 | 0.63±0.03 | 0.63±0.02 | 0.63±0.11 |
| $\chi^2$ | 15.7 | 15.8 | 32.6 | 15.8 |

**Table 5**

**Exponential fit results for 256K pairs**

| Parameter | $\chi^2_C$ | $\chi^2_A$ | PML | $PML_A$ |
|---|---|---|---|---|
| $R_E$ | 4.00±0.24 | 4.01±0.13 | 3.94±0.09 | 4.01±0.19 |
| $\lambda_E$ | 1.34±0.01 | 1.37±0.09 | 1.31±0.05 | 1.37±0.13 |
| $\chi^2$ | 24.5 | 24.0 | 50.9 | 23.9 |

to determine if our data *require* the Gaussian parameterization. As an alternative, we can consider the exponential form

$$C_2(|\vec{q}|) = 1 + \lambda_E e^{-2|\vec{q}|R_E} \quad , \tag{8}$$

where the factor of two appearing in the argument of the exponential has been determined somewhat empirically (see Appendix). This form is motivated by recent observations (see, e.g., [3]) that microscopic models for pion production show a preference for an exponential behavior of the correlation function. The results of fitting to Equation 8 are given in Table 5, and a comparison of the fit to a Gaussian one is shown in Fig. 2. While the $\chi^2$ for the exponential fit is indeed higher than for the fit to the Gaussian distribution (24.0 versus 15.8), these values (and an examination of Fig. 2) indicate that we can exclude the exponential from (true) Gaussian one only at the 90% level, even though we accumulated 256K pairs!

This conclusion is rather unsavory for those who would like to use two-particle interferometry to explore subtle aspects of pion production dynamics. Note in particular that the value of $\lambda$ for the two parameterizations differs by more than a factor of two! However, it is important to remember that all pion pairs are not created equal. Specifically, the 256K pairs used in the studies to this point have been distributed according to their natural phase-space distributions. This necessarily leads to the small number of counts in the lowest $q$-bins. One can get around this

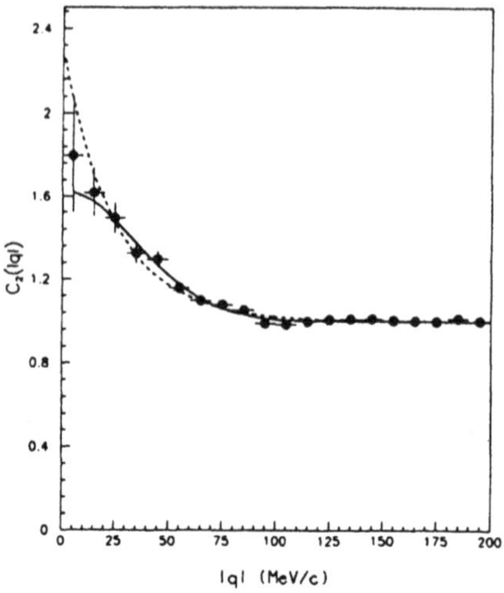


Figure 2: A comparison of fits to the 256K-pair sample for exponential and Gaussian parameterizations. See text for further details.

by *triggering* on the those events with low $q$. In such experiments, one can hope to achieve a uniform distribution of statistics across the region of interest. A typical example from the E802 experiment[4] is shown in Fig. 3. We can take our test data sample of 256K pairs, and rearrange the distribution of errors to be uniform over the interval $0 < |\vec{q}| < 200$ MeV/c. This is the key requirement to be able to make unambiguous statements about the source parameters (for this simple case). As shown in Table 6 and Fig. 4, the improved errors on the points at low-$q$ now establish unambiguously that the Gaussian is the preferred parameterization.

Let me try to summarize the results from this study:

- In the presence of background fluctuations, Principle of Maximum Likelihood fits underestimate the errors.
- For small data sets, the naive $\chi^2_C$ (Eq. 3) overestimates the errors.
- The $\chi^2_A$ parameter (Eq. 4) and the augmented PML (Eq. 7) provide similar (and correct) parameter determination and error estimation for all reasonable data sets.
- For typical data sets, the dominant source of "error" in determining the extracted parameters results not from statistical errors, but from ambiguities in the form of the preferred fitting function.

Thus, even for a well-controlled test-sample, fitting the correlation function is seen to be a somewhat subtle enterprise. It would be naive to think that actual data are any easier to deal with.

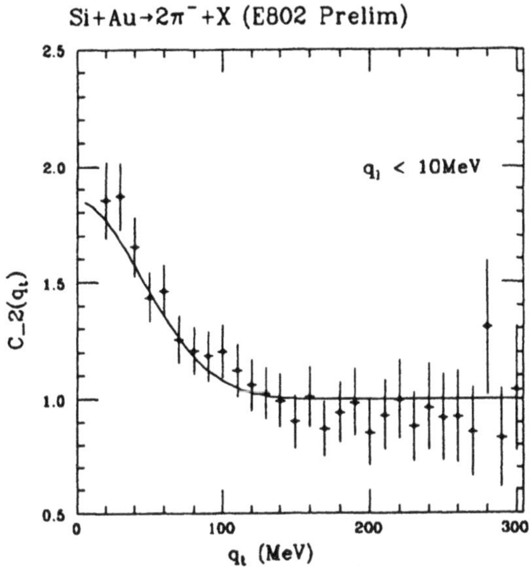


**Figure 3:** The correlation function in $q_T$ in the interval $0 < q_L < 10$MeV/c for Si+Au collisions.

**Table 6**

Exponential vs. Gaussian fit parameters for
256K pairs uniformly distributed in $0 < q < 200$ MeV/c

| Parameter | $\chi^2_C$ | $\chi^2_A$ | PML | $PML_A$ |
|---|---|---|---|---|
| $R$ | 4.03±0.08 | 4.03±0.07 | 4.03±0.03 | 4.03±0.06 |
| $\lambda$ | 0.61±0.01 | 0.13±0.01 | 0.61±0.01 | 0.61±0.01 |
| $\chi^2$ | 15.2 | 15.3 | 31.3 | 15.3 |
| | | | | |
| $R_E$ | 2.88±0.08 | 2.89±0.05 | 2.82±0.04 | 2.87±0.03 |
| $\lambda_E$ | 0.82±0.02 | 0.83±0.01 | 0.81±0.01 | 0.83±0.01 |
| $\chi^2$ | 112.8 | 110.8 | 251.2 | 110.2 |

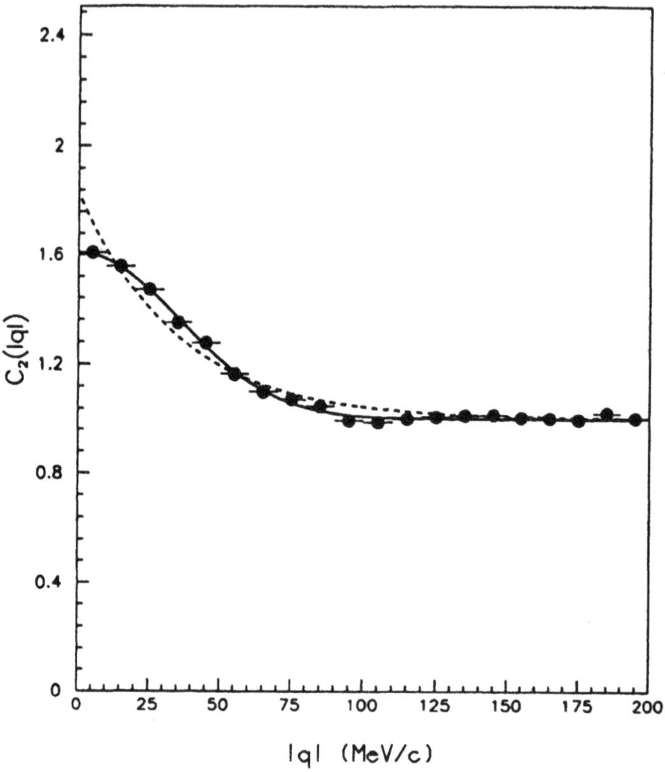


**Figure 4: A comparison of fits to the 256K-pair sample with a uniform distribution of events (roughly corresponding to those for a triggered experiment) for exponential and Gaussian parameterizations. See text for further details.**

# 3 Conclusions

In this paper, the simplest test case has required some effort in order to converge on the (known) correct solution. Let me now provide a brief, non-inclusive list of what sort of effects have *not* been included, the better to allow the reader to judge for him or herself the limitations in fits to *real* data sets:

- The effects of single-particle momentum resolution (which smears the relative momentum determination).
- The effect of two-particle track-finding biases and ambiguities (which can severely distort the value of the low-$q$ points).
- The effect of the residual correlation[1] (which requires weighting and/or correcting the background distribution).
- The correction for the two-pion mutual Coulomb interaction (which strongly affects the low-$q$ bins).
- The subdivision of the data set into many dimensions (which allows determination of source shapes, at the expense of having small statistics in many of the bins).

A natural reaction to such a list is to ask "Is there any hope for HBT measurements?" I believe that the answer is (resoundingly) "**Yes!**", but only with substantial (but worthwhile) effort. As models for the level of rigor required, consider the extraction of QCD-scaling violations or the Weinberg angle. It is a rare interferometry experiment indeed that can be held to the same level of rigor. Mixing that level of rigor with an appropriate amount of vigor can indeed lead us "from statistics to dynamics".

# 4 Acknowledgments

It is a pleasure to take this opportunity to acknowledge Richard Weiner for many years of encouragement and support. In addition, I would like to thank the organizers of CAMP for their support and patience.

## References

[1] W.A. Zajc *et al.* Two pion correlations in heavy ion collisions. *Phys. Rev.*, **C29**:2173, 1984.

[2] K. Kulka. *Bose-Einstein Interferometry in High-Energy Hadron-Hadron Collisions.* PhD thesis, University of Lund, 1987.

[3] T. Csorgo *et al.* Two pion correlations from spacer. *Central Research Institute for Physics, Budapest*, :, 1989.

[4] R. Morse *et al.* to appear in Proceedings of Quark Matter 90, 1990.

[5] G. Goldhbaber *et al.* Influence of Bose-Einstein Statistics on the Antiproton-Proton Annihilation Process. *Phys. Rev.*, **120**:300, 1960.

# A APPENDIX: Exponential vs. Gaussian Parameterizations

Most fits to correlation data have used variations on the Gaussian form presented in Eq. 1. This is motivated more by convenience than physical insight. Of particular usefulness is the factorizability of this form. However, it is important to ask what other forms are plausible, and, in particular, what forms are preferred by the data? Many authors have noted recently that the exponential form of Eq. 8 may better describe the data than the Gaussian one. The fundamental difference, of course, is that the Gaussian approaches the origin with zero slope, while the exponential continues to rise near $q = 0$. Thus, the accuracy of the lowest bins is critically important for the exponential form, while less so for the Gaussian. Since these low bins are the most problematic to both populate and to understand, to this date is hard to point to convincing evidence for an exponential shape. Nonetheless, it may be appropriate to state explicitly the nature of the source implied by an exponential form, and the prospects for measuring this via interferometry.

The source density corresponding to the exponential correlation function of Eq. 8 is given by

$$\rho_E(|\vec{r}|) = \frac{1}{\pi^2}\frac{R_E}{[R_E^2 + |\vec{r}|^2]^2} \quad . \tag{9}$$

We would first like to compare values of $R_E$ with the corresponding Gaussian $R$. The usual prescription of using equivalent root-mean-square radii fails for $\rho_E(|\vec{r}|)$. This is our first hint that the exponential and Gaussian *source* distributions are *very* different.

One might hope to circumvent this by using distributions with equal *most probable* radii. This is shown in Fig. 5, which is hardly convincing, either in coordinate or momentum space! The long tail of the exponential distribution in coordinate space extends well-beyond the relatively restricted region of the Gaussian. An examination of the two corresponding correlation functions (Fig. 5) implies that it would be trivial to discriminate between these two forms in relative momentum space as well.

However, we usually don't have *a prioiri* knowledge of the source distribution. Instead, we fit our correlation functions in order to infer it. If we have a correlation function with the "typical" 10% errors, then we will be unable to distinguish between an exponential form of radius $R_E$, and a Gaussian with radius $\sim \sqrt{8}R$. This is shown in Fig. 6b, which emphasizes that rather similar correlation functions can result from quite different spatial distributions (this effect was noted in the original GGLP paper[5]).

Finally, it is worth inquiring as to what sort of physical processes can result in a $\rho_E(|\vec{r}|)$ of the from given by Eq. 9. Suppose that the pions result from the decay of resonances or clusters, characterized by an exponential lifetime $\tau$. If the cluster has mass $m$ and momentum $p$, then the normalized probability to decay to pions at a radius $r$ is

$$\mathcal{P}_{prod}(p, r) = \frac{1}{8\pi}\left(\frac{m}{p\tau}\right)^3 e^{-mr/p\tau} \quad . \tag{10}$$

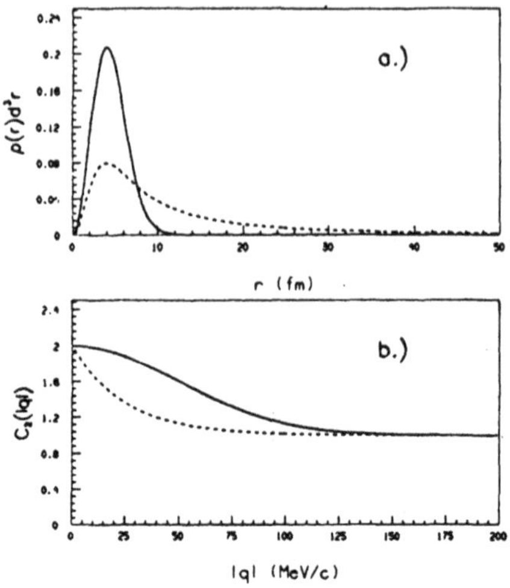


Figure 5: A three-dimensional Gaussian distribution (solid lines) is compared to an exponential one (dashed lines) with the same most probable radius of 4 fm a.) in coordinate space and b.) for the resulting correlation functions.

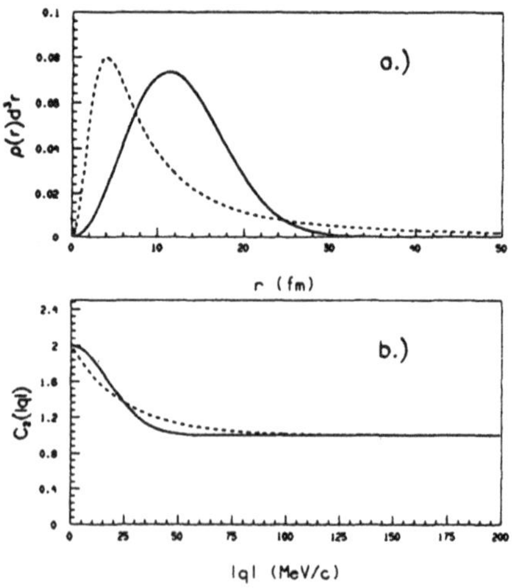


Figure 6: A three-dimensional Gaussian distribution (solid lines) is compared to an exponential one (dashed lines) with the Gaussian radius $R = \sqrt{8}R_E$ a.) in coordinate space and b.) for the resulting correlation functions.

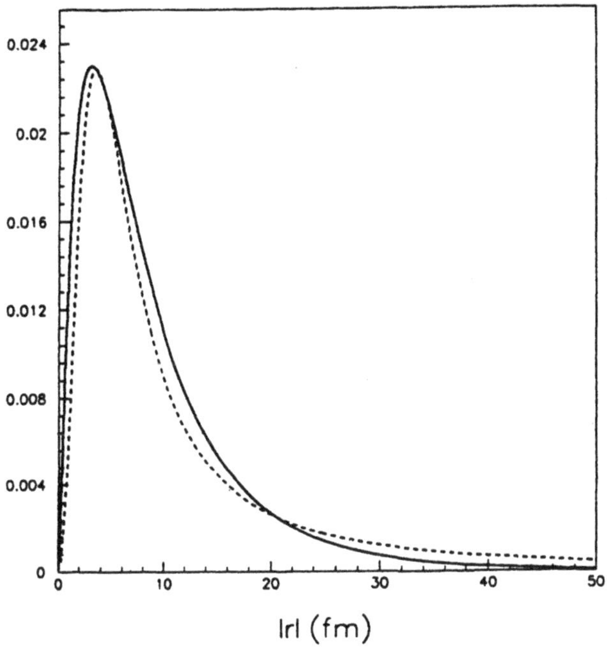


Figure 7: Solid line: The source distribution resulting from convoluting exponential decays with an initial momentum distribution (Eq. 12), for $\alpha m = 1$ and $\tau = 1$ fm. Dashed line: The function $\rho_E(|\vec{r}|)$ evaluated with $R_E = 3.5$ fm.

Further assume that the momentum distribution of these clusters is given by

$$f_p(|\vec{p}|)d^3p = \frac{\alpha^3}{8\pi}e^{-\alpha p}d^3p \quad . \tag{11}$$

(This distribution is an *ansatz* for numerical convenience only.) The probability for a pion to be born at a radius $|\vec{r}|$ is then given by

$$\mathcal{P}(r) = \int_0^\infty \mathcal{P}_{prod}(p,r)f_p(p)d^3p = \frac{1}{8\pi}\left(\frac{\alpha m}{\tau}\right)^3 K_0\left(2\sqrt{\frac{\alpha m r}{\tau}}\right) \quad . \tag{12}$$

As shown in Fig. 7, this simple model produces a source distribution fairly similar to the $\rho_E(|\vec{r}|)$ obtained from the exponential correlation function. However, it should be kept in mind that the real situation is undoubtedly much more complex. There are several different resonances, each with its own characteristic decay time, with (perhaps) differing momentum distributions $f_p$, there is an initial spatial distribution for resonance creation, etc. If this is a correct model of the production dynamics, then it is fortuitous that the end result is simply fit by a simple exponential (or any simple form, for that matter).

# List of Participants

| | |
|---|---|
| ALARD, J.P. | France |
| ANDREEV, I. | U.S.S.R. |
| ARDOUIN, D. | France |
| BARTKE, J. | Poland |
| BARTL, A. | Austria |
| BERGER, F. | Federal Republic of Germany |
| BIALAS, A. | Switzerland |
| BOPP, F.W. | Federal Republic of Germany |
| BOWLER, M. | Great Britain |
| BUSCHBECK, B. | Austria |
| CARRUTHERS, P. | U.S.A. |
| CHEW, C.K. | Republic of Singapore |
| CSERNAI, L. | Norway |
| CSÖRGÖ,T. | Hungary |
| DREMIN, I. | U.S.S.R. |
| DUPIEUX, P. | France |
| EKSPONG, G. | Sweden |
| ERAZMUS, B. | France |
| FERENC, D. | Yugoslavia |
| FINCKE-KEELER, M. | Canada |
| FOWLER, G. | Great Britain |
| FRIEDLANDER, E. | U.S.A. |
| GIOVANNINI, A. | Italy |
| GOLDHABER, G. | U.S.A. |
| GRISHIN, V. | U.S.S.R. |
| GROSSE, E. | Federal Republic of Germany |
| GUSTAFSON, G. | Sweden |
| GUTBROD, H.H. | Federal Republic of Germany |
| HERRERA, G. | Federal Republic of Germany |
| HOFMANN, W. | Federal Republic of Germany |
| HWA, R.C. | U.S.A. |
| JACOBS, P. | U.S.A. |

| | |
|---|---|
| KAGEYA, T. | Japan |
| KITTEL, W. | The Netherlands |
| KOMAROV, V. | U.S.S.R. |
| LAHANAS, M. | Federal Republic of Germany |
| LÖRSTAD, B. | Sweden |
| LUND, B.I. | Federal Republic of Germany |
| LYUBOSHITZ, Y.L. | U.S.S.R. |
| MAKHLIN, A. | U.S.S.R. |
| MORSE, R. | U.S.A. |
| NAVARRA, F. | Brasil |
| ORNIK, U. | Federal Republic of Germany |
| PADULA, S. | Brasil |
| PAN, J.C. | Federal Republic of Germany |
| PEITZMANN, T. | Federal Republic of Germany |
| PLÜMER, M. | Federal Republic of Germany |
| PLUTA, J. | Poland |
| PODGORETZKY, M. | U.S.S.R. |
| RAFELSKI, J. | U.S.A. |
| RAHA, S. | Federal Republic of Germany |
| RIMONDI, F. | Italy |
| RÖHRICH, D. | Federal Republic of Germany |
| ROZENTAL, I.L. | U.S.S.R. |
| SARCEVIC, I. | U.S.A. |
| SEIBERT, D. | U.S.A. |
| SHIH, C.C. | U.S.A. |
| SIEMIARCZUK, T. | Poland |
| SIMAK, V. | Czechoslovakia |
| SINHA, B. | India |
| SINUYKOV, Y. | U.S.S.R. |
| STENLUND, E. | Sweden |
| STEPHANS, G. | U.S.A. |
| STROTTMAN, D. | U.S.A. |
| SUGANO, K. | U.S.A. |

| | |
|---|---|
| SZWED, R. | Poland |
| VERBEURE, F. | Belgium |
| WALKER, R. | U.S.A. |
| WEINER, R. | Federal Republic of Germany |
| WILK, G. | Poland |
| WLODARCZYK, Z. | Poland |
| ZAJC, W.A. | U.S.A. |
| ZHOU, J. | People's Republic of China |
| ZINOVJEV, G. | U.S.S.R. |

www.ingramcontent.com/pod-product-compliance
Lightning Source LLC
LaVergne TN
LVHW060542180926
848329LV00007B/36
* 9 7 8 9 8 1 0 2 0 3 3 1 3 *